LECTURES IN PROBABILITY THEORY AND MATHEMATICAL STATISTICS

Modern Analytic *and* Computational Methods *in* Science *and* Mathematics

A GROUP OF MONOGRAPHS AND ADVANCED TEXTBOOKS

Richard Bellman, EDITOR

Published

1. R. E. Bellman, R. E. Kalaba, and Marcia C. Prestrud, Invariant Imbedding and Radiative Transfer in Slabs of Finite Thickness, 1963
2. R. E. Bellman, Harriet H. Kagiwada, R. E. Kabala, and Marcia C. Prestrud, Invariant Imbedding and Time-Dependent Transport Processes, 1964
3. R. E. Bellman and R. E. Kalaba, Quasilinearization and Nonlinear Boundary-Value Problems, 1965
4. R. E. Bellman, R. E. Kalaba, and Jo Ann Lockett, Numerical Inversion of the Laplace Transform: Applications to Biology, Economics, Engineering, and Physics, 1966
5. S. G. Mikhlin and K. L. Smolitskiy, Approximate Methods for Solution of Differential and Integral Equations, 1967
6. R. N. Adams and E. D. Denman, Wave Propagation and Turbulent Media, 1966
7. R. L. Stratonovich, Conditional Markov Processes and Their Application to the Theory of Optimal Control, 1968
8. A. G. Ivakhnenko and V. G. Lapa, Cybernetics and Forecasting Techniques, 1967
9. G. A. Chebotarev, Analytical and Numerical Methods of Celestial Mechanics, 1967
10. S. F. Feshchenko, N. I. Shkil', and L. D. Nikolenko, Asymptotic Methods in the Theory of Linear Differential Equations, 1967
11. A. G. Butkovskiy, Distributed Control Systems, 1969
12. R. E. Larson, State Increment Dynamic Programming, 1968
13. J. Kowalik and M. R. Osborne, Methods for Unconstrained Optimization Problems, 1968
14. S. J. Yakowitz, Mathematics of Adaptive Control Processes, 1969
15. S. K. Srinivasan, Stochastic Theory and Cascade Processes, 1969
16. D. U. von Rosenberg, Methods for the Numerical Solution of Partial Differential Equations, 1969
17. R. B. Banerji, Theory of Problems Solving: An Approach to Artificial Intelligence, 1969
18. R. Lattès and J.-L. Lions, The Method of Quasi-Reversibility: Applications to Partial Differential Equations. Translated from the French edition and edited by Richard Bellman, 1969
19. D. G. B. Edelen, Nonlocal Variations and Local Invariance of Fields, 1969
20. J. R. Radbill and G. A. McCue, Quasilinearization and Nonlinear Problems in Fluid and Orbital Mechanics, 1970
21. W. Squire, Integration for Engineers and Scientists, 1970
22. T. Parthasarathy and T. E. S. Raghavan, Some Topics in Two-Person Games, 1971
23. T. Hacker, Flight Stability and Control, 1970
24. D. H. Jacobson and D. Q. Mayne, Differential Dynamic Programming, 1970
25. H. Mine and S. Osaki, Markovian Decision Processes, 1970
26. W. Sierpiński, 250 Problems in Elementary Number Theory, 1970
27. E. D. Denman, Coupled Modes in Plasmas, Elastic Media, and Parametric Amplifiers, 1970
28. F. A. Northover, Applied Diffraction Theory, 1971
29. G. A. Phillipson, Identification of Distributed Systems, 1971
30. D. H. Moore, Heaviside Operational Calculus: An Elementary Foundation, 1971
31. S. M. Roberts and J. S. Shipman, Two-Point Boundary Value Problems: Shooting Methods, 1971
32. V. F. Demyanov and A. M. Rubinov, Approximate Methods in Optimization Problems, 1970
33. S. K. Srinivasan and R. Vasudevan, Introduction to Random Differential Equations and Their Applications, 1971
34. C. J. Mode, Multitype Branching Processes: Theory and Applications, 1971
35. R. Tompvić and M. Vukobratović, General Sensitivity Theory, 1971
36. J. G. Krzyż, Problems in Complex Variable Theory, 1971
37. W. T. Tutte, Introduction to the Theory of Matroids, 1971

LECTURES IN PROBABILITY THEORY AND MATHEMATICAL STATISTICS

STEFAN ZUBRZYCKI

University of Wrocław

AMERICAN ELSEVIER PUBLISHING COMPANY, INC.
NEW YORK

AMERICAN ELSEVIER PUBLISHING COMPANY, INC.
52 Vanderbilt Avenue, New York, N.Y. 10017

ELSEVIER PUBLISHING COMPANY
335 Jan Van Galenstraat, P.O. Box 211
Amsterdam, The Netherlands

International Standard Book Number 0-444-00120-4
Library of Congress Catalog Card Number 76-183843

Translated by Robert Bartoszyński from the original Polish *Wykłady z rachunku prawdopodobieństwa i statystyki matematycznej* published by PWN—Polish Scientific Publishers, Warszawa 1970.

COPYRIGHT 1972 BY PAŃSTWOWE WYDAWNICTWO NAUKOWE
WARSZAWA (POLAND), MIODOWA 10

All rights reserved,
No part of this publication may be reproduced,
stored in a retrieval system, or transmitted
in any form or by any means, electronic,
mechanical, photocopying, recording,
or otherwise, without the prior
written permission of the publisher.
American Elsevier Publishing Company, Inc.,
52 Vanderbilt Avenue, New York, N.Y. 10017.

PRINTED IN POLAND

Contents

Preface ix

Chapter I
Events and Probability

§ 1. The Object of Probability Theory 1
§ 2. Events and Sets 1
§ 3. Examples of Events 3
§ 4. The Space of Elementary Events. Events 4
§ 5. Relations Between Operations on Events 9
§ 6. Axiomatic Definition of Probability 13
§ 7. Examples 18

Chapter II
Probabilities in the Most Important Spaces of Elementary Events

§ 8. At Most Countable Spaces of Elementary Events 21
§ 9. Field of Events on the Real Line 22
§ 10. Probability on S_0 25
§ 11. Theorem on Extension of Measures 31
§ 12. Probability Distributions on the Real Line 40
§ 13. Probability Distributions in Multidimensional Euclidean Spaces 42
§ 14. Classification of Probability Distributions on the Real Line 51

Chapter III
Conditional Probability. Stochastic Independence of Events

§ 15. Conditional Probability 54
§ 16. Bayes' Theorem. Probabilities of Causes 59
§ 17. Stochastic Independence of Events 61
§ 18. Independent Trials 64
§ 19. Probabilities in Cartesian Products 66

Chapter IV
Classical Limit Theorems

§ 20. Binomial Distribution 74
§ 21. Poisson Theorem and Poisson Distribution 78
§ 22. The de Moivre–Laplace Theorem 80

Chapter V
Random Variables

§ 23. Random Variable and its Formal Properties	93
§ 24. Expectations of Random Variables which Assume a Finite Number of Values	102
§ 25. Expectation of a Random Variable in the General Case	106
§ 26. Expectation and Integration. Probability Distribution Function and Probability Distribution of a Random Variable	111
§ 27. Variance, Moments, and Other Numerical Characteristics of the Distribution of a Random Variable	115

Chapter VI
Laws of Large Numbers

§ 28. Independence of Random Variables. Additivity of Variance	122
§ 29. Čebyšev Inequality	127
§ 30. Empirical Distribution Function and Sample Characteristics	128
§ 31. Weak Law of Large Numbers. Convergence in Probability and Convergence with Probability One	130
§ 32. Strong Law of Large Numbers	135
§ 33. On the Convergence of Sample Characteristics	142
§ 34. Complements and Examples	148

Chapter VII
Central Limit Theorems of Probability Theory

§ 35. Generating Functions	153
§ 36. Characteristic Functions	158
§ 37. Characteristic Function of Some Important Distributions. Additivity of These Distributions	172
§ 38. Central Limit Theorem of Probability Theory	179
§ 39. Distributions Connected with the Normal Distribution which Occurs in Statistics	187
§ 40. Covariance Matrix of a Probability Distribution in a Multidimensional Euclidean Space	197
§ 41. Characteristic Functions of Multidimensional Random Variables	209
§ 42. Multidimensional Normal Distribution	212
§ 43. Multidimensional Form of the Central Limit Theorem of Probability Theory	225
§ 44. Limit Distributions of Sample Statistics	227

Chapter VIII
Statistical Inference

§ 45. Examples of Tests Related to the Normal Distribution	237
§ 46. Tests Based on Limit Theorems	254
§ 47. Most Powerful and Uniformly Most Powerful Tests	268
§ 48. Problem of Discrimination	282
§ 49. Problem of Estimation. Bayesian Estimates	289
§ 50. Interval Estimation. Confidence Intervals	292
§ 51. Maximum Likelihood Estimators	296
§ 52. Unbiased Estimators with Minimum Variance. Informational Inequality	299
§ 53. Statistical Decision Function. Minimax Estimators	305
References	315
Index	317

Preface

In the last few decades, in particular in the years following the World War II, we have been witnessing a rapid development of methods of probability theory and mathematical statistics, and a considerable evolution of views on the foundations of statistical inference.

One can say that the aim of probability theory is to create a mathematical description of random experiments whose results are not completely determined by the parameters which are controlled by the experimenter. The fundamental theorems of probability theory, namely the so-called laws of large numbers and central limit theorems, try to explain how the statistical independence of observations leads to the phenomenon which in popular terms can be described by stating that a random sample taken out of a large population can be considered as a relatively exact copy of that population.

The rapid development of mathematical statistics, dating from the last years of the 19th century, originated from a series of discoveries concerning probability distributions of sample statistics, which served for the so-called significance tests. As a prototype, one can think here about the problem of verifying on the basis of series of tosses of a coin the hypothesis that the coin in question is good. For the solution, it is necessary to know what is the probability distribution of the number of heads in a series of tosses of a good coin.

In the thirties, together with the construction of the mathematical foundations of probability theory, came the formulation of the basic principles of mathematical statistics as well-defined problems of finding extrema. It was pointed out that—to use an example—it is important to know not only the probability distribution of the number of heads in tossing a good coin, i.e. a coin satisfying the tested null hypothesis, but also it is important to know what are the alternative hypotheses. For instance, it is essential to decide whether the alternative hypotheses assert only that heads may appear more often than tails, or the difference between the probabilities of heads and tails may be of an arbitrary sign. The knowledge of the probability distribution of the test statistic corresponding to the null hypothesis allows us to construct a test which leads to the rejection of the null hypothesis when it is true with the given probability (called the level of significance). One also requires, however, that the probability of rejection of the null hypothesis should be maximal when this hypothesis is false, that is to say, one requires the test to be the most powerful. The fundamental lemma of Neyman and Pearson proved in 1930 gives the explicit form of the most powerful test of a single null hypothesis against a single alternative. This lemma is the cornerstone of the theory of testing statistical hypotheses, one of the most important branches of mathematical statistics.

There have been analogous investigations concerning estimations. To use again an example, the problem lies not in verifying the hypothesis that the coin is good, but in the most exact estimation of the probability of heads for this coin.

In testing hypotheses, as well as in giving an estimation, the basic idea is to construct methods of inference which take into account all the admissible hypotheses but do not utilize the *a priori* probabilities, in particular the so-called Bayes' rule. To use an example; the problem lies in constructing methods of inference about the probability of heads for a given coin which do not utilize the information asserting, say, that that coin was chosen at random from a sack whose content consists in 50% of good coins, and in 50% of coins with probability of heads equal to 3/4. In cases where the *a priori* probability is not known, the Bayes' rule suggests the construction of an inference method based on the assumption that the parameter which is studied has a uniform distribution. It has been pointed out that the *a priori* distribution of the parameter is, in general, unknown, and in many cases the very concept of the *a priori* distribution of the parameter is meaningless. The milestone of mathematically correct methods of statistical inference independent of Bayes' rule and *a priori* distributions is the concept of confidence intervals, introduced by Neyman.

The years 1940–1950 brought new reflections concerning the foundations of statistical inference, connected with the newly developed mathematical theory of games. The object of the theory of games is to provide a rational basis for decisions in cases where we do not know the values of the parameters which influence the results of our actions: we know only the set of possible values of those parameters, and we know how they influence the results of our actions. The most important case occurs when the unknown parameters are at the disposal of our opponent. It has been noted that statistical problems can be treated as games between the statistician who suggests a method of inference in the form of a test or estimator, and his opponent, conveniently called Nature, who selects the unknown value of the parameter. The basic underlying idea here is the pragmatic view asserting that statistical inference ought to serve as the foundation of a certain decision about an action, and that a given method of inference ought to be evaluated in terms of risk, that is, the expected loss resulting from actions inadequate to external parameters. The risk depends, on the one hand, on the method of inference, and on the other hand, on external parameters. This approach has led to the creation of the theory of statistical decision functions, whose foundations are presented in a book by Wald (1950). This theory permits a uniform approach to most statistical problems. Moreover, what is probably more important, it stresses the role of effects of statistical procedures in terms of risks.

This in turn has led to intensive studies, lasting till today, aimed at a revision of statistical methods from the above point of view. These studies have resulted in a partial rehabilitation of the Bayes' rule: it has turned out that in many important cases the choice of a statistical decision function can be restricted to decisions which are optimal against a certain *a priori* distribution of the parameter, i.e. it has turned out that a reasonable choice of a statistical decision function is essentially equivalent to the choice of a certain *a priori* distribution.

The aim of this book is to give an outline of the present state of probability theory and mathematical statistics. The main idea is to introduce the reader into the theory, acquainting him with the motives which decided about the way of posing the problems. Thus, the book starts from the measure-theoretical foundations of probability theory, and the culminating points of the first part of the book are §§ 33 and 44, which discuss the theorems on the convergence of sample characteristics to population characteristics, first in the sense of the laws of large numbers, and then in the sense of limit distributions. The nine sections of the last chapter, treated in a less detailed way than the rest of the book, give the basic methods of treating statistical problems.

Stefan Zubrzycki

CHAPTER I

Events and Probability

§ 1. THE OBJECT OF PROBABILITY THEORY

In everyday language one often uses the term "probable". Here are some examples: we say "it will probably rain tomorrow", "Peter is probably ill". We would use the first of these sentences if we see, for instance, dark clouds at sunset. In such a case the meaning of this sentence can be expressed as follows: although dark clouds at sunset do not necessarily imply rain on the following day, nevertheless a sunset like that is more often followed by a rainy day than by a day without rain. In this case the word "probably" conveys our observations concerning the frequency of days with and without rain following sunsets with dark clouds. A similar frequency interpretation can be extracted from the second sentence, when this sentence is used, say, on the occasion of Peter's absence at work. In this case we can interpret this sentence as conveying the information that "up to now, Peter has been absent at work more often because of illness than for other reasons". The precise meaning of such frequential, or statistical, use of the term "probable" constitutes the object of probability theory. The sources of success of this theory lie primarily in this restriction of the use of the term "probable". Probability theory does not deal with the use of the term "probable" as an evaluation of the degree of faith concerning the truth of a statement, even though some investigations in this direction (the so-called *multi-valued logics*) are currently under way. The latter meaning is clear in the sentence "Fermat's last theorem is probably true".

§ 2. EVENTS AND SETS

As a rule, we connect the concept of probability with the results of some observations or experiments, real or conceptual. This takes place when we speak of throwing dice, of the sex of a newborn baby, of the number of telephone calls, or of the length of a nail produced by a machine.

Note that sometimes we do not know in advance what are the possible results of a given experiment. We usually agree without hesitation that the result of tossing a coin will be either head or tail. In practice, however, the coin may rest standing on its edge. Similarly, when we speak of the length of a nail, we agree without hesitation that the result of the experiment can be expressed by a positive number, giving the length of the nail, say, in inches. Is it true, however, that every positive number can represent the length of a nail? Could it be, say, the number 1,000,000, when the machine is designed to produce nails of the length of about 2 inches, and we expect only small deviations from this prescribed length? Thus, the first thing we must do when starting the theoretical investigation of these various problems is to specify what are the possible results of the experiment. In probability theory, such a possible result of an experiment whose probability we want to discuss is called an *event*. Thus, the first thing we must do when building the theory is to specify the concept of the event.

Let us now consider, using the example of throwing dice, what relations can hold between events. When throwing a die, we speak for instance of the event of throwing an odd number of points. Note that this event may occur in various ways: it will occur when we throw 1, when we throw 3, or when we throw 5. On the other hand, the event "six" can occur in only one way, that is, by throwing 6. We see, therefore, that events may be classified into those which can occur in many ways, and those which can occur in only one way. In our example, the difference between the first and the second event is such as the difference between the three element set $\{1, 3, 5\}$ consisting of 1, 3 and 5, and the one-element set $\{6\}$, consisting only of 6.

Next, the two events considered above have the property that they cannot occur simultaneously. If the outcome is odd, "6" does not occur, and vice versa, if "6" occurs, the outcome is not odd. We call such events *disjoint* or *mutually exclusive*. Note that when we say that the events in question are disjoint, it amounts to saying that the sets $\{1, 3, 5\}$ and $\{6\}$ are disjoint, that is, that they do not contain any elements in common.

It is not true that every two events are disjoint. Consider, for instance, the event of throwing an odd number, and the event of throwing a number smaller than 4. These two events can occur simultaneously: for instance, if the outcome is "3". However, it may also happen that only one of them occurs; for example, if the outcome is "5", only the first of these events occurs. The relation between the above events is such as the relation between the sets $\{1, 3, 5\}$ and $\{1, 2, 3\}$: they contain common elements, but each of them contains also elements which do not belong to the other.

It may happen that the occurrence of one event implies the occurrence of another event. In our case this will be the situation with the event "6" and the

I. EVENTS AND PROBABILITY 3

event "even face". If "6" occurs, then also an even number occurs. This relation between events corresponds to the inclusion of set {6} in the set {2, 4, 6}.

We can speak of an event consisting of the simultaneous occurrence of two events; we call it the *product*, or *intersection*, of two events. As an example, consider the events (a) throwing an even number, and (b) throwing a number exceeding 3. The product of these events may occur in two ways, namely by throwing "4" and by throwing "6". Thus, to the product of these two events we may assign the intersection of the sets {2, 4, 6} and {4, 5, 6}: it is equal to the set {4, 6} containing the elements which are common to these two sets.

Similarly, we may speak of an event consisting of the occurrence of at least one of two events. We call it the *sum*, or the *union*, of two events. The sum of the events "an even number" and "a number exceeding 3" is the event consisting of throwing 2, 4, 5 or 6. It corresponds to the sum of sets {2, 4, 6} and {4, 5, 6}, equal to the set {2, 4, 5, 6}, containing all the elements which belong to at least one of the sets {2, 4, 6} and {4, 5, 6}.

From what we have said it should be clear that there exists a close analogy between events and sets. For the experiment of throwing a die we may specify the set consisting of numbers 1, 2, 3, 4, 5 and 6, these numbers representing all the elementary results of a single throw, i.e. the results which cannot be decomposed any further. If we specify such a set, every event concerning throw of a die can be described as a certain subset of this six-element set {1, 2, 3, 4, 5, 6}. We have seen also that relations between events can be expressed as relations between corresponding sets. This analogy between events and sets will serve as a foundation for probability theory. In this construction, elementary events will be identified simply with elements of a certain set, these elements representing single, elementary and non-decomposable results of the experiment in question, and the *events* will be interpreted simply as subsets of this set.

§ 3. EXAMPLES OF EVENTS

Let us now consider some examples which will illustrate the formulations presented at the end of the preceding section.

EXAMPLE 1. *Throw of a die*. As already mentioned in § 2, we can assume that numbers 1, 2, 3, 4, 5, 6 represent all the possible elementary, non-decomposable results of a throw of a die. Thus, there will be $2^6 = 64$ different events involved in a single throw of a die, that is, as many events as there are subsets of the set {1, 2, 3, 4, 5, 6}, including the empty set, containing no elements.

EXAMPLE 2. *Tossing of a coin*. We may assume that the set of all non-decomposable results of a toss of a coin consists of two elements, denoted by H and

T, corresponding to tossing heads (H) or tails (T). There are four possible events involved in a toss of a coin:

{0} empty set, or the impossible event,
{H} the event of tossing heads,
{T} the event of tossing tails,
{H, T} the sure event, consisting of tossing heads or tails.

EXAMPLE 3. *Two tosses of a coin.* Here we must distinguish four possible elementary outcomes: HH—tossing heads on both trials; HT—tossing heads on the first trial and tails on the second; TH—tossing tails on the first trial and heads on the second, and TT—tossing tails on both trials. In this case we have $2^4 = 16$ different events. Here are some of them: {HH, HT}—tossing heads on the first trial; {HT, TT}—tossing tails on the second trial; {HT}—tossing tails for the first time on the second trial; {HH, HT, TH}—tossing heads at least once, {HH, HT, TH, TT}—sure event.

EXAMPLE 4. *Age of couples.* If we denote by x the age of the husband and by y the age of the wife, then the set of all the possible results of observing the age of a couple coincides with the set B of points (x, y) of a plane for which $x \geqslant 16$ and $y \geqslant 16$, provided 16 is the age limit below which marriage is not allowed. Here are examples of events: the event that in the observed couple the husband is older than the wife will be identified with the set of those points in B for which $x > y$. The event "husband over 25" will be identified with the set of those (x, y) in B for which $x > 25$, whatever the value of y. The event that the sum of ages of husband and wife does not exceed 100 will be identified with the set of those (x, y) in B which lie below the line $x+y = 100$.

EXAMPLE 5. *Waiting time for a bus.* Suppose that buses on a certain route arrive at the stop at exactly five-minute intervals. We are interested in the time that elapses between the moment of somebody's arrival at the bus stop and the departure of the first bus that comes along. Thus, the possible results of such an observation coincide with numbers from the interval $0 \leqslant x < 5$. The interval $0 \leqslant x < 2$ will represent the event that the departure of the first bus occurs within two minutes.

§4. THE SPACE OF ELEMENTARY EVENTS. EVENTS

We can now proceed to formulate axioms concerning events.

In probability theory one speaks about the so-called spaces of elementary events. We shall denote the space of elementary events by X, and its elements, called elementary events, will be denoted by the letter e, possibly with subscripts.

I. EVENTS AND PROBABILITY

By the *space of elementary events* we understand the set of elements which constitute elementary results of observations or experiments, i.e. results which cannot be decomposed any further. In probability theory the concept of elementary event is a primitive concept, which is left undefined. In each particular application of probability theory one has to specify some set as the space of elementary events. In the rather typical examples of § 3 we encountered some of these spaces, adapted to the needs of particular problems. For throwing a die, the space was identified with the six-element set $\{1, 2, 3, 4, 5, 6\}$. For the time of waiting for a bus, the space of elementary events consisted of all numbers from the interval $0 \leqslant x < 5$. For the age of married couples, the space consisted of a portion of the plane. The examples of § 3 are rather typical, in the sense that, as the need arises, we shall take as the space of elementary events finite sets, countable sets, Euclidean spaces of a fixed dimension or their subsets, and so on.

Next, in probability theory, the *events* are identified with subsets of the space of elementary events, or, in other words, with sets of elementary events. We shall denote events by capital initial letters of the Latin alphabet, and the set of all events will be denoted by the capital letter S.

Here we must make some comments concerning the set S of all events. As mentioned above, in the general theory the events are identified with sets of elementary events. It turns out, however, that in most general cases one cannot accept every set of elementary events as an event and include it in S. This is connected with the fact, known from analysis, that not all functions are integrable, and that there exist so-called non-measurable sets. We cannot discuss this problem in more detail here; we want only to point out that it is generally impossible to treat every subset of the space of elementary events as an event. Thus, in the general theory, instead of speaking about events simply as of subsets of the space of elementary events, where all such subsets are allowed, we must introduce the set S of all events, and formulate postulates concerning the closure of S under some operations on events.

There is, however, an important exception, when the set S coincides with the class of all subsets of the space of all elementary events. This occurs when the space of elementary events is either finite or countable. Otherwise, one usually indicates some sets of elementary events, requiring them to belong to S, and one takes as S the smallest class of subsets of the space of elementary events containing these sets and closed under operations which we shall presently discuss.

In the sequel, whenever we speak of events, we shall mean sets of elementary events which belong to class S.

In § 2 we have already given examples of operations on events. We shall now present formal definitions. One considers the following operations on events: sum, product, difference, and complement.

The *sum* (or *union*) of events A and B is the event C consisting of all those elementary events which belong to at least one of the events A or B. We shall denote the operation of summation by $\cup$. Thus, equation $A \cup B = C$ means that C is the sum of A and B.

For the throw of a die, the sum of events $\{1, 3, 5\}$ consisting of throwing an odd number of points, and $\{1, 2, 3\}$ consisting of throwing a number of points smaller than 4, is $\{1, 2, 3, 5\}$, consisting of throwing one, two, three, or five points.

The operation of summation extends to an arbitrary number of events. Thus, the sum of n events $A_1, A_2, \ldots, A_n$ is the event

$$C = A_1 \cup A_2 \cup \ldots \cup A_n = \bigcup_{i=1}^{n} A_i$$

consisting of all those elementary events which belong to at least one of the events $A_1, \ldots, A_n$.

In a similar manner one may define the sum of an infinite sequence of events. Thus, the sum of the events $A_1, A_2, \ldots$ is the event

$$C = A_1 \cup A_2 \cup \ldots = \bigcup_{i=1}^{\infty} A_i$$

consisting of those elementary events which belong to at least one of the events $A_1, A_2, \ldots$

The *product* (or *intersection*) of two events A and B is the event C consisting of all those elementary events which belong to A and to B. The product will be denoted by $\cap$. Thus, equation $A \cap B = C$ means that C is the product of the events A and B.

For instance, the product of the events $\{1, 3, 5\}$ and $\{1, 2, 3\}$ discussed above is $\{1, 3\}$, consisting of throwing one or three. Other examples were already given in § 2.

In a similar way one can define the product of a larger number of terms. Thus, the product of the events $A_1, \ldots, A_n$ is the event

$$C = A_1 \cap A_2 \cap \ldots \cap A_n = \bigcap_{i=1}^{n} A_i,$$

consisting of all elementary events which belong to all the events $A_1, \ldots, A_n$.

I. EVENTS AND PROBABILITY

The product of an infinite sequence of events $A_1, A_2, \ldots$ is defined as the event

$$C = A_1 \cap A_2 \cap \ldots = \bigcap_{i=1}^{\infty} A_i$$

consisting of all elementary events which belong to every event A_j.

The *difference* of two events A and B is defined as the event C consisting of all those elementary events which belong to A but not to B. This operation will be denoted by the symbol $-$. Thus, the equality $C = A - B$ means that C is the difference of A and B.

The *complement* of the event A is the event B consisting of all those elementary events which do not belong to A. This operation will be denoted by the symbol $'$. Thus, equality $A' = B$ means that B is the complement of A.

Two events deserve special mention. One of them is the whole space X of all elementary events. We shall call it the *sure event*. The space of all elementary events represents all the possible results of an experiment; thus, when the experiment is performed, one of the events of X must occur. This justifies the term "sure event". The second particular event is the *impossible event*. It happens sometimes that we speak of an event, and later it turns out that it does not contain any elementary events, because the conditions defining it are inconsistent. To have a convenient way of describing such situations, we introduce the concept of the impossible (or empty) event. We shall denote it by 0; thus, the equality $A = 0$ is read as "A is impossible", and by that we mean that "A does not contain any elementary events".

In § 2 we considered a certain relation between events, expressing the fact that those events cannot occur simultaneously. Now we can easily express formally what it means that two events are disjoint: two events A and B are *disjoint* or exclude each other if they have no elementary events in common. This is equivalent to stating that their intersection is empty, that is, that their product is impossible, or

$$A \cap B = 0.$$

We shall introduce one more symbol. In § 2 we considered a relation between events A and B which holds if the occurrence of A implies the occurrence of B. This condition holds if and only if all the elementary events which are in A are also in B. This relation will be denoted by $\subset$. Thus, the formula

$$A \subset B$$

means that the event A is contained in the event B.

One more convention. So far we have said that elementary events are contained in events, treating events as sets of elementary events. From the formal point of view, we ought to use the same expressions when talking about events containing only one elementary event, thus distinguishing an elementary event e from the event $\{e\}$. Sometimes, however, it will be convenient to neglect this distinction; we shall do so, unless it should lead to a misunderstanding. Thus, for instance, we shall speak of probabilities of elementary events, which are in fact probabilities of events containing only one elementary event.

We shall now formulate the postulates concerning the class S of all events. We shall require the class S to have the following two properties:

(A1) *The complement A' of every event A is an event, that is, if $A \in S$ then $A' \in S$.*

(A2) *The sum of every finite or denumerable sequence of events A_i is an event, i.e. if for every index i from a finite or countable set we have $A_i \in S$, then $\bigcup_i A_i \in S$.*

In the theory of sets, a class of subsets of a given set which for every set contained in it contains also its complement is called *complementative*. A class of sets which contains the union of each two sets contained in it is called *additive*; if it contains the union of every sequence of sets contained in it, it is called *countably additive*, or *σ-additive*. An additive and complementative class of sets is called a *field*. A σ-additive and complementative class of sets is called a *σ-field*. Thus, postulates (A1) and (A2) require S to be a σ-field.

The examples which we discussed in §§ 2 and 3 indicate that we should require the class of all events to be additive: we simply want to reserve the right of speaking of events consisting of the occurrence of at least one of a certain number of events. In the examples considered there there were only a finite number of events; hence a finite additivity was sufficient. The example below will justify the introduction of the requirement of the σ-additivity of the class of all events.

EXAMPLE 1. *Tossing a coin until heads appears.* Consider the space of elementary events for the experiment of tossing a coin until heads appears for the first time. It may happen that heads appears for the first time at the first toss, at the second toss, at the third toss, and so on. For n tosses, the space of elementary events consists of sequences of n letters H and T. These sequences represent the possible results of the experiment: the letter T at the ith place signifies tossing tails at the ith toss. Whatever n we take, it may happen that heads appears for the first time at the $(n+1)$th toss. Thus, it seems natural to take as the space of elementary events for this experiment the set of all infinite sequences of letters H and T. The letter at the ith place of such a sequence would inform us whether heads or tails appeared at the ith toss. The event A_i: "heads appears for the first time at the ith toss" would consist of all infinite sequences of letters H and T which start from $i-1$ letters T and have letter H at the ith place. We can now ask for the event that heads appear for the first time after a finite number of tosses. But what does it mean "after a finite number of tosses?" It means that heads appear either at the first toss, or at the second toss, and so on. Hence the event A that heads appear after a finite number of tosses is the sum of a countable sequence of events $A_1, A_2, \ldots$

In the sequel we shall see that σ-additivity is necessary when we want to speak about the probability that the sum of two random variables does not exceed a given number.

Thus, the introduction of σ-additivity seems justified.

In the next section we shall discuss some simple but important relations between the operations on events which we have just introduced.

§ 5. RELATIONS BETWEEN OPERATIONS ON EVENTS

In this section we shall present some simple consequences of the definitions of operations on events introduced in the last section. We shall write them in the form of equalities, valid for all events. Thus we have:

(5.1) $$A \cup A' = X,$$

i.e., the sum of any event and its complement is a sure event.

In fact, by definition, A' is the set of those elementary events which do not belong to the event A. Thus, every elementary event belongs either to A or to A', which coincides with the assertion (5.1).

Next, we have

(5.2) $$A \cap A' = 0,$$

i.e., any event and its complement are disjoint.

This property is obvious, since the elementary events which are in A' do not belong to A by definition.

Moreover, we have

(5.3) $$A \cup A = A,$$
(5.4) $$A \cap A = A,$$
(5.5) $$(A')' = A.$$

These equalities are evident. We write them down here, since they characterize in a certain way the operations of summation, intersection and complementation. Analogous relations can be written down for an arbitrary number of identical terms in a sum or intersection:

(5.6) $$A \cup A \cup \ldots \cup A = A,$$
(5.7) $$A \cap A \cap \ldots \cap A = A,$$
(5.8) $$A \cup A \cup \ldots = A,$$
(5.9) $$A \cap A \cap \ldots = A.$$

There is one more reason for giving the above formulas. The property (A2) of the set of all events is usually formulated in the following way: for any sequence of events $A_1, A_2, \ldots$ their sum $A_1 \cup A_2 \cup \ldots$ is also an event. From this formulation it follows that the sum of a finite number of events is an event. In fact, equality (5.8) allows us to represent any finite sum of events in the form of an infinite sum of a sequence of events

$$A_1 \cup A_2 \cup \ldots \cup A_n = A_1 \cup A_2 \cup A_3 \cup \ldots \cup A_n \cup A_n \cup \ldots$$

Thus, in condition (A2) we might have omitted the summation of a finite number of events.

The same applies to infinite products.

For the difference of events we have

(5.10) $$A - B = A \cap B'.$$

In fact, every elementary event which belongs to A and does not belong to B, belongs to A and to B', and vice versa; thus, events $A - B$ and $A \cap B'$ consist of the same elementary events.

Equation (5.10) shows that differences of events can be defined in terms of intersections and complementations. However, the introduction of the operation of difference is very convenient, since it leads to simplification of many formulations.

In a similar way, intersection can be defined in terms of summation and complementation. The relations between the operations of summation and intersection are given by laws, which are sometimes called the *laws of de Morgan*. In the sequel we shall frequently use those laws. We present them here in three variants: for two events, for a finite number of events, and for sequences of events. They have the following form:

(5.11) $$(A \cup B)' = A' \cap B', \quad (A \cap B)' = A' \cup B',$$

(5.12) $$(A_1 \cup A_2 \cup \ldots \cup A_n)' = A_1' \cap A_2' \cap \ldots \cap A_n',$$
$$(A_1 \cap A_2 \cap \ldots \cap A_n)' = A_1' \cup A_2' \cup \ldots \cup A_n',$$

(5.13) $$(A_1 \cup A_2 \cup \ldots)' = A_1' \cap A_2' \cap \ldots,$$
$$(A_1 \cap A_2 \cap \ldots)' = A_1' \cup A_2' \cup \ldots$$

I. EVENTS AND PROBABILITY

The proofs of the above equalities are very simple; they reduce to mere checking that the events on both sides of these equations consist of the same elementary events.

We shall give the proof of the first of the equalities (5.11). An elementary event belongs to the sum of events A and B if it belongs to at least one of these events. Therefore, it belongs to the complement $(A \cup B)'$ of the sum of A and B if it is not true that it belongs to at least one of the events A and B. But this means precisely, that it belongs neither to A nor to B, hence it must belong to the complement A' of the event A, and to the complement B' of the event B, and hence to the intersection of these complements. Conversely, if an elementary event belongs to the intersection $A' \cap B'$, it does not belong to A, and it does not belong to B. Thus, it is not true that it belongs to at least one of the events A and B, hence it must belong to the complement $(A \cup B)'$ of their sum. This proves the first of the equalities (5.11). The proofs of the remaining equalities are also very simple and will be omitted.

Replacing in equations (5.11)–(5.13) the events by their complements, and applying (5.5) to the left-hand sides of the resulting equalities, we obtain the following equalities, which may serve as definitions of unions by intersections and complementations, and intersections by unions and complementations:

(5.14)
$$A \cup B = (A' \cap B')',$$
$$A \cap B = (A' \cup B')',$$
$$A_1 \cup A_2 \cup \ldots \cup A_n = (A'_1 \cap \ldots \cap A'_n)',$$
$$A_1 \cap A_2 \cap \ldots \cap A_n = (A'_1 \cup \ldots \cup A'_n)',$$
$$A_1 \cup A_2 \cup \ldots = (A'_1 \cap A'_2 \cap \ldots)',$$
$$A_1 \cap A_2 \cap \ldots = (A'_1 \cup A'_2 \cup \ldots)'.$$

Let us also note some characteristic properties of impossible and sure events:

(5.15) $$A \cup X = X.$$

In fact, every elementary event is, by definition, contained in X, hence (5.15). The relation

(5.16) $$A \subset X,$$

is equivalent to the preceding one. Next, we have

(5.17) $$A \cap 0 = 0.$$

Indeed, as the impossible event contains no elementary events, it cannot have any elementary events in common with any event. Equality (5.17) means that we have

COROLLARY 1. *The impossible event is disjoint with any other event.*

In particular,
$$0 \cap 0 = 0.$$

Moreover, we have

(5.18) $$0 \subset A,$$

(5.19) $$0' = X, \quad X' = 0.$$

We shall now present some consequences of postulates (A1) and (A2) concerning the field of events.

THEOREM 1. *Sure and impossible events belong to S, i.e., $X \in S$ and $0 \in S$.*

This follows from equality (5.1) and from the fact that, by virtue of axiom (A2), the sum of two events is an event. Indeed, let A be an arbitrary event. Then, by (A1) A' is also an event. By (A2) the sum $A \cup A'$ also belongs to S, but by (5.1) we have $A \cup A' = X$, hence X is an event. Since $0 = X'$, we have $0 \in S$ by (A1).

THEOREM 2. *The product of any two events is an event, i.e., if $A \in S$ and $B \in S$, then $A \cap B \in S$.*

This property follows from the second of the equalities (5.14). If A and B are events, then by (A1), A' and B' are also events, hence the same is true also for $A' \cup B'$. Using (A1) again, we see that the same is true for $(A' \cup B')'$, which is equal to $A \cap B$ by the second equality (5.14). This completes the proof.

In a similar manner, using the remaining equalities (5.14), one can prove

THEOREM 3. *A product of a finite or a countable number of events is an event, i.e., if for a finite or countable number of indices i we have $A_i \in S$, then $\bigcap_i A_i \in S$.*

THEOREM 4. *The difference of two events is an event, i.e., if $A \in S$ and $B \in S$, then $A - B \in S$.*

This follows from the fact that by (5.10), the difference can be represented in terms of complements and intersections, and by (A1) and Theorem 2, both the complement and the intersection of two events are events.

THEOREM 5. *The sum of events $A_1, A_2, \ldots$ can always be represented as a sum of disjoint events $E_1, E_2, \ldots$ in such a way that for every n we have*
$$E_1 \cup E_2 \cup \ldots \cup E_n = A_1 \cup A_2 \cup \ldots \cup A_n.$$

For the proof it suffices to put $E_1 = A_1$ and
$$E_i = A_i - (A_1 \cup \ldots \cup A_{i-1}) \quad \text{for} \quad i = 2, 3, \ldots$$

§ 6. AXIOMATIC DEFINITION OF PROBABILITY

As has already been mentioned at the beginning of Section 1, probability theory describes in an abstract mathematical form the phenomena connected with the frequencies of the appearance of results of experiments or observations.

We find empirically that in a long series of tosses of a coin the fraction of heads, i.e. the ratio of the number of heads that appeared to the number of tosses, is close to $1/2$; we also find that the larger the number of tosses, the smaller are the oscillations of this fraction around $1/2$. When we say, speaking of tossing a coin: "the probability of heads equals $1/2$", we have in mind precisely such experiences with tossing coins. It ought to be kept in mind, however, that the sentence "the probability of head equals $1/2$" refers neither to any particular series of tosses which has already been performed nor to any toss that we just want to make, but to the experiment of tossing coins in general. By this sentence we want to express a certain property of tossing coins. In a similar way, the sentence "the probability of throwing "one" in a throw of a die equals $1/6$" refers to throwing dice as an experiment and not to any particular series of throws. We want to express by it sentence our experience gained about this experiment. This reference of probabilities to experiments and their analogies with frequencies will serve as a foundation for an axiomatic definition of probability.

First, however, let us look more closely at the relation between the frequency of results which we observe under a repetition of an experiment (such as tossing a coin) and the probability which is to express certain properties of such experiments. We say that the probability of a head is $1/2$. What is the meaning of this statement? We use the fact that fractions of heads in long series of tosses are close to $1/2$. Yes, but they are rarely equal to $1/2$, and may be different in different series. Thus, why is it $1/2$ and not, say, 0.5001? There were some attempts of defining the probability of heads as the limits of the frequencies of heads in an increasing number of tosses. If, however, we were to accept such definitions literally, we would have to wait with any statement involving the probabilities of heads until we found this limit. One should also add that the results of long series of tosses, reported in literature, are usually inconsistent with the conjecture that the probability of heads is equal to $1/2$.

Thus, we proceed in a different way. As in geometry, theorems concerning figures are merely abstract models of empirically observed relations between lines and figures drawn on paper, in probability theory we want to construct an abstract mathematical model of relations concerning the frequencies which we observe empirically. One can say that the relation between probability and frequency is such as the relation between the straight line considered in geometry and a straight line drawn on paper, or between mass points considered in mechanics, and real distributions of mass in physical bodies. It is only by confronting a theoretical model with the observed empirical data that we gain knowledge about a given experiment. The inconsistency between the results of coin tossing and the conjecture that the probability of a head is 1/2 is due to the fact that a coin is asymmetric; indeed, we discover such asymmetry precisely by observing such inconsistencies. In a similar way, the statistical quality control in industry, by confronting the results of production of a machine with the predictions of a theoretical model, leads to discovering the existence of defects, and helps in locating them.

As we said above, the probabilities will refer to experiments and will be treated as abstract descriptions of their properties. We shall not stop there, however, and we shall continue our efforts leading to the explication of the relations between the probabilities which express properties of experiments, and the frequencies of appearances of various results in long series of repetitions of those experiments. One can say that in explaining this relation lies one of the fundamental tasks of probability theory. The solution is given in the so-called laws of large numbers and limit theorems of probability theory. The exposition of these topics will occupy a large portion of this book.

Let us come again to the meaning of the proposition "the probability of a heads is equal to 1/2". In this sentence, probability is a number; this number imitates frequency: hence it ought to be contained between 0 and 1. This number is connected with an event, namely that of tossing a head.

In a given series of throws of a die, we have well defined fractions of 1's, 2's, 3's, 4's, 5's and 6's. Also, the fraction of "even results" is defined: it is equal to the sum of fractions of 2's, 4's and 6's. This opens the way to the following definition:

Probability is a function which has real numbers as values, and events as arguments, and which satisfies the axioms (A3)–(A5) given below. We shall denote probability by the letter P. Thus, $P(A)$ denotes the probability of the event A. The properties of function P are the following:

(A3) $$0 \leqslant P(A) \leqslant 1.$$

(A4) *The probability of the sure event is equal to 1, i.e.*

$$P(X) = 1.$$

(A5) *The probability is σ-additive, i.e. for a sequence of pairwise disjoint events $A_1, A_2, \ldots$ we have*

$$P(A_1 \cup A_2 \cup \ldots) = P(A_1) + P(A_2) + \ldots$$

The need of introducing axiom (A5), requiring σ-additivity and not merely additivity, is connected with the σ-additivity of events, as required by axiom (A2). Axioms (A3)–(A5) give only characteristic properties of probability; the numerical values are not specified. These values will be selected separately in every specific case, depending on the problem on hand. One can say that these axioms tell us which functions can be called probabilities. Thus, to specify what probability we are discussing, we have to define the set of arguments, i.e. the class S of events, as well as the values assigned to those events. In order to be able to use general theorems of probability theory, one has to verify that the class of events under consideration and the probability defined on this class satisfy axioms (A1)–(A5).

Before giving examples, we derive some simple consequences of axioms (A3)–(A5). First, we have

(6.1) $$P(0) = 0,$$

i.e. the *probability of the impossible event equals zero*.

Indeed, by relation (5.17), the impossible event is disjoint with itself, hence the sequence $0, 0, \ldots$ of impossible events is, formally speaking, a sequence of disjoint events. Thus, we may apply axiom (A5), and write

$$P(0 \cup 0 \cup \ldots) = P(0) + P(0) + \ldots$$

On the other hand, by equation (5.8), we have

$$0 = 0 \cup 0 \cup \ldots,$$

which implies that

$$P(0) = P(0) + P(0) + \ldots$$

This, however, is possible only if $P(0) = 0$.

Now we shall show that axiom (A5) implies the additivity of probability for any finite number of disjoint events.

THEOREM 1. *If the events $A_1, A_2, \ldots, A_n$ are pairwise disjoint, then*

(6.2) $$P(A_1 \cup A_2 \cup \ldots \cup A_n) = P(A_1) + P(A_2) + \ldots + P(A_n).$$

Indeed, we can write

$$A_1 \cup A_2 \cup \ldots \cup A_n = A_1 \cup A_2 \cup \ldots \cup A_n \cup 0$$
$$= A_1 \cup A_2 \cup \ldots \cup A_n \cup 0 \cup 0 \cup \ldots$$

By hypothesis and by relation (5.17), the sequence $A_1, A_2, \ldots, A_n, 0, 0, \ldots$ constitutes a sequence of pairwise disjoint events, hence axiom (A5) applies. By the above equalities and axiom (A5) we obtain

$$P(A_1 \cup \ldots \cup A_n) = P(A_1 \cup A_2 \cup \ldots \cup A_n \cup 0 \cup \ldots)$$
$$= P(A_1) + P(A_2) + \ldots + P(A_n) + P(0) + P(0) + \ldots$$

On the other hand, by (6.1) we obtain

$$P(A_1) + \ldots + P(A_n) + P(0) + P(0) + \ldots = P(A_1) + \ldots + P(A_n).$$

Therefore

$$P(A_1 \cup A_2 \cup \ldots \cup A_n) = P(A_1) + P(A_2) + \ldots + P(A_n).$$

In the above reasoning we have shown how the axiom of additivity for infinite sequences of disjoint events implies additivity for finite systems of disjoint events.

Here are some more properties of probability:

THEOREM 2. *If $A, A_1, A_2, \ldots$ are events such that*

$$A \subset A_1 \cup A_2 \cup \ldots,$$

then

(6.3) $\qquad P(A) \leqslant P(A_1) + P(A_2) + \ldots$

PROOF. In fact, if $A \subset A_1 \cup A_2 \cup \ldots$, then $A = E_1 \cup E_2 \cup \ldots$ where $E_i = A \cap A_i$. Thus, event A can be represented as a sum of a sequence of events; these, however, are not necessarily disjoint. On the other hand, by Theorem 5 of § 5, the event A can also be represented as a sum of disjoint events

$$A = C_1 \cup C_2 \cup \ldots,$$

where $C_i = E_i - (E_1 \cup E_2 \cup \ldots \cup E_{i-1})$. In view of the σ-additivity of probability, we have

(6.4) $\qquad P(A) = P(C_1) + P(C_2) + \ldots$

By definition of events C_i, for every i we have

$$E_i = C_i \cup \big(E_i \cap (E_1 \cup \ldots \cup E_{i-1})\big),$$

and the events on the right-hand side are disjoint, hence

$$P(E_i) = P(C_i) + P\big(E_i \cap (E_1 \cup \ldots \cup E_{i-1})\big).$$

Since probability is non-negative, we get

$$P(C_i) \leqslant P(E_i).$$

The last inequality and formula (6.4) imply already the assertion of Theorem 2.

I. EVENTS AND PROBABILITY

THEOREM 3. *If the events $A_1, A_2, \ldots$ form an increasing sequence, that is, if*

(6.5) $$A_1 \subset A_2 \subset \ldots,$$

and if A is the sum of these events, i.e. if

(6.6) $$A = A_1 \cup A_2 \cup \ldots,$$

then

(6.7) $$P(A) = \lim_{n \to \infty} P(A_n).$$

PROOF. We shall represent the sum of the events A_i as a sum of disjoint events. Putting

$$E_1 = A_1, \quad E_i = A_i - A_{i-1} \quad \text{for} \quad i = 2, 3, \ldots,$$

we get

(6.8) $$A = E_1 \cup E_2 \cup \ldots$$

and

(6.9) $$A_i = E_1 \cup \ldots \cup E_i \quad \text{for} \quad i = 1, 2, \ldots,$$

where the events E_i are disjoint. Now we can write

$$P(A) = P(E_1 \cup E_2 \cup \ldots) = \sum_{i=1}^{\infty} P(E_i) = \lim_{n \to \infty} \sum_{i=1}^{n} P(E_i)$$
$$= \lim_{n \to \infty} P(E_1 \cup \ldots \cup E_n) = \lim_{n \to \infty} P(A_n).$$

The first of these equalities follows from formula (6.8), the second is a consequence of the fact that events E_i are disjoint and of the σ-additivity of probability, the third is a definition of the sum of numerical series, the fourth uses finite additivity of probability, and finally, the fifth uses formula (6.9), thus leading to relation (6.7).

The following simple consequence of Theorem 3 will frequently be used in the sequel:

THEOREM 4. *If the events $A_1, A_2, \ldots$ form an increasing sequence, and their sum equals the whole space X of elementary events, then*

$$\lim_{n \to \infty} P(A_n) = 1.$$

PROOF. It suffices to use Theorem 3, putting X instead of A, and use axiom (A4) according to which $P(X) = 1$.

Now we shall prove the dual version of Theorem 3.

THEOREM 5. *If the events $A_1, A_2, \ldots$ form a decreasing sequence of events, i.e. if*

(6.10) $$A_1 \supset A_2 \supset \ldots,$$

and if A is the product of these events, i.e. if

(6.11) $$A = A_1 \cap A_2 \cap \ldots$$

then

(6.12) $$P(A) = \lim_{n \to \infty} P(A_n).$$

PROOF. Passing to complements, we see that the sequence of events $A'_1, A'_2, \ldots$ is increasing, and $A'_1 \cup A'_2 \cup \ldots = A'$. Thus, by Theorem 3 we have $P(A') = \lim_{n \to \infty} P(A'_n)$. We have, however, $P(A') = 1 - P(A)$, and for every n we have also $P(A'_n) = 1 - P(A_n)$. Thus,

$$1 - P(A) = \lim_{n \to \infty} \left(1 - P(A_n)\right),$$

which is equivalent to (6.12).

The following simple consequence of Theorem 5 will be very useful in the sequel:

THEOREM 6. *If the intersection of a decreasing sequence of events $A_1 \supset A_2 \supset \ldots$ is empty, then*

$$\lim_{n \to \infty} P(A_n) = 0.$$

PROOF. It suffices to take in Theorem 5 the impossible event as A and use the fact that the probability of the impossible event is zero (see (6.1)).

§ 7. EXAMPLES

We shall present several examples. We start from cases in which the space of elementary events is finite.

EXAMPLE 1. *Toss of a coin.* In this case the space of elementary events has two elements: $X = \{H, T\}$, where H stands for heads and T stands for tails. The field of events S contains four elements: X—the sure event, 0—the impossible event, $A = \{H\}$—the event of tossing heads, and $B = \{T\}$—the event of tossing tails.

I. EVENTS AND PROBABILITY

The probabilities of these events will be defined as
$$P(0) = 0, \quad P(A) = \tfrac{1}{2}, \quad P(B) = \tfrac{1}{2}, \quad P(A \cup B) = P(X) = 1.$$
Axioms (A1)–(A5) are obviously satisfied.

EXAMPLE 2. *Throwing dice.* The space of elementary events $X = \{1, 2, 3, 4, 5, 6\}$ consists in this case of six elements. The field S of events consists of 64 elements, that is, as many as there are subsets of the set X. Every event is a sum of a certain number of events $\{1\}, \{2\}, \{3\}, \{4\}, \{5\}, \{6\}$, each containing one elementary event. Thus, it suffices to define the probabilities of these events. We define them as
$$P(\{1\}) = P(\{2\}) = \ldots = P(\{6\}) = \tfrac{1}{6}.$$
In this case the probability of every event equals the ratio of the number of elementary events in it to the total number of elementary events. For instance, the probability of throwing an even number of points equals 1/2, since
$$P(\{2, 4, 6\}) = P(\{2\} \cup \{4\} \cup \{6\})$$
$$= P(\{2\}) + P(\{4\}) + P(\{6\}) = \tfrac{1}{6} + \tfrac{1}{6} + \tfrac{1}{6} = \tfrac{1}{2}.$$

EXAMPLE 3. *Two tosses of a coin.* The field of events for this experiment was already described in Example 3 of § 3. To define the probabilities, assume that all elementary events are equally probable. We have therefore
$$P(\{HH\}) = P(\{HT\}) = P(\{TH\}) = P(\{TT\}) = \tfrac{1}{4}$$
and, according to axiom (A5) on additivity, the probabilities of the remaining events are equal to the product of 1/4 and the number of elementary events which they contain. For instance, the probability of tossing heads at least once equals 3/4, since
$$P(\{HH, HT, TH\}) = P(\{HH\} \cup \{HT\} \cup \{TH\})$$
$$= P(\{HH\}) + P(\{HT\}) + P(\{TH\}) = \tfrac{1}{4} + \tfrac{1}{4} + \tfrac{1}{4} = \tfrac{3}{4}.$$

At this point we should mention the so-called *paradox of d'Alembert*. The problem is to compute the probability of at least one heads in two tosses of a coin, and d'Alembert's reasoning goes as follows: at the first toss, either heads or tails appears. If heads appears, the game ends; in the opposite case we toss again, coming up either with heads or with tails. Thus, there are three possibilities, two of them favourable for the gambler. As all possibilities are equally probable, the probability of at least one heads in two tosses is 2/3, and not 3/4, as computed above.

What is the cause for this difference? The fault of d'Alembert lies in the erroneous description of the space of elementary events: he did not notice that, with two tosses, the event "heads at the first toss" is an event composed of two elementary events. Consequently, he treated as equally probable different events from those we did, thus obtaining different results. As regards the probability of the event under discussion, d'Alembert and we computed it assuming certain events to be equally probable. Such assumptions are, however, sometimes false. One can think of an "asymmetric" coin, for which the probability of at least one heads in two tosses would be 2/3 and not 3/4.

EXAMPLE 4. *Waiting time for a bus.* As has already been mentioned in Example 5 of § 3, the elementary events here are numbers from the interval $0 \leqslant x < 5$. There are, therefore, infinitely many elementary events, and hence we cannot use the postulate of equal probabilities. We shall proceed in a different way. We can require, for instance, that events-intervals of equal lengths should have equal probabilities. This leads to the following definition of probabilities for events-intervals: for each event $\langle a, b \rangle = \{x: a \leqslant x < b\}$, $0 \leqslant a < b \leqslant 5$, we put

(7.1) $$P(\langle a, b \rangle) = \frac{b-a}{5},$$

hence the probability of the event $\langle a, b \rangle$ equals the ratio of the length of the interval $\langle a, b \rangle$ to the length of the interval $\langle 0, 5 \rangle$.

Using the axiom of σ-additivity, we can now compute the probabilities of every other event. For instance, if $\langle a_1, b_1 \rangle$ and $\langle a_2, b_2 \rangle$ are disjoint (this will occur if for instance $0 \leqslant a_1 < b_1 \leqslant a_2 < b_2 \leqslant 5$), then

$$P(\langle a_1, b_1 \rangle \cup \langle a_2, b_2 \rangle) = \frac{(b_1-a_1)}{5} + \frac{(b_2-a_2)}{5}.$$

CHAPTER II

Probabilities in the Most Important Spaces of Elementary Events

§ 8. AT MOST COUNTABLE SPACES OF ELEMENTARY EVENTS

We shall now discuss methods of defining probabilities in the most importa n spaces of elementary events. These will be at most countable spaces and Euclidean spaces. From the point of view of analytical difficulties connected with defining the field of events and probability as a function on this field, the easiest case is that of a finite or countable space of elementary events. If the space of elementary events X has a finite or countable number of elements

$$e_1, e_2, \ldots,$$

then we may take as the field S of events simply the class of all subsets of X, with no exceptions. Axioms (A1) and (A2) concerning the field of events are obviously satisfied: the complement of an event, as well as the sum of a finite or countable number of events, being evidently subsets of X, are therefore events and belong to S. Now, in order to define on the field S the probability P satisfyng the properties described in axioms (A3)–(A5), it suffices to take a sequencet of non-negative numbers $p_1, p_2, \ldots$ such that

(8.1) $$p_1 + p_2 + \ldots = \sum_{i=1}^{\infty} p_i = 1$$

and put

$$P(\{e_i\}) = p_i.$$

Since every event is a sum of a finite or countable number of disjoint elementary events, i.e.,

$$A = \{e_{i_1}, e_{i_2}, \ldots\} = \{e_{i_1}\} \cup \{e_{i_2}\} \cup \ldots,$$

in order to be consistent with axiom (A5) we put

(8.2) $P(A) = P(\{e_{i_1}, e_{i_2}, \ldots\}) = P(\{e_{i_1}\}) + P(\{e_{i_2}\}) + \ldots = p_{i_1} + p_{i_2} + \ldots$

In other words, we define the probabilities of elementary events so that their sum equals one, and we define the probability of every event as equal to the sum of the probabilities of elementary events contained in it.

Simple properties of series with non-negative terms imply that axioms (A3)–(A5) are satisfied. Axiom (A3) is implied by the fact that the numbers p_i are non-negative by assumption, and in view of (8.1), we get $P(A) \leqslant 1$ for every A. Axiom (A4) is satisfied in view of (8.1), while axiom (A5) follows from the fact that the series of non-negative terms can be grouped and arranged in an arbitrary way without affecting its sum; in the case of a finite number of elementary events, this property reduces simply to the fact that addition is associative and commutative. In fact, if the events $A_k = \{e_{k,1}, e_{k,2}, ...\}$, $k = 1, 2, ...$, are disjoint and their sum is $A = A_1 \cup A_2 \cup ... = \{e_{i_1}, e_{i_2}, ...\}$, then by (8.2):

$$P(A_k) = p_{i_{k,1}} + p_{i_{k,2}} + ..., \quad k = 1, 2, ...,$$
$$P(A) = p_{i_1} + p_{i_2} + ...,$$

and the equality

$$P(A) = P(A_1) + P(A_2) + ...$$

follows from the fact that the sum $p_{i_1} + p_{i_2} + ...$ equals the sum $p_{i_{1,1}} + p_{i_{1,2}} + ... + p_{i_{2,1}} + p_{i_{2,2}} + ... + ...$, since both these sums consist of the same non-negative terms, only arranged and grouped in a different way.

Thus, one can say that for finite or countable spaces of elementary events, the investigation of probabilities reduces to the investigation of properties of sequences of non-negative numbers which add to unity. We shall return to this remark in the sequel.

In the case of at most countable spaces of elementary events, the function

$$f(e_i) = P(\{e_i\}) = p_i,$$

defined on the space of elementary events, is sometimes called the *probability distribution*.

§ 9. FIELD OF EVENTS ON THE REAL LINE

Things look different when the space of elementary events is not countable. As has already been mentioned, in this case one cannot in general take as the field of events the field of all subsets, and one has to restrict the considerations to some subsets only.

The most important non-countable space is the real line. We shall now describe the field of events on the real line, and give methods of defining probabilities on it.

As events we shall take sets of a certain class. We shall require namely that all infinite intervals of the form $(-\infty, a) = \{x: x < a\}$ be events, that is, that all sets consisting of numbers smaller than any given number a be events. Geometrically, these sets are half-lines extending to the left from a given point. This requirement is quite natural. Since, together with every event, its complement must also belong to the field of events S (axiom (A1)), we must also include in S the sets of the form $\langle a, \infty) = \{x: a \geqslant x\}$, as $\langle a, \infty) = (-\infty, a)'$. Next, together with every two events, their difference must also belong to S (Theorem 4 of § 5), whence half-open intervals $\langle a, b) = \{x: a \leqslant x < b\}$, closed from the left and open from the right must also belong to S. Indeed, $\langle a, b) = (-\infty, b) - (-\infty, a)$.

Some remarks on notation. As usual, $\langle a, b)$ will denote the interval $\{x: a \leqslant x < b\}$, open from the right and closed from the left. Similarly, (a, b) will denote the open interval $\{x: a < x < b\}$, $\langle a, b \rangle$—the closed interval $\{x: a \leqslant x \leqslant b\}$, and $(a, b\rangle$—the interval closed from the right and open from the left. If $a = b$, then (a, b), $\langle a, b)$ and $(a, b\rangle$ will denote empty sets, while $\langle a, b \rangle$ will be a one-element set $\{x: a = x = b\}$. If $a > b$, then sets (a, b), $\langle a, b\rangle$, $\langle a, b)$ and $(a, b\rangle$ will be empty.

Let us now consider the class S_0 of sums of half-open intervals of the form $\langle a, b)$. We shall include in S_0 also the half-lines $\langle -\infty, a) = (-\infty, a) = \{x: -\infty < x < a\}$ and the half-lines $\langle b, \infty) = \{x: b \leqslant x < \infty\}$. In this section the sets $\langle a, b)$ will be called half-open intervals even if $a = -\infty$, or $b = \infty$.

Remark on symbol ∞. The signs $-\infty$ and ∞ will be used for denoting the fact that the interval considered is unbounded on one side, and these signs will not denote any points of the line. Thus $\langle -\infty, a)$ and $(-\infty, a)$ denote the same half-line $\{x: -\infty < x < a\}$, and $\langle a, \infty \rangle$ and $\langle a, \infty)$ denote the same half-line $\{x: a \leqslant x < \infty\}$.

The class S_0 is a field of sets.

Indeed, a complement of a finite sum of half-open intervals of the form $\langle a, b)$ is again a sum of the same form. For instance, $(\langle 1, 4) \cup \langle 5, 7) \cup \langle 10, \infty))' = \langle -\infty, 1) \cup \langle 4, 5) \cup \langle 7, 10)$. Similarly, a sum of a finite number of such sums is again of this form. For instance, $(\langle 1, 2) \cup \langle 5, 8)) \cup \langle 3, 6) = \langle 1, 2) \cup \langle 3, 8)$.

Thus, if we had needed only a finitely additive field of events, we could take the class S_0 of finite sums of half-open intervals. This class, however, is only finitely additive and not σ-additive, as required by axiom (A2), though some of the sums of a countable number of events from S_0 are in S_0. For instance, S_0 contains the events $A_n = \langle 2n, 2n+1)$, but their sum $A = \bigcup_{n=1}^{\infty} A_n = \bigcup_{n=1}^{\infty} \langle 2n, 2n+1)$ is not a sum of a finite number of half-open intervals, and therefore is not in S_0. On the other hand, the sum of half-open intervals $\left\langle 1 - \frac{1}{n}, 1 - \frac{1}{n+1} \right)$,

$n = 1, 2, \ldots$, equals $\langle 0, 1)$ which is in S_0. Thus, if we want to construct a σ-additive field of events, we shall have to include in S_0 some other sets, which are not equal to sums of finitely many half-open intervals.

DEFINITION. As the field S of events, we take the smallest σ-additive field of events containing half-lines of the form $(-\infty, a)$. In analysis, such a field is called the *field of Borel sets*, and its elements are called *Borel sets*. Thus, we define events as Borel sets.

The question arises of the existence of such a smallest σ-field of events. The proof of the existence is as follows: first, there exists a σ-additive field of sets containing all half-lines of the form $(-\infty, a)$, namely the field of all subsets of the real line.

Second, if S_1 and S_2 are two σ-additive fields of sets, then their intersection $S_1 \cap S_2$, i.e. the class of all sets which belong to both S_1 and S_2, is also a σ-additive field of sets.

In fact, if $A \in S_1 \cap S_2$, then $A \in S_1$ and $A \in S_2$. Since S_1 and S_2 are fields, we have $A' \in S_1$ and $A' \in S_2$, hence $A' \in S_1 \cap S_2$. A similar reasoning extends to the case of sums. It remains to show that if sets $A_1, A_2, \ldots$ belong to $S_1 \cap S_2$, then their sum $A = A_1 \cup A_2 \cup \ldots$ also belongs to $S_1 \cap S_2$. But since sets $A_1, A_2, \ldots$ belong to the product of S_1 and S_2, they must belong to each of these fields separately. Therefore, the sum A belongs to both S_1 and S_2, whence also to $S_1 \cap S_2$. The above reasoning extends to the case of an arbitrary number of fields:

The intersection of an arbitrary number of σ-additive fields of sets is a σ-additive field of sets.

By an intersection of fields of sets we mean the class of sets which belong to all those fields.

The intersection of all σ-additive fields of sets which contain all half-lines of the form $(-\infty, a)$ is the required smallest σ-field of sets containing all half-lines $(-\infty, a)$.

The following theorems on the σ-field S of Borel sets on the real line will prove useful in the sequel:

THEOREM 1. *Open intervals* $(a, b) = \{x: a < x < b\}$ *are Borel sets*.

In fact, the interval (a, b) is the sum of the sequence of half-open intervals $\langle a + \frac{1}{n}, b)$,

$$(a, b) = \bigcup_{i=1}^{\infty} \langle a + \frac{1}{i}, b).$$

THEOREM 2. *One point sets are Borel sets.*

In fact, the one-point set $\{a\}$ equals the intersection of the sequence of half-open intervals $\langle a, a+\frac{1}{n})$:

$$\{a\} = \bigcup_{i=1}^{\infty} \langle a, a+\frac{1}{i}).$$

THEOREM 3. *Countable sets are Borel sets.*

In fact, countable sets are equal to countable unions of one-point sets.

THEOREM 4. *Open sets are Borel sets.*

In fact, every open set on the real line, as known from analysis, can be represented as the sum of at most countably many open intervals.

THEOREM 5. *Closed sets are Borel sets.*

Closed sets are complements of open sets.

§ 10. PROBABILITY ON S_0

We shall now try to define probability on events from the class S, i.e. on Borel sets on the real line; in other words, we shall try to define a real function on S which satisfies axioms (A3)–(A5). We shall proceed in two steps. First, we shall show how one can define probability on the field S_0 of finite sums of half-open intervals. Next, we shall show that the probability defined on S_0 can be extended to S in a unique way, provided we require axioms (A3)–(A5) to hold.

Let $F(x)$ be a real function on the line, satisfying the following properties:

(a) $F(x)$ is non-decreasing, i.e.

$$F(x_1) \leqslant F(x_2) \quad \text{if} \quad x_1 \leqslant x_2.$$

(b) $F(x)$ has the limit 0 in $-\infty$ and the limit 1 in $+\infty$, i.e.

$$\lim_{x \to -\infty} F(x) = 0, \quad \lim_{x \to +\infty} F(x) = 1.$$

(c) $F(x)$ is continuous from the left, i.e. for every a

$$\lim_{x \to a-} F(x) = F(a).$$

Using the function $F(x)$, we shall define a function $P(A)$ first on the class of half-open intervals, and then on S_0, and we shall show that $P(A)$ satisfies axioms (A3)–(A5), i.e. that it is non-negative, equal to 1 for the sure event, and σ-additive.

By definition, put

(10.1) $$P(\langle a, b \rangle) = F(b) - F(a),$$

if $a < b$.

Formula (10.1) is meaningful for finite a and b. To extend this definition to the case of infinite intervals, let us write formally the property (b) of function $F(x)$ in the form $F(-\infty) = 0$ and $F(+\infty) = 1$. Then we shall have

(10.2) $\quad P(\langle -\infty, a \rangle) = F(a), \quad P(\langle b, \infty \rangle) = 1 - F(b), \quad P(\langle -\infty, \infty \rangle) = 1.$

Equality (10.1) gives the values of the function P for half-open intervals. Before we extend this definition to other sets from the class S_0, we shall prove that the function P restricted to the class of half-open intervals has certain properties.

THEOREM 1. *If $A_1, ..., A_n$ are disjoint half-open intervals contained in the half-open interval A_0, then*

$$\sum_{i=1}^{n} P(A_i) \leqslant P(A_0).$$

PROOF. Let $A_i = \langle a_i, b_i \rangle$ and let $a_1 \leqslant a_2 \leqslant ... \leqslant a_n$. Since the intervals A_i, $i = 1, 2, ..., n$, are disjoint and contained in A_0, we have

$$a_0 \leqslant a_1 \leqslant b_1 \leqslant a_2 \leqslant ... \leqslant a_n \leqslant b_n \leqslant b_0.$$

Using the fact that the function $F(x)$ is monotone, we obtain

$$\sum_{i=1}^{n} P(A_i) = \sum_{i=1}^{n} (F(b_i) - F(a_i))$$

$$\leqslant \sum_{i=1}^{n} (F(b_i) - F(a_i)) + \sum_{i=1}^{n-1} (F(a_{i+1}) - F(b_i)) = F(b_n) - F(a_n)$$

$$\leqslant F(b_0) - F(a_0) = P(A_0).$$

THEOREM 2. *If a closed interval $A_0 = \langle a_0, b_0 \rangle$ is covered by a finite number of open intervals $E_i = (a_i, b_i)$, $i = 1, ..., n$, then*

$$F(b_0) - F(a_0) \leqslant \sum_{i=1}^{n} (F(b_i) - F(a_i)).$$

PROOF. Suppose first that a_0 and b_0 are finite. Let k_1 be the number of the interval which covers a_0. If $b_{k_1} \leqslant b_0$, let k_2 be the number of the interval which covers b_{k_1}; if $b_{k_2} \leqslant b_0$, let k_3 be the number of the interval which covers b_{k_3}, and so forth. We must finally arrive at k_m such that E_{k_m} covers b_0. Without loss of generality, assume that $m = n$ and $E_{k_i} = E_i$. This can be achieved by removing certain intervals and changing the numbering. In other words, we assume that $a_1 < a_0 < b_1$ and $a_n < b_0 < b_n$, and if $n > 1$, we assume that $a_{i+1} < b_i < b_{i+1}$, $i = 1, 2, \ldots, n-1$. Hence

$$F(b_0) - F(a_0) \leqslant F(b_n) - F(a_1)$$
$$= \bigl(F(b_n) - F(b_{n-1})\bigr) + \bigl(F(b_{n-1}) - F(b_{n-2})\bigr) + \ldots + \bigl(F(b_1) - F(a_1)\bigr)$$
$$\leqslant \sum_{i=1}^{n} \bigl(F(b_i) - F(a_i)\bigr).$$

If $a_0 = -\infty$, by an interval covering a_0 we mean any interval unbounded from the left. Instead of the inequality $a_1 < a_0$ we shall then have $a_1 = a_0 = -\infty$. If $b_0 = \infty$, by an interval covering b_0 we mean any interval unbounded from the right. Instead of the inequality $b_0 < b_n$ we shall then have $b_0 = b_n = \infty$. With these changes, the proof carries over to the case of unbounded intervals A_0.

THEOREM 3. *If $A_0, A_1, \ldots$ is a sequence of half-open intervals such that*

$$A_0 \subset \bigcup_{i=1}^{\infty} A_i,$$

then

(10.3) $$P(A_0) \leqslant \sum_{i=1}^{\infty} P(A_i).$$

PROOF. Let $A_i = \langle a_i, b_i \rangle$, $i = 0, 1, \ldots$ For $a_0 = b_0$, the theorem is trivially true. Suppose now that $-\infty < a_0 < b_0 < \infty$. Let us fix an arbitrary positive number $\varepsilon < b_0 - a_0$, and a positive number δ.

Consider the closed interval $F_0 = \langle a_0, b_0 - \varepsilon \rangle$ and open intervals $E_i = (a_i - \varepsilon_i, b_i)$, $i = 1, 2, \ldots$, where positive numbers ε_i are chosen in such a way that

$$F(a_i) - F(a_i - \varepsilon_i) \leqslant \delta/2^i.$$

This is possible because $F(x)$ is, by assumption, continuous from the left. Then

$$F_0 \subset \bigcup_{i=1}^{\infty} E_i,$$

and by the Heine–Borel theorem, known from analysis, we can find n such that

$$F_0 \subset \bigcup_{i=1}^{n} E_i.$$

By Theorem 2 we obtain

$$F(b_0-\varepsilon)-F(a_0) = \big(F(b_0)-F(a_0)\big)-\big(F(b_0)-F(b_0-\varepsilon)\big)$$

$$= P(A_0)-\big(F(b_0)-F(b_0-\varepsilon)\big) \leqslant \sum_{i=1}^{n}\big(F(b_i)-F(a_i-\varepsilon_i)\big)$$

$$\leqslant \sum_{i=1}^{n}\left(F(b_i)-F(a_i)+\frac{\delta}{2^i}\right) \leqslant \sum_{i=1}^{\infty}\big(F(b_i)-F(a_i)\big)+\delta,$$

whence

$$P(A_0)-\big(F(b_0)-F(b_0-\varepsilon)\big) \leqslant \sum_{i=1}^{\infty} P(A_i)+\delta.$$

Since $\varepsilon > 0$ and $\delta > 0$ are arbitrary, by the continuity of $F(x)$ from the left, the difference $F(b_0-\varepsilon)-F(b_0)$ tends to zero as $\varepsilon \to 0$, and we obtain the assertion of the theorem.

If $b_0 = \infty$, we take an arbitrary $\varepsilon > 0$, and instead of $b_0-\varepsilon$ we take $b' > a_0$ such that $F(b_0)-F(b') = 1-F(b') < \varepsilon$, which is possible because of property (b) of function $F(x)$. Then we apply the above reasoning to the interval $F_0 = \langle a_0, b' \rangle$. In this manner we obtain

$$P(A_0)-\varepsilon = \big(1-F(a_0)\big)-\varepsilon$$

$$\leqslant \big(1-F(a_0)\big)-\big(1-F(b')\big) = F(b')-F(a_0)$$

$$\leqslant \sum_{i=1}^{\infty}\big(F(b_i)-F(a_i)\big)+\delta = \sum_{i=1}^{\infty} P(A_i)+\delta.$$

Again, in view of the fact that ε and δ are arbitrary, the inequality (10.3) follows.

Finally, if $a_0 = -\infty$, we take an arbitrary $\eta > 0$ and find such an $a' < b_0$ that $F(a') < \eta$, and apply the above reasoning to the interval $\langle a', b_0 \rangle$. This leads to the inequality

$$P(A_0)-\varepsilon-\eta \leqslant \sum_{i=1}^{\infty} P(A_i)+\delta,$$

valid for arbitrary ε, η and δ; this proves the assertion (10.3) for $a_0 = -\infty$, and completes the proof of the theorem.

THEOREM 4. *Function P defined for half-open intervals by formula (10.1) is countably additive in the class of half-open intervals.*

PROOF. Indeed, let $A_1, A_2, \ldots$ be a sequence of disjoint half-open intervals, such that their sum $A = A_1 \cup A_2 \cup \ldots$ is also a half-open interval. By Theorem 1, we have for every n

$$\sum_{i=1}^{\infty} P(A_i) \leqslant P(A).$$

Therefore

$$\sum_{i=1}^{\infty} P(A_i) \leqslant P(A),$$

while by Theorem 3, we have the reverse inequality

$$P(A) \leqslant \sum_{i=1}^{\infty} P(A_i).$$

It follows that

$$P(A) = \sum_{i=1}^{\infty} P(A_i).$$

Until now, we discussed the properties of function $P(A)$ defined by (10.1) on half-open intervals. We shall now prove

THEOREM 5. *Function P defined on half-open intervals by formula* (10.1) *can be extended in a unique way to the field S_0 of finite sums of half-open intervals. The function so extended is countably additive on S_0.*

PROOF. By definition, every element of S_0 can be represented as the sum of a finite number of disjoint half-open intervals on which the function P is defined by (10.1). If

$$A = A_1 \cup \ldots \cup A_n$$

is such a representation, i.e. if $A_1, \ldots, A_n$ are disjoint half-open intervals, then if we want to preserve the additivity property, we must put

$$P(A) = P(A_1) + \ldots + P(A_n).$$

Such a representation, however, is not unique. For instance

$$\langle 1, 3) \cup \langle 5, 9) = \langle 1, 2) \cup \langle 2, 3) \cup \langle 5, 8) \cup \langle 8, 9).$$

If

$$A = E_1 \cup \ldots \cup E_m$$

is another representation of the set A, then, in accordance with the additivity property, we would also have

$$P(A) = P(E_1) + \ldots + P(E_m).$$

The question arises whether such definitions are consistent, i.e. whether in such cases
$$P(A_1)+ \ldots +P(A_n) = P(E_1)+ \ldots +P(E_m).$$
The answer to this question is affirmative, which follows from the additivity of function P in the class of half-open intervals. Indeed, for every $i = 1, 2, \ldots, n$ we can write
$$A_i = \bigcup_{j=1}^{m} (A_i \cap E_j).$$
This is a representation of the half-open interval A_i as the sum of m disjoint half-open intervals. We have therefore
$$P(A_i) = \sum_{j=1}^{m} P(A_i \cap E_j),$$
whence
$$\sum_{i=1}^{n} P(A_i) = \sum_{i=1}^{n} \sum_{j=1}^{m} P(A_i \cap E_j).$$
In a similar way we can write
$$E_j = \bigcup_{i=1}^{n} (A_i \cap E_j),$$
whence also
$$\sum_{j=1}^{m} P(E_i) = \sum_{j=1}^{m} \sum_{i=1}^{n} P(A_i \cap E_j).$$
The above double sums differ only in the order of summation, whence they are equal to one another. This proves that the definition of function P on elements of S_0 according to the additivity postulate is unique. It is also evident that the function so defined is additive on S_0. We shall show that it is also countably additive.

Let $A_1, A_2, \ldots$ be disjoint elements of S_0 such that their sum $A_0 = A_1 \cup A_2 \cup \ldots$ is again an element of S_0. Each of the sets A_i, $i = 0, 1, 2, \ldots$, can be represented as a sum of a finite number of disjoint half-open intervals
$$A_i = \bigcup_{j=1}^{m_i} E_{ij}.$$
Half-open intervals E_{ij}, $i = 1, 2, \ldots$, are disjoint, and there are countably many of them. Every interval E_{0i}, $i = 1, 2, \ldots, m_0$, whose sum equals A_0, is itself a sum of a finite or countable number of them. Let us denote by E_{ij}^* those in-

tervals whose sum is E_{0i}. In view of countable additivity in the class of half-open intervals, we have

$$P(E_{0i}) = \sum_j P(E_{ij}^*).$$

Since P is additive on S_0, it follows that

$$P(A) = \sum_{i=1}^{m_0} P(E_{0i}) = \sum_{i=1}^{m_0} \sum_j P(E_{ij}^*).$$

After changing the order of summation in the last double sum, we obtain

$$P(A) = \sum_{i=1}^{\infty} \sum_{j=1}^{m_i} P(E_{ij}) = \sum_{i=1}^{\infty} P(A_i),$$

which shows that P is countably additive in the field S_0. The proof of Theorem 5 is completed.

§ 11. THEOREM ON EXTENSION OF MEASURES

In the last section we showed how to construct a function which satisfies axioms (A3)–(A5); this function, however, was defined on a field of events which was not countably additive. We showed that this function is countably additive in that field. The question arises whether we can extend this function to other sets, so as to obtain a countably additive function defined on a σ-field? The answer to this question is in the affirmative. We shall obtain it from the *general theorem on extension of measures*, which we shall prove in this section.

THEOREM 1. *Suppose that S_0 is a field of subsets of a space X, and that $P(A)$ is a finite measure defined on S_0, i.e. that P is non-negative, satisfies the condition*

(11.1) $$P(X) < \infty,$$

and

(11.2) $$P(A_1 \cup A_2 \cup \ldots) = P(A_1) + P(A_2) + \ldots,$$

whenever sets $A_1, A_2, \ldots$ are disjoint and belong to S_0, and their sum also belongs to S_0. Then P can be extended to a countably additive function on the smallest σ-field S containing S_0, and the extension is unique.

If, in addition, $P(X) = 1$, then the extended function satisfies all axioms of probability.

PROOF. The proof will consist of several steps. First, using function $P(A)$ we shall construct another function $P^*(A)$ defined on the field of all subsets of the space X. Next, using $P^*(A)$, we shall define a certain class of subsets of the space X, and we shall show that this class is a σ-field containing S_0. Then we shall show that the function P^*, when considered on this class only, is countably additive. Function P^* restricted to the smallest σ-field containing S_0 will give the desired extension.

DEFINITION 1. For every subset E of the space X let $P^*(E)$ denote the greatest lower bound of the sums

$$P(A_1)+P(A_2)+\ldots,$$

where $A_1, A_2, \ldots$ are sets from the field S_0, such that their sum covers the set E. In a more formal way, we may write

$$P^*(E) = \inf\left\{\sum_{i=1}^{\infty} P(A_i): A_i \in S_0, E \subset \bigcup_{i=1}^{\infty} A_i\right\}.$$

The function P^* is sometimes called the *outer measure* induced by P.

The following properties of the function P^* will be used in the sequel.

(a) *The function P^* is an extension of P, that is, if $A \in S_0$, then*

(11.3) $$P^*(A) = P(A).$$

In fact, as

$$A \subset A \cup 0 \cup 0 \cup \ldots,$$

we have

$$P^*(A) \leqslant P(A)+P(0)+P(0)+\ldots = P(A).$$

On the other hand, if $A, A_1, A_2, \ldots$ are elements of S_0, and $A \subset A_1 \cup A_2 \cup \ldots$, then by Theorem 2 of § 6 we have

$$P(A) \leqslant P(A_1)+P(A_2)+\ldots,$$

or

$$P(A) \leqslant P^*(A).$$

This completes the proof of equality (11.3).

The property (a) implies in particular that $P^*(0) = 0$.

(b) *If A and B are subsets of X such that $A \subset B$, then*

$$P^*(A) \leqslant P^*(B).$$

Indeed, as every sum of sets from S_0 which covers B must also cover A, we have

$$\left\{\sum_{i=1}^{\infty} P(A_i): A_i \in S_0, B \subset \bigcup_{i=1}^{\infty} A_i\right\} \subset \left\{\sum_{i=1}^{\infty} P(A_i): A_i \in S_0, A \subset \bigcup_{i=1}^{\infty} A_i\right\}.$$

Since the greatest lower bound of a larger set of numbers can only be smaller, we obtain $P^*(A) \leqslant P^*(B)$.

(c) *If $A, A_1, A_2, \ldots$ are subsets of X such that*

(11.4) $$A = A_1 \cup A_2 \cup \ldots,$$

then

(11.5) $$P^*(A) \leqslant P^*(A_1) + P^*(A_2) + \ldots$$

Indeed, condition (11.4) implies the inclusion

$$A \subset A_1 \cup A_2 \cup A_3 \cup \ldots$$

Let us fix $\varepsilon > 0$. Let

$$A_i \subset E_{i1} \cup E_{i2} \cup \ldots,$$

where E_{ij} are sets in S_0 such that

(11.6) $$P(E_{i1}) + P(E_{i2}) + \ldots \leqslant P^*(A_i) + \frac{\varepsilon}{2^i}.$$

Such sets exist by the definition of function P^*. There are countably many sets E_{ij} and their sum covers A. Thus, in view of inequality (11.6) we get

$$P^*(A) \leqslant \sum_{i=1}^{\infty} \sum_{j=1}^{\infty} P(E_{ij}) \leqslant \sum_{i=1}^{\infty} \left(P^*(A_i) + \frac{\varepsilon}{2^i} \right),$$

whence

$$P^*(A) \leqslant \varepsilon + \sum_{i=1}^{\infty} P^*(A_i).$$

Since $\varepsilon > 0$ is arbitrary, inequality (11.5) follows.

We shall now define, using the function P^*, the class of measurable sets.

DEFINITION 2. A subset A of the space X will be called *measurable with respect to the outer measure P^* induced by P*, or shortly, *measurable*, if for every set E of space X

$$P^*(E) = P^*(E \cap A) + P^*(E \cap A').$$

The class of measurable sets will be denoted by $\overline{S}$.

This definition is somewhat artificial, and it is difficult to show a natural way leading to it. We use this definition, because it gives a simple proof of our theorem. We shall now prove some properties of measurable sets.

(d) *Complements of measurable sets are measurable sets*, i.e. if $A \in \overline{S}$, then $A' \in \overline{S}$.

We must prove that if for every set E of the space X we have
$$P^*(E) = P^*(E \cap A) + P^*(E \cap A'),$$
then also
$$P^*(E) = P^*(E \cap A') + P^*(E \cap (A')').$$
Since $(A')' = A$ (see equality (5.5)), the right-hand sides of these equalities are equal, which proves property (d).

(e) *Sums of measurable sets are measurable*, i.e. if $A \in \overline{S}$, and $B \in \overline{S}$, then $A \cup B \in \overline{S}$.

Let E be an arbitrary subset of the space X. Since A is measurable, we have

(11.7) $\qquad P^*(E) = P^*(E \cap A) + P^*(E \cap A').$

Next, since B is measurable, we get

(11.8) $\qquad P^*(E \cap A) = P^*(E \cap A \cap B) + P^*(E \cap A \cap B'),$

(11.9) $\qquad P^*(E \cap A') = P^*(E \cap A' \cap B) + P^*(E \cap A' \cap B').$

Substituting (11.8) and (11.9) to (11.7), we obtain

(11.10) $\qquad P^*(E) = P^*(E \cap A \cap B) + P^*(E \cap A \cap B')$
$$+ P^*(E \cap A' \cap B) + P^*(E \cap A' \cap B').$$

This formula is valid for every subset E of the space X. Now, let us take instead of E the set $E \cap (A \cup B)$. Since

$$(E \cap (A \cup B)) \cap A \cap B = E \cap A \cap B,$$
$$(E \cap (A \cup B)) \cap A \cap B' = E \cap A \cap B',$$
$$(E \cap (A \cup B)) \cap A' \cap B = E \cap A' \cap B,$$
$$(E \cap (A \cup B)) \cap A' \cap B' = 0,$$
$$P^*(0) = 0,$$

we obtain

(11.11) $\quad P^*(E \cap (A \cup B)) = P^*(E \cap A \cap B) + P^*(E \cap A \cap B') + P^*(E \cap A' \cap B),$

where the right-hand side coincides with the first three terms of the right-hand side of (11.10). Next, since $(A \cup B)' = A' \cap B'$, we have

$$P^*(E \cap (A \cup B)') = P^*(E \cap A' \cap B').$$

The last equality and formulas (11.10) and (11.11) imply that

$$P^*(E) = P^*(E \cap (A \cup B)) + P^*(E \cap (A \cup B)'),$$

which proves that the sum $A \cup B$ is measurable.

(f) *The function P^* is additive in the class of measurable sets*, i.e. *if measurable sets A and B are disjoint, then*

$$P^*(A \cup B) = P^*(A) + P^*(B).$$

If the sets A and B are disjoint, we get $E \cap A \cap B = 0$ for any set E. Equality (11.11) yields in this case

$$P^*\big(E \cap (A \cup B)\big) = P^*(E \cap A \cap B') + P^*(E \cap A' \cap B).$$

Substituting $A \cup B$ in place of E and using the equalities

$$(A \cup B) \cap (A \cup B) = A \cup B,$$
$$(A \cup B) \cap A \cap B' = A,$$
$$(A \cup B) \cap A' \cap B = B,$$

valid for disjoint sets, we obtain the assertion of theorem (f).

By induction, we infer from (e) that:

(e_1) *Sums of finite numbers of measurable sets are measurable.*

We shall prove that

(g) *Sums of sequences of measurable sets are measurable*, i.e. *if $A_i \in \overline{S}$, $i = 1, 2, \ldots$, and $A = A_1 \cup A_2 \cup \ldots$, then $A \in \overline{S}$.*

(h) *The function P^* is countably additive in the class of measurable sets, that is, if measurable sets $A_1, A_2, \ldots$ are disjoint, and $A = A_1 \cup A_2 \cup \ldots$, then*

$$P^*(A) = P^*(A_1) + P^*(A_2) + \ldots$$

To prove (g) we show that for any subset E of the space X we have

$$P^*(E) = P^*(E \cap A) + P^*(E \cap A').$$

In view of property (c) of function P^*, we obtain

$$P^*(E) \leqslant P^*(E \cap A) + P^*(E \cap A').$$

Thus, it suffices to prove that

(11.12) $$P^*(E) \geqslant P^*(E \cap A) + P^*(E \cap A').$$

Suppose first that the sets $A_1, A_2, \ldots$ are disjoint.

From equation (11.11) we obtain for disjoint measurable sets A and B the equality

$$P^*\big(E \cap (A \cup B)\big) = P^*(E \cap A) + P^*(E \cap B)$$

(since in this case $A \cap B = 0$, $A \cap B' = A$, $A' \cap B = B$). By induction, this inequality can be generalized to the case of any finite number of disjoint measurable terms; thus, if measurable sets $A_1, \ldots, A_n$ are disjoint, we have for any subset E of the space X:

(11.13) $$P^*(E\cap \bigcup_{i=1}^{n} A_i) = \sum_{i=1}^{n} P^*(E\cap A_i).$$

By property (e_1) a sum of a finite number measurable sets is a measurable set. Thus, the sets

$$F_n = A_1 \cup A_2 \cup \ldots \cup A_n$$

are also measurable. Using the last equality and (11.13), and then the fact that $E \cap F_n' \supset E \cap A'$ combined with property (b), we obtain

$$P^*(E) = P^*(E\cap F_n) + P^*(E\cap F_n') \geqslant P^*(E\cap F_n) + P^*(E\cap A')$$

$$= \sum_{i=1}^{n} P^*(E\cap A_i) + P^*(E\cap A').$$

Since the last equality holds for every n, we get

(11.14) $$P^*(E) \geqslant \sum_{i=1}^{\infty} P^*(E\cap A_i) + P^*(E\cap A').$$

Let us now apply property (c) to sets $E\cap A_1, E\cap A_2, \ldots$ whose sum is obviously equal to $E\cap A$. We obtain

$$P^*(E) \geqslant P^*(E\cap A) + P^*(E\cap A'),$$

which proves the measurability of a sum of a sequence of measurable sets.

Next, every sum of measurable sets can always be represented in the form of a sum of disjoint measurable sets.

Indeed, instead of sets $A_1, A_2, \ldots$ it suffices to take the sets $E_i = A_i - (A_1 \cup \ldots \cup A_{i-1})$ (see Theorem 5 of § 5). By theorems (d) and (e_1) and properties of operations on events, known from § 5, disjoint sets E_i are measurable, provided the same holds for sets A_i.

Now, let us consider inequality (11.14) for disjoint measurable sets A_i, and let us substitute their sum A for E. In view of the equality $A\cap A_i = A_i$, we obtain

$$P^*(A) \geqslant \sum_{i=1}^{\infty} P^*(A_i).$$

Using (c) we get the assertion of theorem (h).

(i) *The field S_0 is contained in $\overline{S}$.*

We have to prove that if $A \in S_0$, then for every subset E of the space X

$$P^*(E) = P^*(E\cap A) + P^*(E\cap A').$$

II. PROBABILITIES OF ELEMENTARY EVENTS

In view of theorem (c) it suffices to show that

(11.15) $$P^*(E) \geq P^*(E \cap A) + P^*(E \cap A').$$

Let us fix an arbitrary positive number ε. Let $A_1, A_2, \ldots$ be a sequence of sets from S_0 such that

$$E \subset A_1 \cup A_2 \cup \ldots = \bigcup_{i=1}^{\infty} A_i$$

and

(11.16) $$\sum_{i=1}^{\infty} P(A_i) \leq P^*(E) + \varepsilon.$$

Since the set A belongs to S_0, we have for every index i

$$P(A_i) = P(A_i \cap A) + P(A_i \cap A'),$$

whence

$$\sum_{i=1}^{n} P(A_i) = \sum_{i=1}^{n} P(A_i \cap A) + \sum_{i=1}^{n} P(A_i \cap A').$$

Therefore, using property (a) we obtain

(11.17) $$\sum_{i=1}^{\infty} P(A_i) = \sum_{i=1}^{\infty} P(A_i \cap A) + \sum_{i=1}^{\infty} P(A_i \cap A')$$

$$= \sum_{i=1}^{\infty} P^*(A_i \cap A) + \sum_{i=1}^{\infty} P^*(A_i \cap A').$$

Applying theorem (c) to the sets $A_i \cap A$ and to the sets $A_i \cap A'$, we obtain

(11.18) $$\sum_{i=1}^{\infty} P^*(A_i \cap A) + \sum_{i=1}^{\infty} P^*(A_i \cap A') \geq P^*\left(\bigcup_{i=1}^{\infty} A_i \cap A\right) + P^*\left(\bigcup_{i=1}^{\infty} A_i \cap A'\right).$$

Next, in view of the relation $E \subset \bigcup_{i=1}^{\infty} A_i$ and theorem (b) we have

(11.19) $$P^*\left(\bigcup_{i=1}^{\infty} A_i \cap A\right) + P^*\left(\bigcup_{i=1}^{\infty} A_i \cap A'\right) \geq P^*(E \cap A) + P^*(E \cap A').$$

Combining inequalities (11.16)–(11.19) we finally obtain for every positive ε

$$P^*(E) + \varepsilon \geq P^*(E \cap A) + P^*(E \cap A'),$$

which proves inequality (11.15). Theorem (i) is therefore proved.

Properties (d) and (e) imply that the class $\overline{S}$ of measurable sets is a field of sets. Theorem (g) shows that this field is countably additive, and theorem (i) asserts that it contains the field S_0. Thus, it must also contain the smallest countably additive field containing S_0. From property (a) it follows that the function P^* is an extension of the function P, and (f) and (h) imply that P^* is countably additive on the σ-field $\overline{S}$. Since the σ-field S is contained in the σ-field $\overline{S}$, the function P^* is σ-additive on the σ-field S. Thus, the function P^* restricted to S is an extension of the measure P to the smallest σ-field S containing S_0.

It remains to show that this extension is unique. Suppose that $P_1(A)$ and $P_2(A)$ are two countably additive functions defined on S such that $P_1(A) = P_2(A)$ for $A \in S_0$. Let M denote the class of sets for which $P_1(A) = P_2(A)$. The class M contains, by definition, the field S_0. To complete the proof, it suffices to show that it contains the field S; in other words, it suffices to show that $S \subset M$. This property will be proved by the following lemmas:

(j) *The complement of every set in M belongs to M, that is, if $A \in M$, then $A' \in M$.*

In fact, by the additivity of functions P_1 and P_2, we have

$$P_1(A) + P_1(A') = P_1(X)$$

and

$$P_2(A) + P_2(A') = P_2(X).$$

In view of the equality $P_1(A) = P_2(A)$ we obtain $P_1(A') = P_2(A')$, which proves the lemma.

(k) *If $A_1, A_2, \ldots$ are in M and $A_1 \subset A_2 \subset \ldots$, then their sum $A = A_1 \cup A_2 \ldots$ also belongs to M.*

Since $A_1, A_2, \ldots$ is an increasing sequence of sets in S, by Theorem 3 of § 6 we get

$$P_1(A) = \lim_{n \to \infty} P_1(A_n),$$

and

$$P_2(A) = \lim_{n \to \infty} P_2(A_n).$$

By assumption, $P_1(A_n) = P_2(A_n)$ for $n = 1, 2, \ldots$, whence also $P_1(A) = P_2(A)$, which shows that $A \in M$.

Denote by M_0 the smallest class of sets which contains S_0 and has the property that the complements of sets from this class belong to this class, and the sums of increasing sequences of sets from this class belong to it. (A sequence of sets is called increasing if $A_1 \subset A_2 \subset \ldots$)

Clearly, $M_0 \subset M$.

To complete the proof, it suffices to show that

(l) *M_0 is a countably additive field of sets.*

II. PROBABILITIES OF ELEMENTARY EVENTS

Indeed, as M_0 contains S_0 by definition, and S is the smallest σ-field containing S_0, we obtain $S \subset M_0$, and in view of the inclusion $M_0 \subset M$, we obtain $S \subset M$.

To prove lemma (l) we shall show that

(m) *Class M_0 contains the sums of every two of its sets.*

For a given set E denote by $K(E)$ the class of all sets A such that $A \cup E$ belongs to M_0.

Note first that

(n) *For every set E, the class $K(E)$ contains the sums of increasing sequences of sets from this class.*

Indeed, if sets $A_1, A_2, \ldots$ belong to $K(E)$ and $A_1 \subset A_2 \subset \ldots$, then the sets $A_1 \cup E, A_2 \cup E, \ldots$ belong to M_0, and $(A_1 \cup E) \subset (A_2 \cup E) \subset \ldots$ By definition of M_0, their sum belongs to M_0. But

$$\bigcup_{i=1}^{\infty}(A_i \cup E) = \left(\bigcup_{i=1}^{\infty} A_i\right) \cup E,$$

whence $\bigcup_{i=1}^{\infty} A_i$ belongs to $K(E)$.

Next, since the sets A and E appearing in the definition of the class $K(E)$ play symmetric roles, the relations $A \in K(E)$ and $E \in K(A)$ are equivalent.

Now, if $E \in S_0$, then $K(E)$ contains all sets of S_0. In fact, if $A \in S_0$, then also $A \cup E \in S_0$, and the functions P_1 and P_2 are equal on S_0 by assumption.

Next, if $E \in M_0$, then the class $K(E)$ is complementative. In fact, if $A \cup E \in M_0$ and $E \in M_0$, then $P_1(A \cup E) = P_2(A \cup E)$ and $P_1(E) = P_2(E)$. Since $M_0 \subset M \subset S$, and $A \cup E = E \cup (A \cap E')$, where the sets E and $A \cap E'$ are disjoint, we get $P_i(A \cup E) = P_i(E) + P_i(A \cap E')$, $i = 1, 2$, whence $P_1(A \cap E') = P_2(A \cap E')$. We can also write

(11.20) $$A' \cup E = (A \cap E')'.$$

We have already proved that $A \cap E'$ belongs to M_0. In view of the last equality, and the definition of M_0, the set $A' \cup E$ also belongs to M_0, which shows that $A' \in K(E)$.

As we saw, for $E \in S_0$, the field S_0 is contained in the class $K(E)$, and the class $K(E)$ contains the complements of its sets and the sums of increasing sequences of its sets; thus for $E \in S_0$ we get $M_0 \subset K(E)$. This means that if $E \in S_0$ and $A \in M_0$, then $A \in K(E)$. But then also $E \in K(A)$. Thus, for $A \in S_0$ and $E \in M_0$, we have $A \in K(E)$, or, for $E \in M_0$ the class $K(E)$ contains S_0. Thus, it must also contain M_0 since, as we know, for $E \in M_0$ the class $K(E)$ contains the complements of its sets and the sums of increasing sequences of its sets. For any two sets A and B from M_0 we have therefore $A \in K(B)$, which means that $A \cup B \in M_0$. This proves lemma (m).

By definition, M_0 contains the complements of its sets, and by lemma (m) it contains the sums of every two of its sets, whence M_0 is a field of sets. Moreover, by definition, M_0 contains the sums of increasing sequences of its sets. We shall infer from the above that M_0 is a σ-additive field of sets. Indeed, let $A_1, A_2, \ldots$ be an arbitrary sequence of sets in M_0. Then the sets $E_i = A_1 \cup \cup A_2 \cup \ldots \cup A_i$, $i = 1, 2, \ldots$, are also in M_0, since M_0 is a field. Moreover, $E_1 \subset E_2 \subset \ldots$ Thus, M_0 contains the sum $\bigcup_{i=1}^{\infty} E_i$. By the definition of sets E_i we have, however, $\bigcup_{i=1}^{\infty} E_i = \bigcup_{i=1}^{\infty} A_i$. Thus, the sum $\bigcup_{i=1}^{\infty} A_i$ belongs to M_0, which proves lemma (n).

This completes the proof of the first part of the theorem. Finally we show that if $P(X) = 1$, then the function P^* restricted to the field S satisfies all axioms of probability.

Indeed, axioms (A1) and (A2) are satisfied, since S is a σ-field of sets. Axiom (A3) is satisfied, since the function P^* satisfies the inequality $0 \leqslant P^*(A) \leqslant 1$ for all sets A in the space X. The first inequality follows from the fact that P^* is, by definition, the greatest lower bound of non-negative values. The inequality $P^*(A) \leqslant 1$ follows from the fact that every set A in space X is covered by the sum of the sequence of sets $X, 0, 0, \ldots$ belonging to S_0; for this sequence we have $P(X) + P(0) + P(0) + \ldots = 1$. Finally, lemma (h) states that P^* restricted to S is countably additive.

The proof of the theorem is now complete.

§ 12. PROBABILITY DISTRIBUTIONS ON THE REAL LINE

Now we are in a position to prove that one can define a function satisfying all axioms of probability on the field of Borel sets of the real line. We shall prove a somewhat stronger theorem, namely we shall characterize the class of all functions defined on the field of Borel sets of the real line, and satisfying the axioms of probability.

In the sequel, it will be convenient to distinguish probability as a function on the field of events from probability as the value of this function on a particular event. Thus, with regard to the first of these meanings, we shall speak of probability distributions. More precisely, we shall accept the following:

DEFINITION 1. A function $P(A)$ defined on the field S of events in the space X, satisfying all axioms of probability, will be called a *probability distribution* in the space X, or shortly, a *probability distribution*, whenever it is clear from the context what is the underlying space X.

In particular, we shall speak of probability distributions on the real line, meaning functions defined on the Borel class of sets of the real line satisfying axioms (A1)–(A5).

In this section we shall prove two theorems.

THEOREM 1. *To every function $F(x)$ satisfying conditions* (a), (b), (c) *of § 10 there corresponds exactly one probability distribution $P(A)$ on the real line such that*

(12.1) $$P((-\infty, a)) = F(a).$$

THEOREM 2. *If $P(A)$ is a probability distribution on the real line, then the function $F(x)$, defined as*

(12.2) $$F(x) = P((-\infty, x)),$$

satisfies conditions (a), (b) *and* (c) *of § 10*.

These theorems show, first of all, that there exist probability distributions on the real line, and that there is a one-to-one correspondence between probability distributions on the real line and functions satisfying conditions (a), (b), (c) of § 10. This justifies the following definition:

DEFINITION 2. For a given probability distribution $P(A)$ on the real line, the function $F(x)$ defined by (12.2) will be called the *distribution function of the distribution $P(A)$*.

Proof of Theorem 1 is contained in the results of the preceding sections. In § 10 we showed how for an arbitrary function $F(x)$ satisfying conditions (a), (b), (c) of § 10 one can define, on the field S_0 of finite sums of half-open intervals, a non-negative function $P(A)$, bounded by one and equal to one for $A = X$, countably additive on S_0 and satisfying (12.1). Thus, this function satisfies all the assumptions of the theorem on extension of measures given in § 11, and therefore, it can be extended in a unique way to the smallest countably additive field S containing S_0. As all half-lines $(-\infty, a)$ are in S_0, the field S is equal to the field of Borel sets of the real line. By theorem of § 11, this extended function gives the desired probability distribution.

PROOF OF THEOREM 2. Theorem 2 is a simple consequence of the properties of probability distributions on the real line. Thus, let $P(A)$ be a probability distribution on the real line, and define $F(x)$ by formula (12.2). We have to show that $F(x)$ satisfies conditions (a), (b), (c) of § 10.

Function $F(x)$ is non-decreasing. Indeed, as for $x_1 < x_2$ we have $(-\infty, x_1) \cup \langle x_1, x_2 \rangle = (-\infty, x_2)$, by additivity and by the fact that probability is non-negative, we get

$$F(x_2) = P((-\infty, x_2)) = P((-\infty, x_1)) + P(\langle x_1, x_2 \rangle)$$
$$\geqslant P((-\infty, x_1)) = F(x_1).$$

Next, if $x_1 < x_2 < \ldots$ is a sequence increasing to infinity, we can represent the whole real line as the sum of an increasing sequence of events

$$(-\infty, \infty) = \bigcup_{i=1}^{\infty} (-\infty, x_i).$$

By Theorem 4 of § 6 we obtain therefore

$$\lim_{n \to \infty} F(x_n) = 1.$$

Similarly, if $x_1 > x_2 > x_3 > \ldots$ is a sequence decreasing to $-\infty$, then $(-\infty, x_1), (-\infty, x_2), \ldots$ is a decreasing sequence of sets with an empty intersection. Thus, in view of Theorem 6 of § 6, we get

$$\lim_{n \to \infty} F(x_n) = 0.$$

This proves that $\lim_{x \to -\infty} F(x) = 0$.

Finally, if $x_1 < x_2 < \ldots$ is a sequence increasing to the limit x, then $(-\infty, x)$ is equal to the sum of the increasing sequence of sets $(-\infty, x_1), (-\infty, x_2), \ldots$, whence by Theorem 3 of § 6 we have $F(x) = \lim_{n \to \infty} F(x_n)$. This shows that $F(x)$ is continuous from the left, and completes the proof of Theorem 2.

§ 13. PROBABILITY DISTRIBUTIONS IN MULTIDIMENSIONAL EUCLIDEAN SPACES

It often happens that we must take as the space of elementary events a multidimensional Euclidean space, or a subset of it. We already encountered such situations when we considered the age of couples. In this example, each observation yielded a pair of numbers, the age of the husband and the age of the wife. Another example is provided by studying samples of products with respect to several characteristics. Thus, the result of studying a sample of corn may be expressed by, say, two numbers, one giving the dampness of the corn, and the other giving the mass of a thousand corns. These two numbers represent a point in a two-dimensional Euclidean space. It is natural, therefore, that we shall study methods of defining probability distributions in such Euclidean spaces.

Since all essential differences as compared with the case of real line are already apparent in the case of the Euclidean plane, we shall restrict the discussion to the case of two dimensions only. The passage to a large number of dimensions does not require any significant changes.

Thus, suppose that the space of elementary events is the plane R_2, that is, the set of pairs (x, y) of real numbers.

The first question to arise is this: what should we take as the class of events? It seems natural to include in the class of events all sets of the form

(13.1) $$\{(x, y): x < a, y < b\},$$

where a and b are arbitrary real numbers. These sets, however, do not constitute even a field, while we want the class of events to be a σ-field. We accept therefore the following definition:

DEFINITION 1. As the *field S of events in the plane* R_2 we take the smallest countably additive class of sets which contains all sets of the form (13.1).

As known from analysis, S is the class of Borel sets on the plane. Class S contains not only the rectangles (the requirement that rectangles should be in S is as natural as the requirement that all sets (13.1) should be in S), but also all open and closed sets on the plane. One could define S in an equivalent way, by requiring that S should be the smallest countably additive field containing all open sets.

The second question to arise is: how can one define functions P satisfying axioms (A3)–(A5) on the class S of Borel sets of the plane? The answer can be found in a way analogous to that in the case of the real line. In a similar way, one can also characterize the class of all possible probability distributions on the plane in terms of two-dimensional distribution functions.

We shall formulate the corresponding theorems and sketch their proofs, trying to point out the differences as compared with the case of the real line considered in § 12.

Let $F(x, y)$ be a function of two variables satisfying the following conditions:

(a_2) The function $F(x, y)$ is non-decreasing in two variables, that is, for any two points (x_1, y_1) and (x_2, y_2) such that $x_1 \leqslant x_2$ and $y_1 \leqslant y_2$ we have

$$F(x_2, y_2) - F(x_1, y_2) - F(x_2, y_1) + F(x_1, y_1) \geqslant 0.$$

(b_2) For every x_0 and for every y_0 we have

$$\lim_{y \to -\infty} F(x_0, y) = 0, \quad \lim_{x \to -\infty} F(x, y_0) = 0.$$

Moreover,

$$\lim_{\substack{x \to \infty \\ y \to \infty}} F(x, y) = 1.$$

(c_2) The function $F(x, y)$ is continuous from the left in two variables, i.e. if $x_1 \leqslant x_2 \leqslant \ldots$ is a sequence tending to x, and $y_1 \leqslant y_2 \leqslant \ldots$ is a sequence tending to y, then

$$\lim_{n \to \infty} F(x_n, y_n) = F(x, y),$$

where the last relation holds if x or y equals $+\infty$.

We have the following

THEOREM 1. *To every function $F(x, y)$ satisfying conditions (a_2), (b_2) and (c_2), there corresponds exactly one probability distribution $P(A)$ on the plane such that*

(13.2) $P(\{(x, y): x < a, y < b\}) = F(a, b).$

The proof of Theorem 1 is similar to the proof of Theorem 1 of § 12. First, using the function $F(x, y)$, we define the value of $P(A)$ on half-open rectangles of the form

(13.3) $\{(x, y): a_1 \leqslant x < a_2, b_1 \leqslant y < b_2\},$

where $a_1 \leqslant a_2$ and $b_1 \leqslant b_2$ are arbitrary real numbers. By definition, we put

(13.4) $P(\{(x, y): a_1 \leqslant x < a_2, b_1 \leqslant y < b_2\})$
$= F(a_2, b_2) - F(a_1, b_2) - F(a_2, b_1) + F(a_1, b_1).$

In view of property (a_2) of the function $F(x, y)$, the numbers assigned to the rectangles by formula (13.4) are non-negative. We shall extend definition (13.4) also to the case of infinite half-open rectangles, by admitting the possibility that a_1 and b_1 may assume value $-\infty$, and a_2 and b_2 may assume value $+\infty$; here we shall treat the inequalities $-\infty \leqslant x$ and $-\infty < x$ as equivalent. We shall speak of infinite rectangles, meaning the sets of the form (13.3) when one of the values a_1, a_2, b_1, b_2 is "improper", that is, equal to $\pm \infty$. In particular, $R_2 = \{(x, y): -\infty \leqslant x < \infty, -\infty \leqslant y < \infty\}$. In order to extend the definition (13.4) to the case of infinite rectangles, let us agree to write for every x and y

$F(x, -\infty)$	instead of	$\lim_{y \to -\infty} F(x, y),$
$F(-\infty, y)$	instead of	$\lim_{x \to -\infty} F(x, y),$
$F(\infty, y)$	instead of	$\lim_{x \to \infty} F(x, y),$
$F(x, \infty)$	instead of	$\lim_{y \to \infty} F(x, y),$
$F(\infty, \infty)$	instead of	$\lim_{x, y \to \infty} F(x, y).$

In view of property (b$_2$), for every x and y we have $F(x, -\infty) = 0$, $F(-\infty, y) = 0$. In particular, we have $F(-\infty, -\infty) = 0$. Thus, according to definition (13.4) we have

$$P(\{(x, y): x < a, y < b\})$$
$$= P(\{(x, y): -\infty \leqslant x < a, -\infty \leqslant y < b\})$$
$$= F(a, b) - F(a, -\infty) - F(-\infty, b) + F(-\infty, -\infty) = F(a, b),$$

which proves the consistency of definition (13.4) with condition (13.2). In particular,

(13.5) $\quad P(R_2) = P(\{(x, y): -\infty \leqslant x < \infty, -\infty \leqslant y < \infty\})$
$$= F(\infty, \infty) - F(-\infty, \infty) - F(\infty, -\infty) + F(-\infty, -\infty) = 1.$$

Next, one shows that the function $P(A)$ defined on the class of half-open finite and infinite rectangles is countably additive in this class. This can be shown by theorems analogous to Theorems 1–4 of § 10 for half-open rectangles.

The next step is to prove that the function $P(A)$ defined by (13.4) on the class of finite and infinite half-open rectangles can be extended in a unique way to the smallest field S_0 containing this class. In view of this uniqueness, we shall from now on denote by $P(A)$ the function so extended. One can show that the function $P(A)$ is countably additive on S_0, non-negative, and—in view of equality (13.5)—equal to 1 for $A = R_2$. To complete the proof of Theorem 1 it suffices now to use the theorem on extension of measures from § 11.

The converse theorem is also true:

THEOREM 2. *If $P(A)$ is a probability distribution on the plane (i.e. a function defined on the class of Borel sets of the plane satisfying axioms* (A3)–(A5)), *then the function $F(x, y)$ defined by the equality*

(13.6) $\qquad F(a, b) = P(\{(x, y): x < a, y < b\})$

has properties (a$_2$), (b$_2$) *and* (c$_2$).

PROOF. The function $F(x, y)$ has the property (a$_2$). Let $x_1 \leqslant x_2$ and $y_1 \leqslant y_2$. We have to show that $F(x_2, y_2) - F(x_1, y_2) - F(x_2, y_1) + F(x_1, y_1) \geqslant 0$.

```
                (x₁, y₂)  (x₂, y₂)
         ─────────────┬─────────
              J₂      │   J₁
         ─────────────┤         │ (x₂, y₁)
              (x₁, y₁)│         
              J₄      │   J₃
```

Fig. 1

We shall show that the left-hand side of the last inequality equals the probability of the event $\{(x, y): x_1 \leq x < x_2, y_1 \leq y < y_2\}$; this probability is non-negative by axiom (A3), which explain the probabilistic meaning of property (a_2). Let

$$J_1 = \{(x, y): x_1 \leq x < x_2, y_1 \leq y < y_2\},$$
$$J_2 = \{(x, y): -\infty \leq x < x_1, y_1 \leq y < y_2\},$$
$$J_3 = \{(x, y): x_1 \leq x < x_2, -\infty \leq y < y_1\},$$
$$J_4 = \{(x, y): -\infty \leq x < x_1, -\infty \leq y < y_1\}$$

(see Fig. 1). J_1, J_2, J_3 and J_4 are disjoint events. By the additivity of probability and definition (13.6) we have

$$F(x_1, y_1) = P(J_4),$$
$$F(x_2, y_1) = P(J_4 \cup J_3) = P(J_4) + P(J_3),$$
$$F(x_1, y_2) = P(J_4 \cup J_2) = P(J_4) + P(J_2),$$
$$F(x_2, y_2) = P(J_1 \cup J_2 \cup J_3 \cup J_4) = P(J_1) + P(J_2) + P(J_3) + P(J_4).$$

Thus

$$F(x_2, y_2) - F(x_1, y_2) - F(x_2, y_1) + F(x_1, y_1)$$
$$= P(J_1) + P(J_2) + P(J_3) + P(J_4) - \bigl(P(J_4) + P(J_2)\bigr)$$
$$- \bigl(P(J_4) + P(J_3)\bigr) + P(J_4) = P(J_1) \geq 0.$$

Now we shall show that the function $F(x, y)$ has property (b_2).

Let us fix x_0 (possibly $x_0 = -\infty$ or $+\infty$). We shall show that if $y_1 > y_2 > \ldots$ is a sequence decreasing to $-\infty$, then

(13.7) $$\lim_{n \to \infty} F(x_0, y_n) = 0.$$

Let

$$A_i = \{(x, y): -\infty \leq x < x_0, -\infty \leq y < y_i\}.$$

We easily note that, as y_i decrease monotonically to $-\infty$, we have

$$A_1 \supset A_2 \supset \ldots$$

and the product of the events $A_1, A_2, \ldots$ is empty, i.e.

$$\bigcap_{i=1}^{\infty} A_i = 0.$$

By Theorem 3 of § 6 we have therefore

(13.8) $$\lim_{n \to \infty} P(A_n) = 0.$$

According to formula (13.6) we can write
$$P(A_n) = P(\{(x,y): -\infty \leqslant x < x_0, -\infty \leqslant y < y_n\})$$
$$= P(\{(x,y): x < x_0, y < y_n\}) = F(x_0, y_n).$$
Together with (13.8), the above formulas imply the equality (13.7).
In a similar way we show that for every y_0
$$\lim_{n \to -\infty} F(x_n, y_0) = 0.$$
It remains to show that
(13.9)
$$\lim_{\substack{x \to \infty \\ y \to \infty}} F(x,y) = 1.$$

Suppose that $x_1 \leqslant x_2 \leqslant \ldots$ and $y_1 \leqslant y_2 \leqslant \ldots$ are two sequences which increase monotonically to infinity:
$$\lim_{n \to \infty} x_n = \lim_{n \to \infty} y_n = \infty.$$
Denote by A_i the event
$$\{(x,y): x < x_i, y < y_i\}.$$
Clearly, the events $A_1, A_2, \ldots$ form an increasing sequence
$$A_1 \subset A_2 \subset \ldots$$
and their sum equals the whole plane R_2:
$$A_1 \cup A_2 \cup \ldots = R_2.$$
Thus, by Theorem 4 of § 6 we have
$$\lim_{n \to \infty} P(A_n) = 1.$$
Since
$$P(A_i) = P(\{(x,y): x < x_i, y < y_i\}) = F(x_i, y_i),$$
it follows that
$$\lim_{n \to \infty} F(x_n, y_n) = 1,$$
which proves formula (13.9).

It remains to show that the function $F(x,y)$ has property (c_2). This is a simple consequence of Theorem 5 of § 6. In fact, if $x_1 \leqslant x_2 \leqslant \ldots$ is a sequence tending to x_0, and $y_1 \leqslant y_2 \leqslant \ldots$ is a sequence tending to y_0, then the events
$$A_i = \{(x,y): x < x_i, y < y_i\}$$
form an increasing sequence of events whose sum equals
$$A = \{(x,y): x < x_0, y < y_0\}$$

(x_0 and y_0 may assume the values $+\infty$; our reasoning applies to this case. In particular, if $x_0 = y_0 = +\infty$, we again obtain the proof of (13.9)).

By Theorem 5 of § 6 and definition (13.6) we have

$$\lim_{n\to\infty} F(x_n, y_n) = \lim_{n\to\infty} P(\{(x,y): x < x_n, y < y_n\}) = \lim_{n\to\infty} P(A_n)$$
$$= P(A) = P(\{(x,y): x < x_0, y < y_0\}) = F(x_0, y_0),$$

which shows that the function $F(x, y)$ has property (c_2). The proof of Theorem 2 is thus complete.

Theorems 1 and 2 show that there exists a one-to-one correspondence between probability distributions on the plane and functions of two variables satisfying conditions (a_2), (b_2) and (c_2). This justifies the following definition:

DEFINITION 1. For every probability distribution $P(A)$ on the plane, the function of two variables defined as

$$F(a, b) = P(\{(x, y): x < a, y < b\})$$

will be called the *probability distribution function of the distribution $P(A)$*. In general, functions of two variables satisfying conditions (a_2), (b_2) and (c_2) will be called *two-dimensional* or *bivariate distribution functions*.

In view of the one-to-one correspondence, established above, between probability distributions on the plane and bivariate distribution functions, the problem of the existence of at least one probability distribution on the plane reduces to the problem of the existence of at least one bivariate distribution function. We show by an example that bivariate distribution functions exist: the function

$$F(x, y) = \begin{cases} 0 & \text{for } x < 0 \text{ or } y < 0, \\ xy & \text{for } 0 \leqslant x \leqslant 1, 0 \leqslant y \leqslant 1, \\ x & \text{for } 0 \leqslant x \leqslant 1, 1 \leqslant y, \\ y & \text{for } 1 \leqslant x, 0 \leqslant y \leqslant 1, \\ 1 & \text{for } 1 \leqslant x, 1 \leqslant y \end{cases}$$

is a bivariate distribution function. The proof is left to the reader.

We pass to the case of n dimensions. Theorems 1 and 2 remain valid in this case: there exists a one-to-one correspondence between probability distributions in the n-dimensional Euclidean space and n dimensional distribution functions.

DEFINITION 2. A function $F(x_1, ..., x_n)$ on n variables is called an *n-dimensional distribution function* if it satisfies the following conditions:

II. PROBABILITIES OF ELEMENTARY EVENTS

(a_n) $F(x_1, \ldots, x_n)$ is non-decreasing in n dimensions (the definition of this concept will be given below).

(b_n) For every i, for arbitrary $x_1, \ldots, x_{i-1}, x_{i+1}, \ldots, x_n$, we have

$$\lim_{x_i \to -\infty} F(x_1, \ldots, x_n) = 0,$$

and

$$\lim_{i \to \infty} F(x_1^{(i)}, \ldots, x_n^{(i)}) = 1,$$

provided $x_1^{(1)} \leqslant x_1^{(2)} \leqslant \ldots, x_2^{(1)} \leqslant x_2^{(2)} \leqslant \ldots, x_n^{(1)} \leqslant x_n^{(2)} \leqslant \ldots$ are sequences increasing monotonically to infinity.

(c_n) The function $F(x_1, \ldots, x_n)$ is continuous from the left in n variables, that is

$$\lim_{i \to \infty} F(x_1^{(i)}, \ldots, x_n^{(i)}) = F(x_1^{(0)}, \ldots, x_n^{(0)}),$$

provided $x_j^{(1)} \leqslant x_j^{(2)} \leqslant \ldots, j = 1, 2, \ldots, n$, are non-decreasing sequences tending to $x_j^{(0)}$.

Properties (b_n) and (c_n) are direct generalizations of (b) and (c) of § 10 for the case of n dimensions. We shall now define property (a_n) and explain its probabilistic sense.

In the case of the real line, we required the function $F(x)$ to be monotone, in order that the probability assigned to the intervals $\langle a, b \rangle$ by the relation $P(\langle a, b \rangle) = F(b) - F(a)$ be non-negative. In the case of the plane, we assigned probabilities to the rectangles $\{(x, y): a_1 \leqslant x < a_2, b_1 \leqslant y < b_2\}$. To satisfy the axiom of additivity (see proof of property (a) in Theorem 1) it was necessary to assign to such a rectangle the probability

$$F(a_2, b_2) - F(a_1, b_2) - F(a_2, b_1) + F(a_1, b_1).$$

This explains why we required that expressions of this type should be non-negative, as stated in condition (a_2).

In the case of n dimensions, we put by definition

$$P(\{(x_1, \ldots, x_n): x_1 < a_1, \ldots, x_n < a_n\}) = F(a_1, \ldots, a_n).$$

Now, it turns out that if we want to define probabilities for the n-dimensional interval

$$A = \{(x_1, \ldots, x_n): a_1 \leqslant x_1 < a_1 + h_1, \ldots, a_n \leqslant x_n < a_n + h_n\}$$

according to the additivity postulate, we must put

$$P(A) = F(a_1 + h_1, \ldots, a_n + h_n) - F(a_1, a_2 + h_2, \ldots, a_n + h_n)$$
$$- F(a_1 + h_1, a_2, a_3 + h_3, \ldots, a_n + h_n) - \ldots$$
$$- F(a_1 + h_1, \ldots, a_{n-1} + h_{n-1}, a_n) + \ldots + (-1)^n F(a_1, \ldots, a_n).$$

The number on the right-hand side of this equality, equal to the sum of 2^n terms, is called the *nth difference of the function F* at the point $(a_1, ..., a_n)$ corresponding to the increments $h_1, ..., h_n$ of the first, ..., nth variable. To satisfy the axiom of non-negativeness of probability, we must require that all such nth differences be non-negative. Thus, we say that $F(x_1, ..., x_n)$ is *non-decreasing in n dimensions* if all its nth differences are non-negative. The following theorems hold:

THEOREM 3. *To every n dimensional distribution function $F(x_1, ..., x_n)$ there corresponds exactly one distribution in n-dimensional Euclidean space such that*

$$P(\{(x_1, ..., x_n): x_1 < a_1, ..., x_n < a_n\}) = F(a_1, ..., a_n).$$

THEOREM 4. *If $P(A)$ is a probability distribution in the n-dimensional Euclidean space then the function $F(x_1, ..., x_n)$ defined as*

$$F(a_1, ..., a_n) = P(\{(x_1, ..., x_n): x_1 < a_1, ..., x_n < a_n\})$$

is an n-dimensional distribution function.

We shall prove certain theorems which give the connection between multi-dimensional distribution functions and distribution functions in a smaller number of dimensions.

THEOREM 5. *If $F(x, y)$ is a bivariate distribution function, then F is monotone in the usual sense, i.e.*

$$F(x_1, y_1) \leqslant F(x_2, y_2)$$

provided $x_1 \leqslant x_2$ and $y_1 \leqslant y_2$; infinite values for x_1, x_2, y_1, y_2 are allowed.

PROOF. It suffices to show that

$$F(x_1, y_0) \leqslant F(x_2, y_0)$$

where $x_1 \leqslant x_2$. By definition (13.4) and property (b) we have

$$P(\{(x, y): x_1 \leqslant x < x_2, -\infty \leqslant y < y_0\})$$
$$= F(x_2, y_0) - F(x_1, y_0) - F(x_2, -\infty) + F(x_1, -\infty)$$
$$= F(x_2, y_0) - F(x_1, y_0) \geqslant 0,$$

which completes the proof.

THEOREM 6. *If $F(x, y)$ is a bivariate distribution function, then functions of one variable defined as $F(x, +\infty)$ and $F(+\infty, y)$ have all the properties of one-dimensional distribution functions.*

II. PROBABILITIES OF ELEMENTARY EVENTS

PROOF. It suffices to consider the function $F(x, +\infty)$. It is monotone by Theorem 5. The relations

$$\lim_{x \to -\infty} F(x, +\infty) = 0, \quad \lim_{x \to \infty} F(x, +\infty) = 1$$

follow from property (b_2) of function $F(x, y)$. The left continuity of $F(x, +\infty)$ is a particular case of property (c_2) of the function $F(x, y)$, which completes the proof.

This theorem justifies the following definition:

DEFINITION 3. If $F(x, y)$ is a bivariate distribution function, then functions $F(x, +\infty)$ and $F(+\infty, y)$ will be called its *marginal distribution functions*, and the corresponding probability distributions on the real line will be called the *marginal distributions* of the distribution on the plane corresponding to the distribution function $F(x, y)$.

Theorem 6 extends to the case of n dimensions:

THEOREM 7. *If $F(x_1, \ldots, x_n)$ is an n-dimensional distribution function, then functions of k variables obtained by substituting the value $+\infty$ for the arbitrary $n-k$ variables have all the properties of k dimensional distribution functions.*

This, in turn, justifies the following definition:

DEFINITION 4. If $F(x_1, \ldots, x_n)$ is an n-dimensional distribution function, then all distribution functions of lower dimensions which can be obtained from it in the manner described in Theorem 7 will be called its *marginal distribution functions*, and all k-dimensional distributions ($k < n$) determined by these distribution functions in the k-dimensional Euclidean space will be called *marginal distributions* of the distribution in the n-dimensional Euclidean space corresponding to $F(x_1, \ldots, x_n)$.

§ 14. CLASSIFICATION OF PROBABILITY DISTRIBUTIONS ON THE REAL LINE

In this section we shall study in some detail the probability distributions on the real line, and we shall distinguish two important classes of such distributions, namely *discrete distributions* and *continuous distributions*.

Let $P(A)$ be a probability distribution on the real line $(-\infty, +\infty)$ and let $F(x)$ be the distribution function of $P(A)$. For every x we shall denote by $F(x+)$ the right-hand side limit of the function F at the point x.

THEOREM 1. *For every a the probability of the event $\{a\}$ is equal to the difference between the right-hand limit of $F(x)$ at the point a and the value of F at a, that is*

(14.1) $$P(\{a\}) = F(a+) - F(a).$$

PROOF. The event $A = \{a\}$ is equal to the product of the decreasing sequence of events $A_n = \langle a, a+\frac{1}{n} \rangle$. By Theorem 5 of § 6 we have therefore

$$P(\{a\}) = \lim_{n \to \infty} P\left(\langle a, a+\frac{1}{n} \rangle\right).$$

But $P\left(\langle a, a+\frac{1}{n} \rangle\right) = F\left(a+\frac{1}{n}\right) - F(a)$, hence

$$P(\{a\}) = \lim_{n \to \infty} F\left(a+\frac{1}{n}\right) - F(a) = \lim_{n \to \infty} F\left(a+\frac{1}{n}\right) - F(a) = F(a+) - F(a),$$

which proves Theorem 1.

If $F(a+) - F(a) > 0$, we say that F has a *jump* at the point a. This jump equals $P(\{a\})$.

THEOREM 2. *The set of points at which a distribution function has jumps is at most countable.*

PROOF. Since probability never exceeds one, we have $P(\{a\}) \leqslant 1$ for every a. Let A_n denote the set of points a for which $\frac{1}{n} \geqslant P(\{a\}) > \frac{1}{n+1}$. The number of elements of A_n is finite and does not exceed $n+1$, since otherwise we would have $P(A_n) > 1$. Every point a at which $F(x)$ has a jump belongs to one and only one of the sets A_n. Thus, the set A of points at which F has jumps equals to the sum of the sets A_n, i.e. $A = A_1 \cup A_2 \cup \ldots$ As a sum of countably many finite sets, A is at most countable (of course, A can be finite or empty).

The set A, even though it is at most countable, can be everywhere dense on the line. For instance, rational numbers form a countable set and lie everywhere densely on the real line.

It may happen that the sum of all jumps equals one. We then have

$$P(A) = \sum_{a \in A} P(\{a\}) = 1, \quad P(A') = P((-\infty, \infty) - A) = 0.$$

Thus, the whole probability is concentrated on a countable set A. We could take A as the space of elementary events, and we would then have an at most countable space of elementary events. Usually, we do not do that, adding to A

the remaining part of the real line as a set of probability zero, and the distribution for which $P(A) = 1$ is treated as a special case of probability distribution on the real line. The reasons for such a construction are the following: first, we have one space of elementary events suitable for a description of many phenomena, and second, we can formulate general theorems, concerning all probability distributions, without the necessity of specifying those subsets of real line on which the probability is actually concentrated.

Such extensions are quite common. Thus, we shall often deal with probability distributions on the real line for which $P(\langle 0, \infty)) = 1$, i.e. for which one could take the positive half-line as the space of elementary events. Nevertheless, we shall treat such distributions as special cases of distributions on the real line.

We shall now introduce an important definition.

DEFINITION 1. If the sum of all jumps of a distribution function equals one, the probability distribution is called *discrete*.

Note that if the distribution function has no jump at the point a, it must be continuous at this point. Indeed, we then have $F(a+) = F(a)$, and since the distribution function is continuous from the left, we have also $F(a-) = F(a)$, where $F(a-)$ denotes the left-hand limit of F at a. The equality $F(a-) = F(a) = F(a+)$ proves the continuity at a.

Thus, if the distribution function has no jumps, it is continuous. In the class of distributions with continuous distribution functions we shall distinguish a smaller class by the following definition:

DEFINITION 2. If the distribution function $F(x)$ of the distribution $P(A)$ on the real line can be represented for every x as the integral

$$F(x) = \int_{-\infty}^{x} f(a)\,da$$

of a non-negative function $f(x)$, then the distribution $P(A)$ will be called *continuous* (or: *of the continuous type*) and $f(x)$ will be called the *density of the distribution* $P(A)$.

CHAPTER III

Conditional Probability. Stochastic Independence of Events

§ 15. CONDITIONAL PROBABILITY

We shall now introduce one of the most important concepts of probability theory, namely that of conditional probability.

Let us start from an example. Suppose that in a certain population we have $w = w_s + w_n$ women, out of which w_s women smoke cigarettes and w_n women do not smoke, and $m = m_s + m_n$ men, out of which m_s men smoke and m_n men do not smoke. Thus, the whole population consists of $w + m$ people, out of whom $w_s + m_s$ smoke.

We want to describe the experiment consisting of choosing a person from this population. We may take as the space of elementary events the four-point space consisting of points W_s, W_n, M_s and M_n denoting respectively woman-smoker, woman non-smoker, man-smoker and man non-smoker. $W = \{W_s, W_n\}$ is the event consisting of the selection of a woman, $S = \{W_s, M_s\}$ is the event consisting of selecting a smoking person, $M = \{M_s, M_n\}$—the event consisting of selecting a man, and $N = \{W_n, M_n\}$—that of selecting a non-smoker. As the probabilities of the various events we may take the fractions of the persons with the corresponding properties. Thus, for instance, $P(\{W_s\}) = w_s/(w+m)$, $P(\{M\}) = m/(w+m)$, and so on. In particular, the probability that the person selected is a smoker equals the fraction of smokers

$$P(S) = \frac{w_s + m_s}{w + m}.$$

Now, what is the probability that a selected woman smokes? We answer easily: it is equal to the fraction of smoking women, that is, w_s/w. If we were pedantic, we could formulate our question in a slightly different way, namely: what is the probability that a selected person smokes, provided it is a woman? Thus, we ask for the probability of the event S under the condition that the event W has occurred. Such probability will be called *conditional* and will be denoted

III. CONDITIONAL PROBABILITY. STOCHASTIC INDEPENDENCE

by $P(S|W)$. Now, our aim is to define the conditional probability in terms of "unconditional" probabilities, which we have considered until now. The example above will suggest a suitable definition.

The basic idea in defining conditional probabilities lies in restricting the space of elementary events and the field of events to those elementary events which are contained in the event which constitutes the condition, and in defining the probability on this new smaller class of events.

Let us come back to our example. We said that the probability $P(S|W)$ should be equal to the fraction w_s/w of women who smoke. But the fraction of smoking women equals the ratio of persons who smoke and are of female sex to the number of persons who are of female sex. The event of choosing a smoking woman is the product of two events: that of selecting a woman W and that of selecting a smoking person S, i.e., $\{W_s\} = W \cap S$. Next, $P(\{W_s\}) = P(W \cap S) = w_s/(w+m)$ and $P(W) = w/(w+m)$. Thus, we can represent the conditional probability $P(S|W)$ as the ratio of the probability of the product of the event in question and the event which constitutes the condition to the probability of the event which constitutes the condition. This leads to the required definition:

DEFINITION 1. If $P(B) > 0$, then the *conditional probability* of the event A under the condition that B occurred, to be denoted by $P(A|B)$, will be defined as the ratio of the probabilities of the event $A \cap B$ to the probability of the event B. Thus,

$$(15.1) \qquad P(A|B) = \frac{P(A \cap B)}{P(B)}.$$

In more general terms, the conditional probability informs us what portion of the event B is occupied by the event A.

EXAMPLE 1. What is the probability of an even result of a throw of a die, provided that the result is smaller than 4?

The elementary events here are numbers 1, 2, 3, 4, 5, 6, and the events are all the subsets of this six-element set. Probability is defined by the postulate of equal probabilities for all one-element events; hence the probability of every event is equal to the product of $1/6$ and the number of elements of that event.

The event A consisting of throwing an even number of points, equals $\{2, 4, 6\}$. The event B, consisting of throwing the number of points smaller that 4, equals $\{1, 2, 3\}$. By Definition 1

$$P(A|B) = \frac{P(A \cap B)}{P(B)} = \frac{P(\{2\})}{P(\{1, 2, 3\})} = \frac{\frac{1}{6}}{\frac{3}{6}} = \frac{1}{3}.$$

Note that the conditional probability of the event A is different from the unconditional probability of the same event, as

$$P(A) = P(\{2, 4, 6\}) = \tfrac{3}{6} = \tfrac{1}{2}.$$

EXAMPLE 2. What is the conditional probability of the event A consisting in throwing an even number of points on a die on condition that the event $B = \{1, 2\}$ has occurred.

By Definition 1 we have

$$P(A|B) = \frac{P(\{2,4,6\} \cap \{1,2\})}{P(\{1,2\})} = \frac{P(\{2\})}{P(\{1,2\})} = \frac{\tfrac{1}{6}}{\tfrac{2}{6}} = \tfrac{1}{2}.$$

In this case $P(A|B) = P(B)$.

We shall now prove

THEOREM 1. *The conditional probability $P(A|B)$ satisfies all axioms of probability.*

PROOF. For every event A we have $0 \leqslant P(A|B) \leqslant 1$.

Indeed, non-negativeness follows from the fact that $P(A \cap B) \geqslant 0$. Since $P(A \cap B) \leqslant P(B)$, we have also $P(A|B) \leqslant 1$.

Next, $P(X|B) = 1$, since $X \cap B = B$, and

$$P(X|B) = \frac{P(B)}{P(B)} = 1.$$

Finally, $P(A|B)$ is countably additive. In fact, if $A_1, A_2, \ldots$ are disjoint and their sum is A, then $A_1 \cap B, A_2 \cap B, \ldots$ are also disjoint and their sum is $A \cap B$. By the countable additivity of probability we have

$$P(A \cap B) = P(A_1 \cap B) + P(A_2 \cap B) + \ldots$$

By dividing both sides by $P(B)$, we have, in virtue of Definition 1,

$$P(A|B) = P(A_1|B) + P(A_2|B) + \ldots,$$

which completes the proof of Theorem 1.

EXERCISE. The reader will verify that if the probability $P(A)$ is defined on the field S of events, then the conditional probability $P(A|B)$ may be considered as the usual probability defined on the field S_B of events of the form $A \cap B$, where $A \in S$. The role of the space of elementary events is played here by the event B.

We shall now prove a theorem which shows that unconditional probability may be defined in terms of conditional probabilities and the probabilities of the conditions.

III. CONDITIONAL PROBABILITY. STOCHASTIC INDEPENDENCE

THEOREM 2. *If $A_1, \ldots, A_n$ are disjoint and such that*

(15.2) $A_1 \cup \ldots \cup A_n = X$ *and* $P(A_i) > 0$ *for* $i = 1, 2, \ldots, n$,

then for every event A

(15.3) $P(A) = P(A_1)P(A|A_1) + \ldots + P(A_n)P(A|A_n).$

The formula (15.3) is sometimes called the *formula for absolute probability*.

PROOF. Since the events $A_1, \ldots, A_n$ are disjoint, the events $A \cap A_1, A \cap A_2, \ldots, A \cap A_n$ are also disjoint. In view of (15.2) we can write

$$(A \cap A_1) \cup \ldots \cup (A \cap A_n) = A \cap (A_1 \cup \ldots \cup A_n) = A \cap X = A.$$

Hence

(15.4) $P(A) = P(A \cap A_1) + \ldots + P(A \cap A_n).$

By Definition 1 we have for $i = 1, 2, \ldots, n$

$$P(A \cap A_i) = P(A_i)\frac{P(A \cap A_i)}{P(A_i)} = P(A_i)P(A|A_i).$$

Substituting this in (15.4), we obtain equality (15.3).

EXAMPLE 3. *Drawing balls from two urns*. Suppose we have two urns, the first containing 30 white and 20 black balls, and the second containing 7 white and 3 black balls. Suppose also that we draw balls from these two urns in such a way that we first choose an urn at random, and then draw a ball at random from the urn selected. What is the probability that we shall choose a white ball?

To construct a probabilistic model of our experiment, we take a four-element set as the space of elementary events: $X = \{w_1, w_2, b_1, b_2\}$; here, w_1 and w_2 denote the events of drawing white balls from the first urn and the second urn respectively, similarly, b_1 and b_2 denote the events of drawing black balls from the first urn and the second urn, respectively. Let $U_1 = \{w_1, b_1\}$ denote the event of drawing from the first urn, and let $U_2 = \{w_2, b_2\}$ denote the event of drawing from the second urn. Further, let $W = \{w_1, w_2\}$ denote the event of drawing a white ball, and $B = \{b_1, b_2\}$—the event of drawing a black ball. Now, choosing the urns at random may be interpreted as equal probabilities of the events U_1 and U_2, whence $P(U_1) = P(U_2) = 1/2$, since U_1 and U_2 are disjoint and $U_1 \cup U_2 = X$. Next, from the compositions of the urns it follows that the probabilities of drawing white balls are

$$P(W|U_1) = \frac{30}{50} = \frac{3}{5}, \quad P(W|U_2) = \frac{7}{10}.$$

By Theorem 2 we have
$$P(W) = P(U_1)P(W|U_1) + P(U_2)P(W|U_2)$$
$$= \frac{1}{2} \cdot \frac{3}{5} + \frac{1}{2} \cdot \frac{7}{10} = \frac{1}{2}\left(\frac{6}{10} + \frac{7}{10}\right) = \frac{13}{20}.$$

The importance of Theorem 2 lies in the fact that in practical situations it is often easier to determine the conditional probabilities of various events than their "ordinary" or unconditional probabilities. Such a situation is illustrated by Example 3.

One more remark. We have talked here of "unconditional" and conditional probabilities. In fact, this distinction is not necessary, since we always speak of conditional probabilities. For instance, we could speak of the probability of the event that a Pole we shall meet has blue eyes. With respect to this event, the probability that an inhabitant of Wrocław whom we shall meet has blue eyes is a conditional probability. But also the initial probability can be treated as conditional, with respect to, say, the event that a European we shall meet has blue eyes. Thus, there is no essential difference between conditional and unconditional probability, since the very conditions of the observations or the experiment to which the probability applies cause this probability to be conditional.

After the above remark we shall return to the theory.

In this section we have defined conditional probability for cases in which the event constituting the condition has positive probability. In the sequel we shall encounter conditional probabilities for cases in which the condition has probability zero. In those cases we shall have to overcome various analytic difficulties, similar to those which we encountered when passing from countable to uncountable spaces of elementary events.

One can prove a theorem somewhat stronger than Theorem 2. Its proof is analogous to that of Theorem 2, with the only difference that one has to use the axiom of countable additivity.

THEOREM 3. *If $A_1, A_2, \ldots$ are disjoint and such that*
$$A_1 \cup A_2 \cup \ldots = X,$$
and $P(A_i) > 0$ for $i = 1, 2, \ldots$, then for every event A
$$P(A) = P(A_1)P(A|A_1) + P(A_2)P(A|A_2) + \ldots$$

Let us also mention a simple consequence of Definition 1, which will prove useful in the sequel:

III. CONDITIONAL PROBABILITY. STOCHASTIC INDEPENDENCE

THEOREM 4. *If $P(B) > 0$, then*

$$P(A \cap B) = P(B)P(A|B).$$

This formula follows upon multiplying (15.1) by $P(B)$.

§ 16. BAYES' THEOREM. PROBABILITIES OF CAUSES

In this section we shall prove the simplest case of the so-called *Bayes'* * *theorem*. This theorem gives a key to a wide class of problems connected with inductive inference on the basis of observations, and constitutes the starting point for various branches of mathematical statistics.

THEOREM 1 (Bayes). *If $A_1, A_2, \ldots, A_n$ are disjoint and such that*

$$A_1 \cup A_2 \cup \ldots \cup A_n = X,$$

and $B_1, B_2, \ldots, B_m$ are also disjoint and such that

$$B_1 \cup B_2 \cup \ldots \cup B_m = X,$$

and if $P(A_i) > 0$ for $i = 1, 2, \ldots, n$ and $P(B_j) > 0$ for $j = 1, 2, \ldots, m$ (so that conditional probabilities exist), then for every i and j

$$(16.1) \qquad P(B_j|A_i) = \frac{P(B_j)P(A_i|B_j)}{P(B_1)P(A_i|B_1) + \ldots + P(B_m)P(A_i|B_m)}.$$

This theorem shows that if one knows the probabilities of events B_j and the conditional probabilities of events A_i when B_j's constitute the conditions, then one can compute the conditional probabilities of events B_j when A_i's constitute the conditions.

PROOF. By Definition 1 of § 15

$$(16.2) \qquad P(B_j|A_i) = \frac{P(A_i \cap B_j)}{P(A_i)}.$$

By Theorem 4 of § 15 we can write

$$(16.3) \qquad P(A_i \cap B_j) = P(B_j)P(A_i|B_j).$$

Since the events $B_1, \ldots, B_m$ satisfy the assumptions of Theorem 2 of § 15, we have

$$(16.4) \qquad P(A_i) = P(B_1)P(A_i|B_1) + \ldots + P(B_m)P(A_i|B_m).$$

By substituting (16.3) and (16.4) in (16.2), we obtain (16.1), which completes the proof.

---

* Thomas Bayes, (dec. 1763).

As we saw in Example 3 of § 15, it is sometimes easy to determine the conditional probabilities. In Bayes' theorem, the events $B_1, \ldots, B_m$ are sometimes called *causes*, and the events $A_1, \ldots, A_n$ are sometimes called *results*. The probabilities of the events $B_1, \ldots, B_m$ are called the *a priori probabilities of the causes*, and the conditional probabilities $P(B_j|A_i)$ are called the *a posteriori probabilities of the causes*. By *a posteriori* it is meant that one refers to the probabilities of the causes when the results are already known. The theorem of Bayes states that one can compute the *a posteriori* probabilities provided one knows the *a priori* probabilities of the causes and the conditional probabilities of the results for every given cause.

We shall illustrate by a simple example how this theorem is related to inductive inference.

EXAMPLE 1. Suppose that we have a lot of 100 screws. We draw one screw at random, and it turns out to be good (not defective). What can we say about the number of defective screws in the whole lot? One thing is certain, namely that it contains no more that 99 defective screws. However, we are not interested in this type of statement. We want to know the answer to questions such as: what is the probability that the lot contains, say, 50 defective items?

In this example, the drawing of a good or a defective screw can be treated as a result whose probability depends on a cause, the cause being the real number of defective items in the lot. The event that the lot contains i defective items will be denoted by B_i, $i = 0, 1, \ldots, 100$. Denote by A_1 the event that the screw selected is good, and by A_2—the event that it is defective. If the lot contains i bad items, then the probability of selecting a bad item equals $i/100$, and the probability of selecting a good item equals $(100-i)/100$. These are the conditional probabilities of the results for given causes; we can write this as follows:

$$P(A_1|B_i) = \frac{i}{100}, \quad P(A_2|B_i) = \frac{100-i}{100}.$$

If we know the *a priori* probabilities, $P(B_i)$, then by Bayes' theorem for the known result (in our case equal to A_2) we could compute the *a posteriori* probability of B_{50} that our lot contains 50 defective items, that is $P(B_{50}|A_2)$. These *a posteriori* probabilities contain all our knowledge about the number of defective items in the lot which can be gathered from the fact that the item selected was good. But in order to be able to draw such an inference, we must know something *a priori*, before sampling, namely we must know the *a priori* probabilities of different causes. How can we obtain such knowledge in cases similar to that of our example? Can one infer something even without such knowledge? These are the questions that arise immediately. This example shows that essentially the problem reduces to that of inference about a phenomenon on the basis of observations. One can say that mathematical statistics is devoted entirely to this problem.

Theorem 1 can easily be generalized to the case of countably many results and causes.

THEOREM 2. *If* $A_1, A_2, \ldots$ *are disjoint and such that*

$$A_1 \cup A_2 \cup \ldots = X,$$

and $B_1, B_2, \ldots$ *are disjoint and such that*

$$B_1 \cup B_2 \cup \ldots = X,$$

and if $P(A_i) > 0$ *for* $i = 1, 2, \ldots,$ $P(B_j) > 0$ *for* $j = 1, 2, \ldots,$ *then for every* i *and* j

$$P(B_j|A_i) = \frac{P(B_j)P(A_i|B_j)}{\sum_{j=1}^{\infty} P(B_j)P(A_i|B_j)}.$$

The proof differs from that of Theorem 1 only at the point where countable additivity instead of finite additivity must be used.

§ 17. STOCHASTIC INDEPENDENCE OF EVENTS

In the example of § 15 we saw that the conditional probability of an event may be equal to its unconditional probability or may be different from it. If the conditional probability of A in condition that B has occurred differs from the unconditional probability of A, then, generally speaking, the occurrence of B tells us something about A, consequently, A and B are, in a sense, connected, or *dependent* on each other. If the conditional probability of A provided that B has occurred is the same as the unconditional probability, the occurrence of B implies nothing about the event A. In this case we say that A is *independent* of B.

Thus, the event A is independent of the event B if

$$P(A|B) = P(A).$$

By definition of conditional probability it means that

$$P(A|B) = \frac{P(A \cap B)}{P(B)} = P(A),$$

hence

(17.1) $$P(A \cap B) = P(A)P(B),$$

that is, the probability of the product of events equals to the product of the probabilities of those events. If (17.1) holds, we have also

$$(17.2) \qquad P(B|A) = \frac{P(A \cap B)}{P(A)} = \frac{P(A)P(B)}{P(A)} = P(B),$$

whence the conditional probability of B given that A occurred equals the unconditional probability of the event B. This shows that if the event A is independent of the event B, then, conversely, the event B is also independent of the event A. Thus, we can simply speak of the independence of the events A and B.

In the above reasoning we considered conditional probabilities, and the corresponding events had to have positive probabilities, since otherwise conditional probability is meaningless. Such a restriction is no longer necessary when we consider relation (17.1), which is meaningful even if the events A and B have probability zero. Thus, we shall use this relation as the formal definition of independence:

DEFINITION 1. Events A and B are called *independent* if the probability of their product equals to the product of their probabilities, i.e., if relation (17.1) holds.

Relation (17.2) implies that if A and B are independent, and $P(B) > 0$, then $P(A|B) = P(A)$.

Here are some consequences of Definition 1.

THEOREM 1. *An arbitrary event A and the sure event X are always independent*

In fact, we have $P(A \cap X) = P(A) = P(A) \cdot 1 = P(A) \cdot P(X)$, since $A \cap X = A$ and $P(X) = 1$.

THEOREM 2. *An arbitrary event A and the impossible event 0 are always independent.*

In fact, $P(A \cap 0) = P(0) = P(0)P(A) = 0$, since $A \cap 0 = 0$ and $P(0) = 0$.

These two theorems imply, in particular, that sure and impossible events are independent.

THEOREM 3. *If the events A and B are independent, then the same is true for the pairs of events*

(a) A and B',
(b) A' and B,
(c) A' and B'.

We shall prove assertion (a). As the events $A \cap B$ and $A \cap B'$ are disjoint and their sum is A, we have
$$P(A) = P(A \cap B) + P(A \cap B'),$$
hence
(17.3) $$P(A \cap B') = P(A) - P(A \cap B).$$
Since A and B are independent, we have
$$P(A \cap B) = P(A)P(B) \quad \text{and} \quad P(B') = 1 - P(B);$$
therefore
$$P(A) - P(A \cap B) = P(A) - P(A)P(B) = P(A)(1 - P(B)) = P(A)P(B').$$
The last equality combined with (17.3) implies that
$$P(A \cap B') = P(A)P(B'),$$
which means that A and B' are independent.

Assertions (b) and (c) can be proved in a similar manner. Thus, we see that independence is preserved under complementation.

THEOREM 4. *If the events A and C and B and C constitute two pairs of independent events, and if A and B are disjoint, then $A \cup B$ and C are independent.*

PROOF. We have
$$P((A \cup B) \cap C) = P((A \cap C) \cup (B \cap C)) = P(A \cap C) + P(B \cap C)$$
$$= P(A)P(C) + P(B)P(C) = (P(A) + P(B))P(C)$$
$$= P(A \cup B)P(C),$$
which was to be proved.

This theorem asserts that independence is preserved under the addition of disjoint events. We shall show that the assumption of disjointness cannot be omitted.

EXAMPLE 1. For the experiment of throwing a die, consider the events $A = \{1, 3, 4\}$, $B = \{1, 5, 6\}$, and $C = \{1, 2\}$. We have $P(A) = P(B) = 1/2$, $P(C) = 1/3$, $P(A \cap C) = 1/6$, $P(B \cap C) = 1/6$, $P((A \cup B) \cap C) = 1/6$. Thus, the events A and C are independent, and the same holds for the events B and C, but the events $A \cup B$ and C are not independent. This example shows that one cannot omit the assumption of disjointness in Theorem 4.

We shall leave the proof of the following theorem to the reader:

THEOREM 5. *If $A_1, A_2, \ldots$ are disjoint and their sum is A, and if for every $i = 1, 2, \ldots$ the events A_i and B are independent, then A and B are independent.*

The concept of independence can be generalized to the case of more than two events.

DEFINITION 2. Let $A_1, A_2, \ldots$ be a finite or infinite system of events. We shall say that this system is *n-wise independent* if for every sequence of distinct events $A_{i_1}, \ldots, A_{i_k}$ of the length at most equal to n we have

(17.4) $\qquad P(A_{i_1} \cap A_{i_2} \cap \ldots \cap A_{i_k}) = P(A_{i_1})P(A_{i_2}) \ldots P(A_{i_k}).$

We shall say that the events of this system are *independent en bloc*, or simply: *independent*, if they are *n*-wise independent a) for every natural n if the system is infinite, and b) for n equal to the number of events in the system if the system is finite.

We present here an example of three events which are pairwise independent but are not independent *en bloc*.

EXAMPLE 2. Imagine drawing balls from an urn which contains four balls, of which one is red, one is green, one is white, and one is painted with all these three colours. Let A be the event of drawing a ball with the red colour on it, B—the event of drawing a ball with the green colour on it, and C—the event of drawing a ball with the white colour on it. Assuming equal probabilities for every ball, we have $P(A) = P(B) = P(C) = 1/2$, as every colour appears on two balls. Next, we have $P(A \cap B) = P(A \cap C) = P(B \cap C) = 1/4$, which shows that the events A, B and C are pairwise independent. These events, however, are not independent, since $P(A \cap B \cap C) = 1/4 \neq 1/8 = P(A)P(B)P(C)$; indeed, there is only one ball marked with all three colours.

This example justifies Definition 2.

§ 18. INDEPENDENT TRIALS

Before now we discussed probability distributions which describe a single observation or experiment. When we spoke of two tosses of a coin, or two throws of a die, we treated two tosses as a single experiment. In many cases it is relatively easy to describe the probability distribution for a single trial, single observation or experiment, and it is difficult to speak of the probability distribution for a whole series of trials, experiments or observations. For instance, it is easy to describe a single throw of a die, and the problem becomes complicated when we speak of series of throws. But for a series of throws of a die, we may assume that the results of first throws do not influence the results of further throws, as all throws are performed in identical conditions. In other words, we may assume that the events concerning particular throws are independent.

III. CONDITIONAL PROBABILITY. STOCHASTIC INDEPENDENCE 65

When we deal with series of trials, each of them performed in the same conditions and independently of the results of other trials, we speak of *independent trials*. The question is: can one start from probabilities concerning single trials, and describe the probability distribution for the whole series of trials? The answer is in the affirmative, and will be discussed presently. Before we give general definitions and theorems, we shall analyse two examples in some detail.

EXAMPLE 1. *Two throws of a die*. The result of a throw of a die can be characterized by the number of points scored. Thus, for a single throw we take as the space of elementary events the set $\{1, 2, 3, 4, 5, 6\}$ consisting of six numbers. The field of events coincides with the class of all subsets of this set, and we define probability by assuming all elementary events to be equally probable. We say that such a space of elementary events, field of events and probability distribution describes the throw of a die.

The result of two throws may be described by a pair of numbers: the number resulting from the first throw, and the number resulting from the second throw. Thus, for the space of elementary events in this case one may take the set of 36 pairs (i,j), $i,j = 1, 2, 3, 4, 5, 6$, take the class of all subsets of this set as the field of events, and define probability by requiring that all pairs (i,j) have equal probabilities. Then we shall have for each pair $P(\{(i,j)\}) = 1/36$. Denoting by A_i and B_j the events consisting of throwing i points at the first throw and j points at the second throw, respectively, we shall have $P(A_i) = P(B_j) = 6/36 = 1/6$. Since $A_i \cap B_j = \{(i,j)\}$, and $P(\{(i,j)\}) = 1/36$, we see that the events A_i and B_j are independent for every i and j. From Theorem 4 of § 17 we infer that any events concerning the first throw are independent of any events concerning the second throw.

Let us look more closely at the space of elementary events describing the experiment of two throws of a die. This space has the form of the set of ordered pairs (i,j) where both i and j are elements of the set $\{1, 2, 3, 4, 5, 6\}$, i.e. elements of the space of elementary events for a single throw of a die. Let us consider the event consisting of throwing "six" at the first throw. This event consists of six pairs $(6, 1)$, $(6, 2)$, $(6, 3)$, $(6, 4)$, $(6, 5)$ and $(6, 6)$, in which 6 appears at the first place, and at the second place appears an element of the set $\{1, 2, 3, 4, 5, 6\}$. In the space corresponding to a single throw, the same event, i.e. throwing "six", consisted of one point only.

EXAMPLE 2. *Sampling without replacement*. Suppose we have n objects $a_1, a_2, \ldots, a_n$ and we draw k of them without replacement. What is the probabilistic description of such an experiment? One can proceed as follows: the result of our experiment is described completely by indicating an ordered subset of k

elements of the set in question. Thus, as the space of elementary events we may take the set of all ordered k-tuples of elements of our set. From combinatorial analysis it is known that there are $n(n-1) \ldots (n-k+1)$ such systems. It remains to define probability in our space of elementary events. But the word "sampling" means that all systems of elements have equal probabilities, hence to each elementary event we assign the probability

$$\frac{1}{n(n-1) \ldots (n-k+1)}.$$

The same sampling without replacement could be described in a different manner, by splitting it, as it were, into independent actions. Imagine that our objects are ordered. If we want to draw k objects, we must draw the first one. This can be done in n different ways, by indicating the number of the object which is to be drawn. When the first object has already been drawn, the second can be drawn in $n-1$ different ways. To make the second drawing independent of the first, let us agree that it is performed by showing the number of one of the remaining objects which is to be drawn. Proceeding in this manner, we shall be able to draw the third object in $n-2$ different ways, and so on. The result of sampling will be determined when we indicate a k-element sequence $n_1, n_2, \ldots, n_k$ where $1 \leqslant n_i \leqslant n-i+1$, $i = 1, 2, \ldots, k$. The ith number is drawn from the set consisting of $n-i+1$ elements, so that we can assume that the probability of drawing any specified number n_i equals $1/(n-i+1)$. By the independence of successive drawings, the probability of drawing the numbers $n_1, n_2, \ldots, n_k$ will be equal to

$$\frac{1}{n(n-1)(n-2) \ldots (n-k+1)}.$$

These examples illustrate the general construction which will be presented in the next section.

§ 19. PROBABILITIES IN CARTESIAN PRODUCTS

After the examples of § 18, the following general definitions should appear quite natural.

DEFINITION 1. By the *Cartesian product of the spaces of elementary events* X_1 *and* X_2 we shall mean the space X consisting of all ordered pairs (e', e'') such that $e' \in X_1$ and $e'' \in X_2$. The Cartesian product will be denoted by $X_1 \times X_2$.

DEFINITION 2. By the *Cartesian product of the events* $A_1 \subset X_1$ *and* $A_2 \subset X_2$ where X_1 and X_2 are two elementary spaces, we shall mean the event $A = A_1 \times A_2$

II. CONDITIONAL PROBABILITY. STOCHASTIC INDEPENDENCE 67

in the space $X_1 \times X_2$ equal to the set of all pairs (e', e'') such that $e' \in A_1$ and $e'' \in A_2$.

DEFINITION 3. By the *Cartesian product of two fields of events*, S_1 and S_2, in the spaces of elementary events X_1 and X_2, respectively, we shall mean the smallest countably additive field S of subsets of the Cartesian product $X_1 \times X_2$ containing all sets of the form $A_1 \times A_2$ with $A_1 \in S_1$ and $A_2 \in S_2$. This field will be denoted by $S_1 \times S_2$. In particular, $S_1 \times S_2$ contains all sets of the form $A_1 \times X_2$ with $A_1 \in S_1$, and all sets of the form $X_1 \times A_2$ with $A_2 \in S_2$.

The meaning of these definitions is such that if a space X_1 and a field S_1 describe one experiment, and a space X_2 and a field S_2 describe another experiment, then $X_1 \times X_2$ and $S_1 \times S_2$ describe these two experiments jointly. The events defined in terms of the first experiment only are described as those events in $S_1 \times S_2$ which are of the form $A_1 \times X_2$ with $A_1 \in S_1$. Similarly, the events defined in terms of the second experiment only are described as those events in $S_1 \times S_2$ which are of the form $X_1 \times A_2$ with $A_2 \in S_2$. The products of two events, one described in terms of the results of the first experiment and the other described in terms of the results of the second experiment have the form $A_1 \times A_2 \in S_1 \times S_2$ with $A_1 \in S_1, A_2 \in S_2$.

If our experiments are independent, in the sense that the result of one of them does not influence the conditions of the other, then the events $A_1 \times X_2$ and $X_1 \times A_2$ should be independent, i.e. the probability of the product of these events should be equal to the product of the probabilities of these events. The question arises, whether this postulate can always be satisfied?

Let us start from a simple particular case, where the space of elementary events is at most countable. As we know, in order to define probability in such a space it suffices to define the probabilities for elementary events. We have the following

THEOREM 1. *If* $X_1 = \{e'_1, e'_2, \ldots\}$ *and* $X_2 = \{e''_1, e''_2, \ldots\}$ *are two at most countable spaces of elementary events, and the probabilities are defined by equalities*

(19.1) $\qquad P_1(\{e'_i\}) = p'_i, \quad P_2(\{e''_j\}) = p''_j,$

and if we define probability in the Cartesian product $X_1 \times X_2$ *by*

(19.2) $\qquad P(\{(e'_i, e''_j)\}) = p'_i p''_j,$

then for every event $A_1 \subset X_1$ *and* $A_2 \subset X_2$ *we have*

(19.3) $\qquad P(A_1 \times A_2) = P_1(A_1) P_2(A_2).$

This theorem states that relations (19.2) determine the probability distribution in $X_1 \times X_2$ such that all events defined in terms of the results of the first trial are independent of the events defined in terms of the results of the second trial.

PROOF. Relation (19.3) results from the following equations, the most important of which, the second from the end, is based on the properties of numerical series:

$$P(A_1 \times A_2) = \sum_{\substack{e'_i \in A_1 \\ e''_j \in A_2}} (\{e'_i, e''_j\}) = \sum_{\substack{e'_i \in A_1 \\ e''_j \in A_2}} P_1(\{e'_i\}) P_2(\{e''_j\})$$

$$= \Big[\sum_{e'_i \in A_1} P_1(\{e'_i\})\Big] \cdot \Big[\sum_{e''_j \in A_2} P_2(\{e''_j\})\Big] = P_1(A_1) P_2(A_2).$$

In the general case we have the following theorem:

THEOREM 2. *Let X_1 and X_2 be two spaces of elementary events, let S_1 and S_2 be two σ-fields of events in the spaces X_1 and X_2, and let P_1 and P_2 be the probabilities defined on S_1 and S_2 respectively. Then there exists only one probability P on $S_1 \times S_2$ such that*

(19.4) $P(A_1 \times A_2) = P_1(A_1) P_2(A_2), \quad \text{if} \quad A_1 \in S_1, \quad A_2 \in S_2.$

The proof of this theorem is somewhat complex and will be omitted.

DEFINITION 4. The probability P which appears in Theorem 2 is called the *Cartesian product of the probabilities* P_1 and P_2 and will be denoted by $P_1 \times P_2$.

We shall now present the definition of marginal distribution.

DEFINITION 5. If P is a probability distribution defined on the σ-field $S_1 \times S_2$ where S_1 and S_2 are σ-fields of events in spaces X_1 and X_2 of elementary events, then the *marginal distributions* P_1 and P_2 of P are defined on S_1 and S_2 respectively as

$$P_1(A_1) = P(A_1 \times X_2) \quad \text{for} \quad A_1 \in S_1$$

and

$$P_2(A_2) = P(X_1 \times A_2) \quad \text{for} \quad A_2 \in S_2.$$

According to Definition 5, the probability distribution $P_1 \times P_2$ in Theorem 2 equals the Cartesian product of its marginal distributions.

We shall now discuss probability distributions on the plane. If we have two experiments, each of them producing a number as a result, then the pair of experiments is described by a pair of real numbers, that is, by a point of the plane. According to Definition 1, the Euclidean plane equals to the Cartesian product of two real lines.

When discussing probability distributions on the plane in § 13, we introduced the notion of the field of Borel sets of the plane. What is the relation between this field and the Cartesian product of the Borel fields of sets of the real line? The answer is given by the following theorem:

THEOREM 3. *If $X_1 = R_1$ and $X_2 = R_1$, i.e. if X_1 and X_2 are real lines, while S_1 and S_2 are fields of Borel sets in X_1 and X_2, then $X_1 \times X_2 = R_2$, i.e. $X_1 \times X_2$ is the Euclidean plane, while $S_1 \times S_2$ is the field of Borel sets of the plane.*

PROOF. $X_1 \times X_2$ is the Euclidean plane, since by the definition of the Cartesian product, $X_1 \times X_2$ equals to the set of pairs of real numbers, which equals the Euclidean plane. It remains to prove that $S_1 \times S_2 = S$, where S is the class of Borel sets of the plane. We easily note that $S \subset S_1 \times S_2$. Indeed, by definition, S is the smallest countably additive class of sets containing all sets of the form $\{(x, y): x < a, y < b\}$. But every such set is the Cartesian product of the sets $\{x: x < a\}$ and $\{y: y < b\}$; since the latter are Borel sets on the real line, we have the inclusion $S \subset S_1 \times S_2$.

To prove the reverse inclusion, we shall show that the class K of sets of the form $A_1 \times A_2$ with $A_1 \in S_1$ and $A_2 \in S_2$ is contained in the field of Borel sets of the plane, that is, $K \subset S$. This, however, is quite clear. The field S, as the smallest σ-field containing all sets of the form $\{(x, y): x < a, y < b\}$, contains also the smallest σ-field which contains the sets of the form $A_1 \times A_2$, where A_1 is any set of the form $\{x: x < a\}$ and A_2 is any set of the form $\{y: y < b\}$. But this means that S contains all sets of the form $A_1 \times A_2$ with $A_1 \in S_1$ and A_2 of the form $\{y: y < b\}$. If now A_1 is a fixed set in S_1, then, by the above reasoning, S must contain the smallest σ-field which contains all sets of the form $A_1 \times \{y: y < b\}$. But this implies that S contains all sets of the form $A_1 \times A_2$ with $A_2 \in S_2$. Since $A_1 \in S_1$ is arbitrary, we have $K \subset S$. Therefore, the smallest σ-additive field of sets which contains the class K, i.e. the σ-field $S_1 \times S_2$, is contained in the smallest σ-field S' containing S. But, by definition, S is a σ-field, hence $S' = S$. Thus, $S_1 \times S_2 \subset S$, which was to be proved.

We shall now prove a theorem which characterizes the distribution functions of those probability distributions on the plane which are equal to the products of their marginal distributions.

THEOREM 4. *If X_1, X_2 are two real lines, S_1 and S_2 are Borel fields of sets in X_1 and X_2, P is a probability distribution on $S_1 \times S_2$ and P_1, P_2 are its marginal distributions, defined on S_1 and S_2 respectively, then P is the Cartesian product of the distributions P_1 and P_2, i.e., $P = P_1 \times P_2$ if and only if the distribution function $F(x, y)$ of the distribution P is of the form*

(19.5) $$F(x,y) = F_1(x)F_2(y),$$

where $F_1(x)$ and $F_2(y)$ are one-dimensional distributions.

PROOF. If $P = P_1 \times P_2$, then for every set of the form $A_1 \times A_2$ with $A_1 \in S_1$, $A_2 \in S_2$, we have

(19.6) $$P(A_1 \times A_2) = P_1(A_1)P_2(A_2).$$

In particular, we may put $A_1 = \{x: x < a\}$ and $A_2 = \{y: y < b\}$. Then $A_1 \times A_2 = \{(x,y): x < a, y < b\}$; we have, however, $P_1(A_1) = F_1(a)$, $P_2(A_2) = F_2(b)$, $P(A_1 \times A_2) = F(a,b)$, where F_1 and F_2 are one-dimensional distribution functions of the distributions P_1 and P_2. Substituting these equalities in (19.6), we obtain (19.5), which proves the necessity.

Conversely, suppose that $F_1(x)$, $F_2(y)$ are one-dimensional distribution functions. Then the function $F(x,y) = F_1(x)F_2(y)$ is a joint distribution function, which can easily be verified by checking, and F_1, F_2 are its marginals. The distribution functions F_1 and F_2 determine the marginal distributions P_1 and P_2. By Theorem 2 there exists exactly one probability distribution P on $S_1 \times S_2$ such that $P(A_1 \times A_2) = P_1(A_1)P_2(A_2)$ for $A_1 \in S_1$ and $A_2 \in S_2$. This distribution equals to the Cartesian product $P_1 \times P_2$ of the distributions P_1 and P_2. Thus, the distribution function $F(x,y)$ equals the probability distribution function of a probability distribution equal to the Cartesian product of its own marginals, which proves the sufficiency of condition (19.5).

What we have discussed here for the case of the Cartesian products of two spaces of elementary events extends to the case of the Cartesian products of any finite number of spaces. Here are generalizations of Definitions 1–3.

DEFINITION 6. Given spaces $X_1, X_2, \ldots, X_n$, the set of all ordered n-tuples $(e_1, e_2, \ldots, e_n)$ with $e_i \in X_i$, $i = 1, 2, \ldots, n$, will be called the *Cartesian product of the spaces* $X_1, X_2, \ldots, X_n$, and will be denoted by $X_1 \times X_2 \times \ldots \times X_n$.

DEFINITION 7. Given spaces $X_1, X_2, \ldots, X_n$ and events $A_1 \subset X_1$, $A_2 \subset X_2, \ldots, A_n \subset X_n$, the event $A = A_1 \times A_2 \times \ldots \times A_n \subset X_1 \times X_2 \times \ldots \times X_n$ consisting of all systems $(e_1, e_2, \ldots, e_n)$ such that $e_i \in A_i$, $i = 1, 2, \ldots, n$, will be called the *Cartesian product of the events* $A_1, A_2, \ldots, A_n$.

DEFINITION 8. Given fields $S_1, S_2, \ldots, S_n$ of events in spaces $X_1, X_2, \ldots, X_n$ of elementary events, the smallest σ-field of events in the space $X_1 \times X_2 \times \ldots \times X_n$ which contains all events of the form $A_1 \times A_2 \times \ldots \times A_n$ with $A_i \in S_i$, $i = 1, 2, \ldots, n$, will be called *Cartesian product of fields* and will be denoted by $S = S_1 \times S_2 \times \ldots \times S_n$.

Some remarks concerning Definition 6 seem to be in order. If we have three spaces X_1, X_2 and X_3, then, according to Definition 1, we can form first the

III. CONDITIONAL PROBABILITY. STOCHASTIC INDEPENDENCE

Cartesian product of X_1 and X_2, and then the Cartesian product of the spaces $X_1 \times X_2$ and X_3. In this way we will obtain the space $(X_1 \times X_2) \times X_3$, whose elements will be pairs $((x_1, x_2), x_3)$, the first element in each pair being the pair (x_1, x_2). Similarly, if we formed first the Cartesian product of the spaces X_2 and X_3, and then the Cartesian product of the spaces X_1 and $X_2 \times X_3$, we would obtain the space $X_1 \times (X_2 \times X_3)$ whose elements would be pairs $(x_1, (x_2, x_3))$. Thus, the spaces $X_1 \times X_2 \times X_3$, $(X_1 \times X_2) \times X_3$ and $X_1 \times (X_2 \times X_3)$ are, formally speaking, all different. We shall use, however, the fact that the elements of these spaces can be put in a natural way into a one-to-one correspondence, and—at the cost of a slight logical error—we shall identify the elements

$$(x_1, x_2, x_3), \quad ((x_1, x_2), x_3), \quad (x_1, (x_2, x_3)).$$

According to this convention, the three spaces mentioned above will be treated as identical. The same identification will be used also for the case of a larger number of dimensions. Thus, for instance, we shall treat as identical the elements $((x_1, x_2), (x_3, x_4))$ and $((x_1, x_2, x_3), x_4)$ of spaces $(X_1 \times X_2) \times (X_3 \times X_4)$ and $(X_1 \times X_2 \times X_3) \times X_4$.

One more remark. Sometimes we shall speak of the Cartesian product $X_1 \times \ldots \times X_n$ as of an n-dimensional space; here the spaces X_i will be one-dimensional axes of the space $X_1 \times \ldots \times X_n$. In statements of this kind we shall mean only that the space $X_1 \times \ldots \times X_n$ is built from its components $X_1, \ldots, X_n$ in the manner described above; the spaces X_i may be multidimensional in the usual sense; for instance, they can be Euclidean planes (see Definition 10).

The above remarks apply equally to Cartesian products of events and fields.

According to these remarks, the product $X_1 \times \ldots \times X_n$ can be treated as a multiple repetition of a product of two spaces, for instance

$$X_1 \times \ldots \times X_n = (\ldots (X_1 \times X_2) \times \ldots) \times X_n.$$

Thus, we can use induction and prove the theorems which we have proved for the products of two spaces in the case of the products of n spaces. Here are generalizations of Theorems 1 and 2.

THEOREM 5. *If $X_1, X_2, \ldots, X_n$ are at most countable spaces of elementary events, and $P_1, P_2, \ldots, P_n$ are probabilities in these spaces, then setting for every*

$$(e_1, e_2, \ldots, e_n) \in X_1 \times X_2 \times \ldots \times X_n$$

(19.7) $\qquad P(\{(e_1, \ldots, e_n)\}) = P_1(\{e_1\}) P_2(\{e_2\}) \ldots P_n(\{e_n\})$

we define in $X_1 \times X_2 \times \ldots \times X_n$ a probability measure satisfying the property that for $A_i \in S_i$, $i = 1, 2, \ldots, n$, we have

(19.8) $\qquad P(A_1 \times A_2 \times \ldots \times A_n) = P_1(A_1) P_2(A_2) \ldots P_n(A_n).$

THEOREM 6. *If $X_1, \ldots, X_n$ are spaces of elementary events, $S_1, \ldots, S_n$ are fields of events in these spaces, and $P_1, \ldots, P_n$ are probabilities on $S_1, \ldots, S_n$ respectively, then there exists exactly one probability P defined on $S_1 \times \ldots \times S_n$ such that for $A_i \in S_i$, $i = 1, \ldots, n$, we have*

$$P(A_1 \times A_2 \times \ldots \times A_n) = P_1(A_1) P_2(A_2) \ldots P_n(A_n).$$

DEFINITION 9. The probability P which appears in Theorem 6 is called the *Cartesian product of the probabilities* $P_1, \ldots, P_n$ and will be denoted by $P_1 \times \times \ldots \times P_n$.

DEFINITION 10. If X and S are respectively the Cartesian product of the spaces of elementary events $X_1, \ldots, X_n$ and fields $S_1, \ldots, S_n$ of events in these spaces, and P is a probability distribution defined on S, then the distributions $P_1, \ldots, P_n$ defined on $S_1, \ldots, S_n$ respectively by the relation

$$P_i(A_i) = P(X_1 \times \ldots \times X_{i-1} \times A_i \times X_{i+1} \times \ldots \times X_n)$$

will be called the *marginal distributions* of P.

Clearly, the distribution $P_1 \times \ldots \times P_n$ equals the product of its marginal distributions.

In a similar manner one can generalize Theorems 3 and 4 dealing with Euclidean spaces:

THEOREM 7. *If $X_1, \ldots, X_n$ are real lines, and $S_1, \ldots, S_n$ are Borel fields of subsets of $X_1, \ldots, X_n$ respectively, then $S_1 \times \ldots \times S_n$ is the Borel field of subsets of $X_1 \times \times \ldots \times X_n$, i.e. of n-dimensional Euclidean space.*

THEOREM 8. *The probability distribution P in the n-dimensional Euclidean space equals the product of its marginal distributions if and only if its probability distribution function can be represented in the form of a product of n one-dimensional distributions.*

From what we said above it follows that if the probability distribution P_1 describes a certain experiment, hence to describe n independent repetitions of this experiment one should use the distribution

$$P = \underbrace{P_1 \times P_1 \times \ldots \times P_1}_{n \text{ times}}.$$

Finally, let us mention that the theorem on the product of finitely many probabilities can be extended to the case of a sequence of such distributions. Namely, we have

THEOREM 9. *Let $X_1, X_2, \ldots$ be spaces of elementary events with distinguished fields of events $S_1, S_2, \ldots$ Let $P_1, P_2, \ldots$ be the probabilities defined on $S_1, S_2, \ldots$*

respectively. Finally, let S be the smallest σ-field of events in the product

$$X = X_1 \times X_2 \times \ldots$$

(which, by definition, consists of all sequences $(e_1, e_2, \ldots)$ such that $e_i \in X_i$), which contains all events of the form

$$A_1 \times A_2 \times \ldots \times A_n \times X_{n+1} \times X_{n+2} \times \ldots,$$

for arbitrary n and $A_i \in S_i$, $i = 1, 2, \ldots, n$. Then there exists exactly one probability P on S such that

$$P(A_1 \times A_2 \times \ldots \times A_n \times X_{n+1} \times X_{n+2} \times \ldots) = P_1(A_1) P_2(A_2) \ldots P_n(A_n)$$

for every n and $A_i \in S_i$, $i = 1, 2, \ldots, n$.

The probability P will be called the *Cartesian product of the probabilities* $P_1, P_2, \ldots$

CHAPTER IV

Classical Limit Theorems

§ 20. BINOMIAL DISTRIBUTION

Let us consider the following example.

EXAMPLE 1. Suppose that a deck of cards contains 12 red cards and 12 black cards. Imagine an experiment consisting of n successive repetitions of the following procedure: drawing of one card from the deck, noting its colour, and replacing it back to the deck, so that each time we draw from the entire deck. What is the probability that we shall draw a red card k times?

Our experiment is an n-fold repetition of the experiment of drawing one card, and all repetitions are performed in the same conditions. One experiment can be described in terms of a two-point space of elementary events $X_1 = \{0, 1\}$ where 0 denotes drawing of a black card, and 1 denotes drawing of a red card. To describe the whole series of n drawings we take the n-fold Cartesian product of the space X_1, that is, the space

$$X = \underbrace{X_1 \times X_1 \times \ldots \times X_1}_{n \text{ times}},$$

which consists of all systems $(i_1, i_2, \ldots, i_n)$ with each i_j equal 0 or 1. Thus, the space X contains 2^n elements.

Next, in each drawing the probability of drawing a black card is 1/2, since the colours are symmetric. Thus, $P_1(\{1\}) = P_1(\{0\}) = 1/2$. It follows that the probability

$$P = \underbrace{P_1 \times P_1 \times \ldots \times P_1}_{n \text{ times}}$$

in the space X will be equal for all elementary events, and, consequently, for every system $(i_1, i_2, \ldots, i_n)$ we shall have

$$P(\{(i_1, \ldots, i_n)\}) = \frac{1}{2^n}.$$

The event A_k consisting of drawing a red card k times consists of all systems $(i_1, i_2, \ldots, i_n)$ in which number 1 appears exactly k times. From combinatorics we know that the number of such systems equals

$$\binom{n}{k} = \frac{n!}{k!(n-k)!}.$$

Consequently,

$$P(A_k) = \binom{n}{k} \frac{1}{2^n}.$$

EXAMPLE 2. What is the probability of the event A_k of drawing k red cards in n drawings if the deck contains 12 red cards, and 24 black ones?

To describe this experiment we may use the same space of elementary events as in Example 1. The difference will consist in a different probability of drawing a red card in a single experiment: now it is equal to $12/36 = 1/3$ and not $1/2$. We have, therefore, $P_1(\{1\}) = 1/3$, $P_1(\{0\}) = 2/3$. Thus, in the space X

$$P(\{(i_1, i_2, \ldots, i_n)\}) = \left(\frac{1}{3}\right)^k \left(\frac{2}{3}\right)^{n-k},$$

if the system $(i_1, i_2, \ldots, i_n)$ contains k numbers 1 and $n-k$ numbers 0. We have, therefore,

$$P(A_k) = \binom{n}{k} \left(\frac{1}{3}\right)^k \left(\frac{2}{3}\right)^{n-k}.$$

In general, we shall use the term *Bernoulli trials** for independent trials with probability P_1 defined on a two-element space of elementary events $X_1 = \{0, 1\}$ as

$$P_1(\{1\}) = p, \quad P_1(\{0\}) = 1-p,$$

where p is an arbitrary number from the interval $\langle 0, 1 \rangle$.

Denote by X the n-fold Cartesian product of the space X_1, as in the above examples, i.e., X consists of systems $(i_1, i_2, \ldots, i_n)$ where i_j, $j = 1, 2, \ldots, n$, is either 1 or 0. Let A_k, $k = 0, 1, \ldots, n$, denote the event consisting of systems $(i_1, i_2, \ldots, i_n)$ which contain exactly k numbers 1. Finally, let P denote the probability $\underbrace{P_1 \times P_1 \times \ldots \times P_1}_{n \text{ times}}$ defined on X. If we call the element 1 of the space $X_1 = \{0, 1\}$ a "success", then A_k means the event "k succeeds in n trials". As before, we easily compute that

$$P(A_k) = \binom{n}{k} p^k (1-p)^{n-k}.$$

Note that the events $A_0, A_1, \ldots, A_k, \ldots, A_n$ are disjoint and their sum equals to the sure event, i.e. $A_0 \cup A_1 \cup \ldots \cup A_n = X$. Thus,

$$P(A_0) + P(A_1) + \ldots + P(A_n) = 1$$

---
* Jacques Bernoulli, 1654–1705.

consequently, the probabilities $P(A_k)$, $k = 0, 1, ..., n$, may serve for defining a probability distribution in an $(n+1)$-element space of elementary events consisting of points $0, 1, ..., n$, in which the event k corresponds to A_k in X. In fact, for the probability of elementary event k we may take $P(A_k)$. This is the probability distribution of the number of successes in n Bernoulli trials.

DEFINITION 1. The discrete probability distribution on the real line defined by the equalities

$$P(\{k\}) = \binom{n}{k} p^k (1-p)^{n-k}, \quad k = 0, 1, ..., n,$$

where p is an arbitrary number from the interval $0 \leq p \leq 1$ will be called the *Bernoulli* or *binomial distribution*; probabilities in this distribution will be denoted by $b(k; n, p)$:

$$b(k; n, p) = \binom{n}{k} p^k (1-p)^{n-k}.$$

This distribution is called binomial, since the numbers $b(k; n, p)$ are equal to the successive terms of the expansion of $(p+q)^n$ where $q = 1-p$.

We shall prove a certain interesting property of the binomial distribution.

THEOREM 1. *The numbers*

$$b(0; n, p), b(1; n, p), ..., b(n; n, p)$$

form a sequence which first increases, and then decreases; the maximal term is $b(m; n, p)$, where m is an integer from the interval $(n+1)p-1 \leq m \leq (n+1)p$; if $(n+1)p$ is an integer, then $b(m; n, p) = b(m+1; n, p)$.

PROOF. Let us consider the ratio of two neighbouring terms of the sequence $b(0; n, p), b(1; n, p), ..., b(n; n, p)$. Writing $q = 1-p$ we have

$$\frac{b(k+1; n, p)}{b(k; n, p)} = \frac{\binom{n}{k+1} p^{k+1} (1-p)^{n-k-1}}{\binom{n}{k} p^k (1-p)^{n-k}}$$

$$= \frac{n!}{(n-k-1)!(k+1)!} \cdot \frac{k!(n-k)!}{n!} \cdot \frac{p}{q} = \frac{n-k}{k+1} \cdot \frac{p}{q}.$$

Let us find numbers k for which this ratio exceeds 1. Solving the inequality

$$\frac{n-k}{k+1} \cdot \frac{p}{q} > 1$$

we obtain successively

$$(n-k)p > (k+1)q,$$

$$np > kp + kq + q,$$

IV. CLASSICAL LIMIT THEOREMS

and finally
$$np-q > k(p+q) = k.$$

Thus, the ratio under consideration exceeds 1 as long as $k < np-q = (n+1)p-1$.

In a similar way we see that the ratio in question is smaller than 1 for k satisfying the inequality $(n+1)p-1 < k$.

This completes the proof of our theorem.

The term $b(m; n, p)$ from the theorem is sometimes called the *central term*, and number m is called the *most likely number of successes*. It should be noted, however, that for large n all terms of the sequence $b(k; n, p)$, including the largest, are small.

We shall now give the definition of a multinomial distribution.

Let P_1 denote a probability distribution in a k-element space of elementary events $X_1 = \{e_1, \ldots, e_k\}$, with $P_1(\{e_i\}) = p_i$. This distribution may be used to describe an experiment which could lead to k different results. A series of n independent observations of such an experiment may be described by the probability distribution $P = \underbrace{P_1 \times P_1 \times \ldots \times P_1}_{n \text{ times}}$, defined in the space $X = X_1 \times X_1 \times \ldots \times X_1$ consisting of systems $(e_{i_1}, \ldots, e_{i_n})$ such that $e_{i_j} \in X_1$, and $P(\{(e_{i_1}, \ldots, e_{i_n})\}) = P_1(\{e_{i_1}\}) \ldots P_1(\{e_{i_n}\})$.

Now let $A_{n_1, n_2, \ldots, n_k}$ denote the event consisting of n_1 occurrences of $e_1, \ldots,$ and n_k occurrences of e_k in n repetitions; the order of occurrences is irrelevant. Thus, the event $A_{n_1, \ldots, n_k}$ consists of systems $(e_{i_1}, \ldots, e_{i_n})$ which contain n_1 elements e_1, n_2 elements $e_2, \ldots, n_k$ elements e_k. The probability of every such system equals $p_1^{n_1} p_2^{n_2} \ldots p_k^{n_k}$, and the total number of such systems, as known from combinatorics, is $n!/n_1! \ldots n_k!$. We have therefore

$$P(A_{n_1, \ldots, n_k}) = \frac{n!}{n_1! \ldots n_k!} p_1^{n_1} p_2^{n_2} \ldots p_k^{n_k}.$$

We can now introduce the following definition:

DEFINITION 2. By a *multinomial distribution* we shall mean a discrete distribution in a k-dimensional Euclidean space defined by the equalities

$$P\{(n_1, \ldots, n_k)\} = \frac{n!}{n_1! \ldots n_k!} p_1^{n_1} \ldots p_k^{n_k},$$

where $p_1, \ldots, p_k$ are non-negative numbers whose sum is 1, and $n_1, \ldots, n_k$ are systems of k non-negative integers whose sum is n.

This distribution is called multinomial because the probabilities assigned to the elementary events are equal to the terms of the expansion of $(p_1 + \ldots + p_k)^n$.

§ 21. POISSON THEOREM AND POISSON DISTRIBUTION

The calculation of binomial probabilities $b(k; n, p)$ directly from the definition is somewhat cumbersome. If we were to construct tables of those probabilities, the tables would have to be rather large, since the distribution depends on two parameters, n and p. It turns out, however, that one can find reasonably simple expressions which approximate the probabilities $b(k; n, p)$ as $n \to \infty$, the approximation being quite good already for small values of n. Our next aim will be to find these approximations.

The first theorem which we shall prove concerns the behaviour of binomial distributions in cases where n is large and np is moderately large. We deal with this case always when the number n of trials is large, while the probability p of success in a single trial is small, and we want to find the probability of the joint number of successes.

THEOREM 1 (Poisson*). *If for $n \to \infty$ the probabilities p_n vary with n in such way that*

(21.1) $$\lim_{n \to \infty} np_n = \lambda,$$

then for every k

(21.2) $$\lim_{n \to \infty} b(k; n, p_n) = \frac{\lambda^k}{k!} e^{-\lambda}.$$

PROOF. Let us fix k. We can write

$$b(k; n, p_n) = \frac{n!}{k!(n-k)!} p_n^k (1-p_n)^{n-k}$$

$$= \frac{1}{k!} n(n-1) \ldots (n-k+1) p_n^k (1-p_n)^{n-k}$$

$$= \frac{1}{k!} \cdot \frac{n}{n} \cdot \frac{n-1}{n} \cdot \ldots \cdot \frac{n-k+1}{n} (np_n)^k (1-p_n)^{n-k}.$$

Clearly, we have

$$\lim_{n \to \infty} \frac{n}{n} \cdot \frac{n-1}{n} \cdot \ldots \cdot \frac{n-k+1}{n} = 1$$

and, in view of condition (21.1) we have also

$$\lim_{n \to \infty} (np_n)^k = \lambda^k.$$

Thus, the theorem will be proved when we show that

$$\lim_{n \to \infty} (1-p_n)^{n-k} = e^{-\lambda}.$$

---
* S. D. Poisson, 1781–1840.

IV. CLASSICAL LIMIT THEOREMS

Since we can write
$$(1-p_n)^{n-k} = (1-p_n)^n(1-p_n)^{-k},$$
and $\lim_{n\to\infty}(1-p_n)^{-k} = 1$ (since we have $\lim_{n\to\infty} p_n = 0$ by (21.1)), it suffices to prove that

(21.3) $$\lim_{n\to\infty}(1-p_n)^n = e^{-\lambda}.$$

We know from analysis, that for every a
$$\lim_{n\to\infty}\left(1-\frac{a}{n}\right)^n = e^{-a},$$
where the convergence is uniform in every finite interval $a_0 \leqslant a \leqslant a_1$. In view of (21.1) all numbers np_n can be restricted to such a finite interval, whence for every $\varepsilon > 0$ we have for $n > n_1(\varepsilon)$:
$$\left|\left(1-\frac{np_n}{n}\right)^n - e^{-np_n}\right| < \varepsilon.$$

Next, in view of convergence (21.1) and the continuity of the function e^{-a}, we can write for $n > n_2(\varepsilon)$:
$$|e^{-np_n} - e^{-\lambda}| < \varepsilon.$$
Denoting by $n(\varepsilon)$ the larger of the numbers $n_1(\varepsilon)$ and $n_2(\varepsilon)$, we have for $n > n(\varepsilon)$:
$$|(1-p_n)^n - e^{-\lambda}| = \left|\left(1-\frac{np_n}{n}\right)^n - e^{-\lambda}\right|$$
$$= \left|\left(1-\frac{np_n}{n}\right)^n - e^{-np_n} + e^{-np_n} - e^{-\lambda}\right|$$
$$\leqslant \left|\left(1-\frac{np_n}{n}\right)^n - e^{-np_n}\right| + |e^{-np_n} - e^{-\lambda}| < 2\varepsilon,$$
which proves relation (21.3) and completes the proof of Theorem 1. Here are some examples illustrating the degree of approximation.

EXAMPLE 1. *Birthdays.* What is the probability p_k that among 500 persons there will be exactly k persons born on New Year's Day? If these persons are selected at random, and if we assume that the dates of birth are uniformly distributed over the year, we could assume that we look for the probability of exactly k successes in 500 Bernoulli trials in which the probability of success in a single trial is 1/365. From the binomial distribution we could compute $p_0 = (364/365)^{500} = 0.2537$, $p_1 = 0.3484$, $p_2 = 0.2388$, $p_3 = 0.1089, \ldots$ If we used the Poisson approximation taking $\lambda = 500 \dfrac{1}{365} = 1.3699$, we would obtain instead of the above values the following numbers: 0.2541, 0.3481, 0.2385, 0.1089, ... As we see, the differences appear at the fourth decimal place.

EXAMPLE 2. *Defective items.* Suppose that a machine produces screws under fixed conditions, and that the probability of producing a defective screw equals 0.015. What is the probability that among 100 screws there will be no defective ones? Applying the scheme of Bernoulli trials with probability of success 0.015 in a single trial, we find that this probability equals $p = (0.985)^{100} = 0.22061$. The Poisson approximation for this probability with $\lambda = 100 \cdot 0.015 = 1.5$ gives $e^{-1.5} = 0.22313$.

Let us note that the numbers

(21.4) $$p(k; \lambda) = \frac{\lambda^k}{k!} e^{-\lambda}$$

have the sum one for every value of λ:

$$\sum_{k=0}^{\infty} p(k; \lambda) = 1.$$

Indeed, we have

$$\sum_{k=0}^{\infty} p(k; \lambda) = \sum_{k=0}^{\infty} \frac{\lambda^k}{k!} e^{-\lambda} = e^{-\lambda}\left(1 + \lambda + \frac{\lambda^2}{2!} + \frac{\lambda^3}{3!} + \dots\right),$$

and the series in parentheses is an expansion of e^λ. Thus, these numbers may be treated as probabilities.

DEFINITION 1. By the *Poisson distribution* we mean a probability distribution on the real line such that for every non-negative integer k we have

$$P(\{k\}) = p(k; \lambda).$$

As we see, the Poisson distribution is a discrete distribution: the whole probability is concentrated on the countable set $\{0, 1, 2, \dots\}$ of non-negative integers. This distribution depends on one parameter λ.

§ 22. THE DE MOIVRE–LAPLACE THEOREM

The second theorem which we announced concerns the limit of the binomial distribution $b(k; n, p)$ for fixed p and increasing n. Thus, we shall investigate the limit distribution of the number of successes in an increasing series of Bernoulli trials.

IV. CLASSICAL LIMIT THEOREMS

THEOREM 1 (Local de Moivre–Laplace theorem*). *For a fixed $p = 1-q$, $0 < p < 1$, and $k = k_n$ changing with n in such a way that there exists the limit*

(22.1) $$\lim_{n\to\infty} \frac{k_n - np}{\sqrt{npq}} = x,$$

we have

(22.2) $$\lim_{n\to\infty} \sqrt{npq}\, b(k_n; n, p) = \frac{1}{\sqrt{2\pi}} e^{-x^2/2},$$

where the convergence is nearly uniform with respect to the variable x, i.e., for every $A > 0$ and $\varepsilon > 0$ there exists an $n(A, \varepsilon)$ such that for $n > n(A, \varepsilon)$ we have

$$\left| \sqrt{npq}\, b(k; n, p) - \frac{1}{\sqrt{2\pi}} e^{-x_k^2/2} \right| < \varepsilon$$

provided that

$$-A < x_k = \frac{k - np}{\sqrt{npq}} < A.$$

PROOF. First we show that the assumptions of the theorem imply relation (22.2).

Using the definition of the probabilities $b(k; n, p)$ we can write

$$\sqrt{npq}\, b(k; n, p) = \sqrt{npq}\, \frac{n!}{k!(n-k)!} p^k q^{n-k}.$$

Applying the Stirling formula, which asserts that for every natural n we have

$$n! = n^n e^{-n} \sqrt{2\pi n}\, e^{\theta/12n}, \quad 0 < \theta \leq 1,$$

we obtain the inequality

$$\sqrt{npq}\, b(k; n, p)$$

$$= \sqrt{npq}\, p^k q^{n-k} \frac{n^n e^{-n} \sqrt{2\pi n}\, e^{\theta_1/12n}}{k^k e^{-k} \sqrt{2\pi k}\, e^{\theta_2/12k} (n-k)^{n-k} e^{-n+k} \sqrt{2\pi(n-k)}\, e^{\theta_3/12(n-k)}}$$

$$= \frac{1}{\sqrt{2\pi}} \sqrt{\frac{np}{k} \cdot \frac{nq}{(n-k)}} \left(\frac{np}{k}\right)^k \left(\frac{nq}{n-k}\right)^{n-k} \exp\left(\frac{\theta_1}{12n} - \frac{\theta_2}{12k} - \frac{\theta_3}{12(n-k)}\right).$$

If we put now

$$r_n(k) = \sqrt{\frac{np}{k} \cdot \frac{nq}{n-k}},$$

---

* Abraham de Moivre, 1667–1754. Pierre Simon Laplace, 1749–1827.

$$s_n(k) = \left(\frac{np}{k}\right)^k \left(\frac{nq}{n-k}\right)^{n-k},$$

$$t_n(k) = \exp\left(\frac{\theta_1}{12n} - \frac{\theta_2}{12k} - \frac{\theta_3}{12(n-k)}\right),$$

we obtain

$$\sqrt{npq}\, b(k;n,p) = \frac{1}{\sqrt{2\pi}}\, r_n(k) s_n(k) t_n(k).$$

To prove the theorem it suffices to investigate the limits of the functions $r_n(k)$, $s_n(k)$ and $t_n(k)$.

Put

$$x = \frac{k-np}{\sqrt{npq}}$$

and let us treat x, and hence also k, as continuous variables. In this case we have

$$k = np + x\sqrt{npq}, \quad n-k = nq - x\sqrt{npq},$$

and

$$R_n(x) = r_n(np + x\sqrt{npq}), \quad S_n(x) = s_n(np + x\sqrt{npq}),$$

$$T_n(x) = t_n(np + x\sqrt{npq}).$$

We shall prove that

(a) *The functions $R_n(x)$ tend to the limit $R(x) = 1$ in every finite interval of the variable x;*

(b) *The functions $S_n(x)$ tend to the limit $S(x) = e^{-x^2/2}$ in every finite interval of the variable x;*

(c) *The functions $T_n(x)$ tend to the limit $T(x) = 1$ in every finite interval of the variable x.*

PROOF OF LEMMA (a). We have

$$R_n(x) = r_n(np + x\sqrt{npq}) = \sqrt{\frac{np}{np + x\sqrt{npq}} \cdot \frac{nq}{nq - x\sqrt{npq}}}$$

$$= \sqrt{\frac{p}{p + x\sqrt{\frac{pq}{n}}} \cdot \frac{q}{q - x\sqrt{\frac{pq}{n}}}},$$

which implies the truth of lemma (a).

IV. CLASSICAL LIMIT THEOREMS

PROOF OF LEMMA (b). We have
$$S_n(x) = s_n(np + x\sqrt{npq}) = \left(\frac{np}{np+x\sqrt{npq}}\right)^k \left(\frac{nq}{nq-x\sqrt{npq}}\right)^{n-k}$$
$$= \left(1+x\sqrt{\frac{q}{np}}\right)^{-k}\left(1-x\sqrt{\frac{p}{nq}}\right)^{-(n-k)}.$$

To prove (b) we shall show that $\log S_n(x)$ tends uniformly in every finite interval of the variable x to the limit $-x^2/2$.

Indeed, suppose that $-A < x < A$. Using the expansion
$$\log(1+x) = x - \frac{x^2}{2} + O(x^3),$$
where $O(t)$ denotes a function of t satisfying for some C the inequality $|O(t)| < Ct$, we obtain
$$\log S_n(x) = -k\left[x\sqrt{\frac{q}{np}} - \frac{x^2}{2}\cdot\frac{q}{np} + O\left(\frac{1}{n^{3/2}}\right)\right]$$
$$-(n-k)\left[-x\sqrt{\frac{p}{nq}} - \frac{x^2}{2}\cdot\frac{p}{nq} + O\left(\frac{1}{n^{3/2}}\right)\right].$$

Ordering these expressions with respect to the powers of x, and using the fact that $kO\left(\frac{1}{n^{3/2}}\right) = O\left(\frac{1}{\sqrt{n}}\right)$ and $(n-k)O\left(\frac{1}{n^{3/2}}\right) = O\left(\frac{1}{\sqrt{n}}\right)$ (since k and $n-k$ do not exceed n), we get

$$\log S_n(x) = -x\left(k\sqrt{\frac{q}{np}} - (n-k)\sqrt{\frac{p}{nq}}\right)$$
$$+\frac{x^2}{2}\left(k\frac{q}{np} + (n-k)\frac{p}{nq}\right) + O\left(\frac{1}{\sqrt{n}}\right)$$
$$= -x\left((np+x\sqrt{npq})\sqrt{\frac{q}{np}} - (nq-x\sqrt{npq})\sqrt{\frac{p}{nq}}\right)$$
$$+\frac{x^2}{2}\left((np+x\sqrt{npq})\sqrt{\frac{q}{np}} + (nq-x\sqrt{npq})\sqrt{\frac{p}{nq}}\right) + O\left(\frac{1}{\sqrt{n}}\right)$$
$$= -x(\sqrt{npq} + xq - \sqrt{npq} + xp)$$
$$+\frac{x^2}{2}\left(q + x\frac{q^{3/2}}{\sqrt{np}} + p - x\frac{p^{3/2}}{\sqrt{nq}}\right) + O\left(\frac{1}{\sqrt{n}}\right)$$
$$= -\frac{x^2}{2} + O\left(\frac{1}{\sqrt{n}}\right).$$

This proves the nearly uniform convergence of $\log S_n(x)$ to $-x^2/2$; "nearly uniform" meaning uniform in every finite interval. Thus, lemma (b) is proved.

PROOF OF LEMMA (c). We have

$$T_n(x) = t_n(np + x\sqrt{npq})$$
$$= \exp\left(\frac{\theta_1}{12n} - \frac{\theta_2}{12(np + x\sqrt{npq})} - \frac{\theta_3}{12(nq - x\sqrt{npq})}\right).$$

From this form of the function $T_n(x)$ we see that $\log T_n(x)$ tends nearly uniformly to zero, which proves lemma (c).

From lemmas (a), (b) and (c) it follows that

(d) *The functions* $\dfrac{1}{\sqrt{2\pi}} R_n(x) S_n(x) T_n(x)$ *tend nearly uniformly to the function* $\dfrac{1}{\sqrt{2\pi}} e^{-x^2/2}$.

In these lemmas the variable x was treated as a continuous variable. Let us now turn back to our binomial distribution, where k is not a continuous variable, since it can assume only integer values. If we denote by x_k the number $(k-np)/\sqrt{npq}$ with integer k, we can write

(22.3) $$\sqrt{npq}\, b(k; n, p) = \frac{1}{\sqrt{2\pi}} R_n(x_k) S_n(x_k) T_n(x_k).$$

By lemma (d) it follows that for every $A > 0$ and $\varepsilon > 0$ there exists an $n(A, \varepsilon)$ such that for $n > n(A, \varepsilon)$

$$\left|\sqrt{npq}\, b(k; n, p) - \frac{1}{\sqrt{2\pi}} e^{-x_k^2/2}\right| < \varepsilon,$$

provided that

$$|x_k| = \left|\frac{k-np}{\sqrt{npq}}\right| < A.$$

But this means precisely that the convergence is nearly uniform in x, as stated in the theorem.

We shall prove one more lemma concerning uniform convergence:

(e) *If the functions* $f_1(x), f_2(x), \ldots$ *are uniformly convergent to a continuous function* $f(x)$, $a \leq x \leq b$, *then for every sequence* x_n *converging to* x_0 *we have*

$$\lim_{n \to \infty} f_n(x_n) = f(x_0).$$

PROOF OF LEMMA (e). Fix $\varepsilon > 0$. Since the functions $f_n(x)$ are uniformly con-

IV. CLASSICAL LIMIT THEOREMS 85

vergent to $f(x)$, there exists an $n_1(\varepsilon)$ such that for all $n > n_1(\varepsilon)$ we have for every x
$$|f_n(x)-f(x)| < \tfrac{1}{2}\varepsilon.$$
In particular, for $n > n_1(\varepsilon)$:
(22.4) $$|f_n(x_n)-f(x_n)| < \tfrac{1}{2}\varepsilon.$$

Next, since $f(x)$ is continuous, we can find $\delta(\varepsilon) > 0$ such that $|x-x_0| < \delta(\varepsilon)$ implies $|f(x)-f(x_0)| < \tfrac{1}{2}\varepsilon$. From the convergence of the sequence x_n to x_0 it follows that there exists an $n_2(\varepsilon)$ such that for $n > n_2(\varepsilon)$ we have $|x_n-x_0| < \delta(\varepsilon)$. Thus, for $n > n_2(\varepsilon)$ we obtain
(22.5) $$|f(x_n)-f(x_0)| < \tfrac{1}{2}\varepsilon.$$

Adding inequalities (22.4) and (22.5) we see that for $n > n(\varepsilon)$, where $n(\varepsilon)$ is equal to the larger of the numbers $n_1(\varepsilon)$ and $n_2(\varepsilon)$, we have
$$|f_n(x_n)-f(x_0)| < \varepsilon,$$
which proves lemma (e).

Applying lemma (e) to the function $\dfrac{1}{\sqrt{2\pi}} R_n(x) S_n(x) T_n(x)$ and the sequence $x_n = (k_n - np)/\sqrt{npq}$ satisfying relation (22.1) we see, using relation (22.3), that (22.2) holds. This completes the proof of Theorem 1.

We shall now prove

THEOREM 2 (Integral de Moivre–Laplace theorem). *For a fixed p, $0 < p < 1$, and $q = 1-p$ the probability*
$$P_n(a, b) = \sum_{np+a\sqrt{npq} < k < np+b\sqrt{npq}} b(k; n, p)$$
that in the series of n Bernoulli trials with the probability of success equal to p the number of successes will be contained in the interval $np+a\sqrt{npq} < k < np+ +b\sqrt{npq}$, tends, as $n \to \infty$, to the limit $\int_a^b \dfrac{1}{\sqrt{2\pi}} e^{-x^2/2}dx$, or

(22.6) $$\lim_{n\to\infty} P_n(a, b) = \int_a^b \frac{1}{\sqrt{2\pi}} e^{-x^2/2}dx.$$

The convergence is uniform with respect to the variables a and b.

PROOF. Let
$$S(a, b) = \int_a^b \frac{1}{\sqrt{2\pi}} e^{-x^2/2} dx,$$

$$x_{nk} = \frac{k-np}{\sqrt{npq}}, \quad K_n = \{k: a < x_{nk} < b\},$$

$$S_n(a, b) = \frac{1}{\sqrt{2\pi}} \cdot \frac{1}{\sqrt{npq}} \sum_{k \in K_n} e^{-x_{nk}^2/2}.$$

We shall prove that

(f) *The sums $P_n(a, b)$ and $S_n(a, b)$ have the same limit,* and

(g) *The sums $S_n(a, b)$ converge to the integral $S(a, b)$.*

Lemmas (f) and (g) imply relation (22.6).

PROOF OF LEMMA (f). We can write

$$P_n(a, b) = \sum_{np+a\sqrt{npq} < k < np+b\sqrt{npq}} b(k; n, p) = \sum_{k \in K_n} b(k; n, p).$$

If a and b are fixed, by the second assertion of Theorem 1 for given $\varepsilon > 0$ we can find $n(\varepsilon)$ such that for $n > n(\varepsilon)$ and $k \in K_n$

$$\left| \sqrt{npq}\, b(k; n, p) - \frac{1}{\sqrt{2\pi}} e^{-x_{nk}^2/2} \right| < \varepsilon.$$

It follows that

$$|P_n(a, b) - S_n(a, b)| = \left| \sum_{k \in K_n} b(k; n, p) - \sum_{k \in K_n} \frac{1}{\sqrt{npq}} \cdot \frac{1}{\sqrt{2\pi}} e^{-x_{nk}^2/2} \right|$$

$$= \left| \sum_{k \in K_n} \left(b(k; n, p) - \frac{1}{\sqrt{npq}} \cdot \frac{1}{\sqrt{2\pi}} e^{-x_{nk}^2/2} \right) \right|$$

$$\leq \sum_{k \in K_n} \left| b(k; n, p) - \frac{1}{\sqrt{npq}} \cdot \frac{1}{\sqrt{2\pi}} e^{-x_{nk}^2/2} \right|$$

$$= \sum_{n \in K_n} \frac{1}{\sqrt{npq}} \left| \sqrt{npq}\, b(k; n, p) - \frac{1}{\sqrt{2\pi}} e^{-x_{nk}^2/2} \right|$$

$$< \sum_{k \in K_n} \frac{1}{\sqrt{npq}} \varepsilon = \frac{\varepsilon}{\sqrt{npq}} l_n,$$

where l_n is the number of elements in the set K_n. Since

$$x_{n,k+1} - x_{nk} = \frac{k+1-np}{\sqrt{npq}} - \frac{k-np}{\sqrt{npq}} = \frac{1}{\sqrt{npq}},$$

we have $l_n \leq (a-b)\sqrt{npq} + 1$. Thus, for $n > n(\varepsilon)$

$$|P_n(a, b) - S_n(a, b)| \leq \varepsilon \left((b-a) + \frac{1}{\sqrt{npq}} \right).$$

IV. CLASSICAL LIMIT THEOREMS

For sufficiently large n the difference becomes arbitrarily small, which proves lemma (f).

PROOF OF LEMMA (g). The integral $S(a, b)$, as an integral of a continuous function, can be represented as a limit of sums of the form

$$\sum_{i=1}^{m-1} (x_{i+1}-x_i)\varphi(x_{i+1}),$$

where $\varphi(x) = \dfrac{1}{\sqrt{2\pi}} e^{-x^2/2}$ and the points $a = x_1 < \ldots < x_m = b$ partition the interval $\langle a, b \rangle$ into $m-1$ disjoint intervals; in the passage to the limit it is required that the number of segments (x_i, x_{i+1}) should increase to infinity, and the length of the longest segment should tend to zero,

We easily note that these conditions are satisfied by the sums

$$(22.7) \quad S'_n(a, b) = (x_{n,k_n}-a)\varphi(x_{n,k_n}) + \frac{1}{\sqrt{npq}} \sum_{k=k_n+1}^{\bar{k}_n} \varphi(x_{n,k}) + (b-x_{n,\bar{k}_n})\varphi(b)$$

$$= (x_{n,k_n}-a)\varphi(x_{n,k_n}) + S_n(a, b) - \frac{1}{\sqrt{npq}} \varphi(x_{n,k_n}) + (b-x_{n,\bar{k}_n})\varphi(b),$$

where x_{n,k_n} denotes the smallest, and $x_{n,\bar{k}_n}$ denotes the largest number from the set K_n.

Thus

$$\lim_{n\to\infty} S'_n(a, b) = S(a, b).$$

From equality (22.7) it follows that the difference between the integral sum $S'_n(a, b)$ and the sum $S_n(a, b)$ tends to zero as $n \to \infty$. For every x we have $\varphi(x) \leqslant 1$, and none of the coefficients appearing in the above difference exceed $1/\sqrt{npq}$, whence they all tend to zero. Thus, we have also

$$\lim_{n\to\infty} S_n(a, b) = S(a, b),$$

which completes the proof of lemma (g), and therefore, completes the proof of relation (22.6).

It remains to prove that the convergence is uniform with respect to the numbers a and b. The proof will be based on the following lemma.

(h) *If*

$$f_1(x), f_2(x), f_3(x), \ldots$$

are non-decreasing, and $g(x)$ is a non-decreasing continuous function, and if for every n

(22.8) $$f_n(\infty) - f_n(-\infty) = g(\infty) - g(-\infty) < \infty,$$

then the condition

(22.9) $$\lim_{n \to \infty} f_n(x) = g(x) \quad \text{for every } x$$

implies that the functions $f_n(x)$ tend to $g(x)$ uniformly.

PROOF OF LEMMA (h). We have to prove that for every $\varepsilon > 0$ there exists an $n(\varepsilon)$ such that for $n > n(\varepsilon)$ we have for every x the inequality

$$|f_n(x) - g(x)| < \varepsilon.$$

Let ε be a fixed positive number. By assumption we have

$$g(\infty) - g(-\infty) < \infty,$$

hence we can find $C > 0$ such that

(22.9a) $$g(\infty) - g(-\infty) - \big(g(C) - g(-C)\big)$$
$$= g(\infty) - g(C) + g(-C) - g(-\infty) < \varepsilon/2.$$

In view of equality (22.8) we have also

(22.9b) $$f_n(\infty) - g(C) + g(-C) - f_n(-\infty) < \varepsilon/2.$$

Let us partition the interval $-C < x < C$ by means of $s+1$ points

$$x_0 = -C < x_1 < \ldots < x_s = C$$

into s intervals in such a way that for every $k = 1, 2, \ldots, s$ we have the inequality

$$g(x_k) - g(x_{k-1}) < \varepsilon/2.$$

In view of condition (22.9) we can find an n' such that for $n > n'$ we have

(22.10) $$|f_n(x_k) - g(x_k)| < \varepsilon/2$$

for all $k = 0, 1, \ldots, s$.

We shall now show that for $n > n'$ the inequality

(22.11) $$|f_n(x) - g(x)| < \varepsilon$$

holds for every x.

In fact, if x is one of the points $x_0, x_1, \ldots, x_s$, the inequality holds. If $-C < x < C$ and x is not equal to any of the points $x_0, x_1, \ldots, x_s$, then there exists a number k among $0, 1, \ldots, s-1$ such that

$$x_k < x < x_{k+1}.$$

We then have the following inequalities

$$f_n(x) - g(x) \leq f_n(x_{k+1}) - g(x_k)$$
$$= f_n(x_{k+1}) - g(x_{k+1}) + g(x_{k+1}) - g(x_k)$$
$$\leq \big(f_n(x_{k+1}) - g(x_{k+1})\big) + \big(g(x_{k+1}) - g(x_k)\big)$$
$$< \tfrac{1}{2}\varepsilon + \tfrac{1}{2}\varepsilon = \varepsilon$$

and

$$f_n(x) - g(x) \geq f_n(x_k) - g(x_{k+1})$$
$$= f_n(x_k) - g(x_k) + g(x_k) - g(x_{k+1})$$
$$\geq \big(f_n(x_k) - g(x_k)\big) - \big(g(x_{k+1}) - g(x_k)\big)$$
$$> -\tfrac{1}{2}\varepsilon - \tfrac{1}{2}\varepsilon = -\varepsilon.$$

Thus, (22.11) is proved for every such x.

Finally, if $x < -C$, then we have, on the one hand,

$$f_n(x) - g(x) \leq f_n(-C) - g(-\infty)$$
$$\leq f_n(-C) - g(-\infty) + \big(g(+\infty) - g(C)\big)$$
$$= \big(g(+\infty) - g(C) + g(-C) - g(-\infty)\big) + \big(f_n(-C) - g(-C)\big)$$
$$\leq \tfrac{1}{2}\varepsilon + \tfrac{1}{2}\varepsilon = \varepsilon;$$

(in the last inequality we have used formula (22.9a) and formula (22.10) applied for $x_0 = -C$), and on the other hand

$$f_n(x) - g(x) \geq f_n(-\infty) - g(-C)$$
$$\geq f_n(-\infty) - g(-C) - \big(f_n(\infty) - f_n(C)\big)$$
$$= -\big(f_n(\infty) - g(C) + g(-C) - f_n(-\infty)\big) - \big(g(C) - f_n(C)\big)$$
$$\geq -\tfrac{1}{2}\varepsilon - \tfrac{1}{2}\varepsilon = -\varepsilon;$$

(in the last inequality we used formula (22.9b) and (22.10) applied for $x_s = C$). The above inequalities imply that (22.11) holds also for $x < -C$. In a similar way we show that this inequality holds also for $x > C$.

This completes the proof of lemma (h).

To complete the proof of Theorem 2 we assume in lemma (h) that:

$$f_n(x) = \begin{cases} P_n(0, x) & \text{for } x \geq 0, \\ -P_n(x, 0) & \text{for } x < 0, \end{cases}$$

$$g(x) = \begin{cases} \int_0^x \frac{1}{\sqrt{2\pi}} e^{-u^2/2} du & \text{for } x \geq 0, \\ -\int_x^0 \frac{1}{\sqrt{2\pi}} e^{-u^2/2} du & \text{for } x < 0. \end{cases}$$

By these definitions and by formula (22.6) we have for every x

$$\lim_{n \to \infty} f_n(x) = g(x).$$

Moreover, functions $f_n(x)$ and $g(x)$ are monotone, and $g(x)$ is continuous. Besides,

$$f_n(+\infty) - f_n(-\infty) = 1,$$

since this quantity expresses the probability of the sure event, and

$$g(+\infty) - g(-\infty) = 1,$$

since

$$g(+\infty) - g(-\infty) = \int_{-\infty}^{\infty} \frac{1}{\sqrt{2\pi}} e^{-u^2/2} du = 1.$$

The relation

(22.12) $$\int_{-\infty}^{\infty} \frac{1}{\sqrt{2\pi}} e^{-u^2/2} du = 1,$$

can be verified as follows. Denoting $I = \int_{-\infty}^{+\infty} e^{-u^2/2} du$ we may write

$$I^2 = \int_{-\infty}^{\infty} e^{-u^2/2} du \int_{-\infty}^{\infty} e^{-v^2/2} dv = \int_{-\infty}^{\infty} \int_{-\infty}^{\infty} e^{-(u^2+v^2)/2} du dv.$$

Passing to polar coordinates by substitution

$$u = r\cos\varphi, \quad v = r\sin\varphi, \quad du\,dv = r\,dr\,d\varphi,$$

we obtain

$$I^2 = \int_{-\infty}^{\infty} \int_{-\infty}^{\infty} e^{-(u^2+v^2)/2} du dv = \int_0^{\infty} \int_0^{2\pi} re^{-r^2/2} d\varphi\,dr$$

$$= 2\pi \int_0^{\infty} re^{-r^2/2} dr = 2\pi[-e^{-r^2/2}]_0^{\infty} = 2\pi.$$

Hence $I = \sqrt{2\pi}$, which proves (22.12).

Thus, all assumptions of lemma (h) are satisfied. For given $\varepsilon > 0$ one can thus find an $n(\varepsilon)$ such that for $n > n(\varepsilon)$ we have for every x

$$|f_n(x) - g(x)| < \varepsilon/2.$$

IV. CLASSICAL LIMIT THEOREMS

On the other hand, by the definition of the functions $f_n(x)$ and $g(x)$ we have for every a and $b > a$

$$P_n(a, b) = f_n(b) - f_n(a),$$
$$S(a, b) = g(b) - g(a).$$

Thus, for $n > n(\varepsilon)$ and arbitrary $a < b$ we have also

$$\begin{aligned}|P_n(a, b) - S(a, b)| &= |(f_n(b) - f_n(a)) - (g(b) - g(a))| \\ &= |(f_n(b) - g(b)) - (f_n(a) - g(a))| \\ &\leqslant |f_n(b) - g(b)| + |f_n(a) - g(a)| \\ &< \tfrac{1}{2}\varepsilon + \tfrac{1}{2}\varepsilon = \varepsilon,\end{aligned}$$

which was to be proved. Theorem 2 is thus proved completely.

As we saw, the function $\dfrac{1}{\sqrt{2\pi}} e^{-x^2/2}$ which appears in Theorems 1 and 2 is a probability density. The probability distribution with this density is a particular case of the so-called normal or Gaussian distribution, and its probability distribution function

$$\Phi(x) = \int_{-\infty}^{x} \frac{1}{\sqrt{2\pi}} e^{-u^2/2} du$$

is sometimes called the *Laplace function*.

Let us consider the function

(22.13) $$f(x) = \frac{1}{\sigma\sqrt{2\pi}} e^{-(x-m)^2/2\sigma^2},$$

where σ is a positive number and m is an arbitrary real number.

We shall prove that this function is a probability density. Obviously, it is non-negative and continuous, whence it suffices to verify that

$$\int_{-\infty}^{\infty} f(x) dx = 1.$$

Indeed, after substituting

$$u = \frac{x-m}{\sigma}, \quad dx = \sigma du,$$

we obtain

$$\int_{-\infty}^{\infty} \frac{1}{\sigma\sqrt{2\pi}} e^{-(x-m)^2/2\sigma^2} dx = \int_{-\infty}^{\infty} \frac{1}{\sqrt{2\pi}} e^{-u^2/2} du = 1$$

(here we used formula (22.12)).

DEFINITION 1. By the *normal* or *Gaussian** *probability distribution* we shall mean the probability distribution on the real line with the density given by (22.13).

The normal distribution depends on two parameters, σ and m, whose interpretation will be given in § 34. In particular, for $\sigma = 1$ and $m = 0$ we obtain the probability density which already appeared in Theorems 1 and 2.

The theorems proved allow us to find the approximate values of $b(k; n, p)$ by using tables of the normal distribution. Thus, Theorem 1 allows to use the approximate formula

$$(22.14) \quad b(k; n, p) \approx \frac{1}{\sqrt{npq}} \cdot \frac{1}{\sqrt{2\pi}} e^{-(k-np)^2/2npq} = \frac{1}{\sqrt{npq}} \varphi\left(\frac{k-np}{\sqrt{npq}}\right),$$

where

$$\varphi(x) = \frac{1}{\sqrt{2\pi}} e^{-x^2/2}.$$

Another excellent approximation can be obtained from Theorem 2. We can use the formula

$$(22.15) \quad b(k; n, p) = \int_a^b \frac{1}{\sqrt{2\pi}} e^{-x^2/2} dx = \Phi(b) - \Phi(a),$$

where

$$a = \frac{k - \frac{1}{2} - np}{\sqrt{npq}}, \quad b = \frac{k + \frac{1}{2} - np}{\sqrt{npq}}.$$

Table 1 illustrates the approximation given by (22.15).

Table 1

$n = 6$, $p = 2/3$

k	$b(k; n, p)$	approximation	absolute error	relative error %
0	0.0878	0.0970	0.0092	10.5
1	0.2634	0.2355	0.0279	10.6
2	0.3292	0.3350	0.0058	1.8
3	0.2195	0.2355	0.0160	7.3
4	0.0823	0.0820	−0.0003	0.3
5	0.0164	0.0139	−0.0025	15.2
6	0.0014	0.0011	−0.0003	21.4

This table shows that even for such a small value of n and a rather asymmetric binomial distribution ($p = 2/3$), which obviously disturbs the approximation, it is nevertheless quite good.

---
* Carl Friedrich Gauss, 1777–1855.

CHAPTER V

Random Variables

§ 23. RANDOM VARIABLE AND ITS FORMAL PROPERTIES

In § 18, when describing the experiment consisting of n tosses of a coin, we took as the space of elementary events 2^n systems $\{i_1, \ldots, i_n\}$ where each i_k was either 0 or 1. When computing the probability of k heads in n tosses, we counted the number of systems $\{i_1, \ldots, i_n\}$ containing exactly k unities; the event A_k consisting of all such systems is precisely the event of tossing k heads in n tosses. We included the system $\{i_1, \ldots, i_n\}$ in A_k if and only if the number of unities in that system was k. Thus, without using the term, we were speaking of a random variable: this random variable was a function defined on all systems $\{i_1, \ldots, i_n\}$ whose values are the numbers of ones in those systems.

Another example is provided by the age of couples, discussed in § 7. In that example, points of the plane constituted the space of elementary events. The age of the husband is here a random variable: it is a function whose arguments are points of the plane and whose values are the first coordinates of those points.

Let us add that these types of random variables are most common in probability theory: one takes the points $(x_1, \ldots, x_n)$ of the n-dimensional Euclidean space as elementary events, and one considers random variables $\xi_1, \ldots, \xi_n$ defined by the equalities

$$\xi_i(x_1, \ldots, x_n) = x_i, \quad i = 1, 2, \ldots, n.$$

In this section we shall present the formal definition of random variable, and we shall prove some elementary properties of random variables.

Let X be the space of elementary events, let S be the field of events in it, and let P be the probability defined on S.

DEFINITION. The term *random variable* will denote every real valued function $\xi = \xi(e)$ defined on the space X of elementary events such that the following property is satisfied:

(M) for every real α, the set of elementary events for which $\xi(e) < \alpha$ is an event; in other words, for every α

$$\{e: \xi(e) < \alpha\} \in S.$$

Note that if the space of elementary events is at most countable, then property (M) is not restrictive. Indeed, in this case all subsets of X are events. The property (M) constitutes a restriction only in the case where not all subsets of X are events.

In measure theory, we call functions satisfying (M) *measurable with respect to the field S*. Thus, every real valued function measurable with respect to the field of events is called a *random variable*.

Below we give some properties of random variables. Roughly, they assert that various functions of random variables are also random variables. These theorems are significant only in cases where not all subsets of the set of elementary events are events.

THEOREM 1. *If $\xi = \xi(e)$ is a random variable, and A is a Borel set on the real line, then the set of elementary events for which $\xi(e) \in A$ is an event*:

$$\{e: \xi(e) \in A\} \in S.$$

PROOF. From the definition of a random variable we have for every α

$$\{e: \xi(e) < \alpha\} \in S.$$

We can write

$$\{e: \xi(e) < \alpha\} = \{e: \xi(e) \in A\}$$

where $A = \langle -\infty, \alpha \rangle$.

Denote by K the class of sets A on the real line for which

$$\{e: \xi(e) \in A\} \in S.$$

To prove that K contains all Borel sets, it suffices to show that

(a) K contains all half-lines of the form $\langle -\infty, \alpha \rangle$,

(b) the class K is complementative,

(c) the class K is countably additive.

Property (a) is equivalent to measurability with respect to the field of events, as noted earlier.

Properties (b) and (c) follow from the corresponding properties of the field of events.

Thus, if $A \in K$, then $\{e: \xi(e) \in A\} \in S$. Since the field S is complementative, we have also $\{e: \xi(e) \in A\}' \in S$. But

$$\{e: \xi(e) \in A\}' = \{e: \xi(e) \in A'\}$$

which proves that $A' \in K$.

V. RANDOM VARIABLES

Similarly, if the sets $A_1, A_2, \ldots$ belong to K, then so does their sum: $A = A_1 \cup \cup A_2 \cup \ldots \in K$.

By assumption, we have for $i = 1, 2, \ldots$

$$\{e: \xi(e) \in A_i\} \in S.$$

It follows that

$$\bigcup_{i=1}^{\infty} \{e: \xi(e) \in A_i\} \in S.$$

Obviously, we have

$$\bigcup_{i=1}^{\infty} \{e: \xi(e) \in A_i\} = \left\{e: \xi(e) \in \bigcup_{i=1}^{\infty} A_i\right\}.$$

The last two relations show that $A \in K$, which completes the proof of Theorem 1.

From Theorem 1 it follows in particular that we have

COROLLARY 1. *If ξ is a random variable, then sets of the form*

$$\{e: \xi(e) \leqslant a\} = \{e: \xi(e) \in \langle -\infty, a \rangle\},$$
$$\{e: a < \xi(e)\} = \{e: \xi(e) \in (a, \infty)\},$$
$$\{e: a < \xi(e) < b\} = \{e: \xi(e) \in (a, b)\}$$

are events, a and b being real numbers.

We suggest that the reader should give the formal proofs himself. We shall use Corollary 1 in proving Theorem 2.

THEOREM 2. *If $\xi = \xi(e)$ is a random variable, then the functions*

$$|\xi|, \quad \xi+a, \quad a\xi, \quad \xi^2,$$

where a is an arbitrary real number, are also random variables.

PROOF. The function $|\xi|$ is a random variable, since for every $\alpha > 0$

$$\{e: |\xi(e)| < \alpha\} = \{e: -\alpha < \xi(e) < \alpha\},$$

and for $\alpha \leqslant 0$

$$\{e: |\xi(e)| < \alpha\} = 0 \in S.$$

The function $\xi+a$ is a random variable, since for every α

$$\{e: \xi(e)+a < \alpha\} = \{e: \xi(e) < \alpha-a\} \in S.$$

The function $a\xi$ is a random variable, since for every α, when $a = 0$ the set $\{e: a\xi(e) < \alpha\}$ is either impossible or certain, whence it always belongs to S; for $a > 0$

$$\{e: a\xi(e) < \alpha\} = \{e: \xi(e) < \alpha/a\} \in S;$$

and for $a < 0$
$$\{e: a\xi(e) < \alpha\} = \{e: \xi(e) > \alpha/a\} \in S.$$

The function ξ^2 is a random variable, since for $\alpha > 0$
$$\{e: \xi^2(e) < \alpha\} = \{e: -\sqrt{\alpha} < \xi(e) < \sqrt{\alpha}\} \in S.$$

Theorem 2 shows that certain functions of random variables are also random variables. We shall now show that this property remains true for the whole class of functions called Borel functions. We recall that a real valued function $f(x)$ of a real variable x (where x belongs to the whole real line, an interval, or generally, a Borel set) is called a *Borel function*, if for every real α the set
$$\{x: f(x) < \alpha\}$$
is a Borel set. We have the following

THEOREM 3. *If $\xi(e)$ is a random variable, and $f(x)$ is a Borel function defined on the real line, then the function*
$$\eta(e) = f(\xi(e))$$
is also a random variable.

PROOF. To prove that $\eta(e)$ is a random variable, it suffices to show that the set of elementary events e for which $\eta(e) < \alpha$ belongs to the field of events.

By Theorem 1, if $\xi(e)$ is a random variable, then for every Borel set A, the set of points e for which $\xi(e) \in A$ is an event:
$$\{e: \xi(e) \in A\} \in S.$$

Now, if we take as A the set
$$A = \{x: f(x) < \alpha\},$$
we can write the following equalities:
$$\{e: \eta(e) < \alpha\} = \{e: f(\xi(e)) < \alpha\} = \{e: \xi(e) \in A\}.$$

Since, as noted before, the last of these sets is an event, the set $\{e: \eta(e) < \alpha\}$ is also an event, which proves Theorem 3.

Since the functions $f(x) = |x|$, $f(x) = x+a$, $f(x) = ax$ and $f(x) = x^2$ are Borel functions, Theorem 2 can be treated as a corollary to Theorem 3.

Up to now we have discussed which functions of one random variable are themselves random variables. Now we shall discuss the problem of determining which functions of several random variables are random variables. First, we shall prove

V. RANDOM VARIABLES

THEOREM 4. *If $\xi = \xi(e)$ and $\eta = \eta(e)$ are random variables, then their sum $\xi + \eta$ is also a random variable.*

PROOF. We have to show that for every α the set of points e such that $\xi(e) + \eta(e) < \alpha$ is an event, that is,

$$\{e: \xi(e) + \eta(e) < \alpha\} \in S.$$

We shall show that

$$\{e: \xi(e) + \eta(e) < \alpha\} = \bigcup_{i=1}^{\infty} \bigcup_{j=1}^{\infty} A_{ij},$$

where

$$A_{ij} = \left\{e: \xi(e) < \frac{j}{i}, \eta(e) < \alpha - \frac{j}{i}\right\}$$

$$= \left\{e: \xi(e) < \frac{j}{i}\right\} \cap \left\{e: \eta(e) < \alpha - \frac{j}{i}\right\}.$$

Clearly, the sets A_{ij} are events; thus, the above equality implies our theorem, since countable summation does not lead outside the field of events. It remains to check the truth of the above equality.

Obviously, for every $e \in A_{ij}$, and hence for every $e \in \bigcup_{i=1}^{\infty} \bigcup_{j=1}^{\infty} A_{ij}$ we have

(23.1) $$\xi(e) + \eta(e) < \alpha.$$

We shall prove that the converse is also true, i.e. if for some e inequality (23.1) holds, then we can find i and j such that

$$e \in A_{ij}.$$

In fact, if (23.1) holds, we can find i such that

(23.1') $$\xi(e) + \eta(e) < \alpha - \frac{1}{i}.$$

Suppose now that j is an integer such that

(23.2) $$\frac{j-1}{i} \leqslant \xi(e) < \frac{j}{i}.$$

Then

$$-\xi(e) \leqslant -\frac{j-1}{i}.$$

Adding this inequality to (23.1') we obtain

$$\eta(e) < \alpha - \frac{j}{i}.$$

This inequality together with the second inequality of (23.2) show that $e \in A_{ij}$, which completes the proof.

Using induction we immediately obtain from Theorems 4 and 2:

COROLLARY 2. *If $\xi_1 = \xi_1(e), \ldots, \xi_n = \xi_n(e)$ are random variables, then their sum*

$$\xi_1 + \ldots + \xi_n$$

and their average

$$\frac{1}{n}(\xi_1 + \ldots + \xi_n)$$

are random variables.

We shall prove now

THEOREM 5. *If $\xi = \xi(e)$ and $\eta = \eta(e)$ are random variables, then their product $\xi\eta$ is also a random variable.*

PROOF. The product of two random variables can be represented in the form

$$\xi\eta = \tfrac{1}{4}((\xi+\eta)^2 - (\xi-\eta)^2).$$

Since on the right-hand side the only operations are adding, squaring and multiplying by a constant, Theorem 5 follows as a corollary to Theorems 2 and 4.

By induction, we obtain from Theorem 5

COROLLARY 3. *If $\xi_1 = \xi_1(e), \ldots, \xi_n = \xi_n(e)$ are random variables, then their product $\xi_1 \xi_2 \ldots \xi_n$ is also a random variable.*

Theorems 4 and 5 together with the corollaries to them give the most important functions of random variables which are themselves random variables. As in the case of one random variable, for the case of more than one random variable one can show a large class of functions with the property that, when random variables are substituted in place of their arguments, we obtain random variables. These are Borel functions of n variables. We recall that a real function $f(x_1, \ldots, x_n)$ of n real variables is a *Borel function* if for every α the set

$$\{(x_1, \ldots, x_n): f(x_1, \ldots, x_n) < \alpha\}$$

is an n dimensional Borel set.

Note first that we have

THEOREM 6. *If $\xi_1 = \xi_1(e), \ldots, \xi_n = \xi_n(e)$ are random variables, and A is any n-dimensional Borel set, then the set*

$$\{e: (\xi_1(e), \ldots, \xi_n(e)) \in A\}$$

is an event.

With obvious formal changes, the proof is the same as the proof of Theorem 1. We note first that the relation

(23.3) $$\{e\colon (\xi_1(e), \ldots, \xi_n(e)) \in A\} \in S$$

holds for all sets of the form

$$A = \{(x_1, \ldots, x_n)\colon x_1 < \alpha_1, \ldots, x_n < \alpha_n\}$$

where $\alpha_1, \ldots, \alpha_n$ are arbitrary real numbers. Next, we have to show that the class K of sets A for which (23.3) holds is complementative and countably additive. It follows that the class K contains all Borel sets in n dimensions.

THEOREM 7. *If $\xi_1 = \xi_1(e), \ldots, \xi_n = \xi_n(e)$ are random variables, and $f(x_1, \ldots, x_n)$ is a Borel function of n variables, then the function*

$$\eta(e) = f(\xi_1(e), \ldots, \xi_n(e))$$

is a random variable.

PROOF. If

$$A = \{(x_1, \ldots, x_n)\colon f(x_1, \ldots, x_n) < \alpha\},$$

then

$$\{e\colon \eta(e) < \alpha\} = \{e\colon f(\xi_1(e), \ldots, \xi_n(e)) < \alpha\}$$
$$= \{e\colon (\xi_1(e), \ldots, \xi_n(e)) \in A\}.$$

By Theorem 6, the set $\{e\colon (\xi_1(e), \ldots, \xi_n(e)) \in A\}$ is an event, which proves the theorem.

Since the functions

$$f(x_1, x_2) = x_1 + x_2, \quad f(x_1, x_2) = x_1 x_2$$

are Borel functions, Theorems 4 and 5 may be treated as corollaries to Theorem 7.

We shall now prove the essentially simple but important

THEOREM 8. *If $\xi_1 = \xi_1(e), \xi_2 = \xi_2(e), \ldots$ is a sequence of random variables, and A is the set of elementary events e for which the numerical sequence $\xi_1(e), \xi_2(e), \ldots$ is convergent, then the function*

$$\xi(e) = \begin{cases} \lim_{n \to \infty} \xi_n(e) & \text{for } e \in A, \\ 0 & \text{otherwise} \end{cases}$$

is a random variable.

Before we present formal proof, here are some remarks. First, the theorem asserts essentially that the limits of sequences of random variables are random variables. Secondly, this theorem is formulated in such a way that it is not required that sequences $\xi_1(e), \xi_2(e), \ldots$ converge for all elementary events e.

Thirdly, it is not important that we put $\xi(e) = 0$ for e in A'; instead of zero we could take any constant, or even allow for $\xi(e)$ not to be constant on A'. It is important only that the sets $\{e: \xi(e) < \alpha, e \in A'\}$ should be events.

PROOF. First we shall prove that the set A is an event (perhaps impossible).

In fact, an elementary event e belongs to the set A if and only if for every $\varepsilon > 0$ there exists an n such that for $r, s \geq n$ we have

$$|\xi_r(e) - \xi_s(e)| < \varepsilon.$$

Indeed, this is a necessary and sufficient condition for the convergence of the sequence $\xi_1(e), \xi_2(e), \ldots$

Next, we note that the last condition can be formulated as follows:

(W) For every $k = 1, 2, \ldots$ there exists an n such that for all $r, s \geq n$ we have

(23.4) $$|\xi_r(e) - \xi_s(e)| < 1/k.$$

It should be noted that the essential change here consists in replacing all positive numbers ε by the sequence $1, \frac{1}{2}, \frac{1}{3}, \ldots$

The question arises whether the set of elementary events e satisfying (W) is an event. We shall show that this set can be constructed by means of countable summation and multiplication of events. Indeed, for $k, r, s = 1, 2, \ldots$ let

$$D_{krs} = \{e: |\xi_r(e) - \xi_s(e)| < 1/k\}.$$

Then

$$C_{kn} = \bigcap_{r=n}^{\infty} \bigcap_{s=n}^{\infty} D_{krs}$$

equals the set of points e for which inequality (23.4) holds starting from $r = n$ and $s = n$. Since the sets D_{krs} are events, the sets C_{kn} are also events.

Next,

$$B_k = \bigcup_{n=1}^{\infty} C_{kn}$$

is the set of points e for which there exists an n such that for $r \geq n$ and $s \geq n$ inequality (23.4) holds. As sums of countably many events, sets B_k are also events.

Clearly, we have

$$A = \bigcap_{k=1}^{\infty} B_k,$$

and the set A, as the product of a sequence of events is also an event.

To prove that e is a random variable, one has to show that for every α the set of points e for which $\xi(e) < \alpha$ is an event. We note first that this set can be represented as the union of two disjoint sets

V. RANDOM VARIABLES

$$\{e: \xi(e) < \alpha\} = (\{e: \xi(e) < \alpha\} \cap A) \cup (\{e: \xi(e) < \alpha\} \cap A').$$

At any rate, the second term is an event, and it suffices to prove that the first is an event.

In order for the limit $\lim_{n \to \infty} \xi_n(e)$ to be smaller than α for a certain point e it is necessary and sufficient that the following condition be satisfied:

(M') There exists a natural number k such that for infinitely many r we have

(23.5) $$\xi_r(e) \leqslant \alpha - \frac{1}{k}.$$

Let Z denote the set of all points e for which condition (M') holds. We shall show that Z is an event. Thus, for $k, r = 1, 2, \ldots$ put

$$E_{kr} = \left\{e: \xi_r(e) \leqslant \alpha - \frac{1}{k}\right\}.$$

Obviously, the sets E_{kr} are events. Now,

$$F_{kn} = \bigcup_{r=n}^{\infty} E_{kr}$$

is the set of points e for which there exists an $r \geqslant n$ such that inequality (23.5) holds.

Next, the set

$$G_k = \bigcap_{n=1}^{\infty} F_{kn}$$

equals the set of all points e for which there exist infinitely many indices r for which (23.5) holds.

Finally

$$Z = \bigcup_{k=1}^{\infty} G_k$$

equals the set of all points e for which condition (M') holds. Since this set is formed by countable addition and multiplication of events, Z is also an event.

Since the set

$$\{e: \xi(e) < \alpha\} \cap A$$

equals the set of all points e for which the limit of the sequence $\xi_1(e), \xi_2(e), \ldots$ exists and is smaller than α, we can write

$$\{e: \xi(e) < \alpha\} \cap A = Z \cap A.$$

The first term, as the product of two events is itself an event, which completes the proof of Theorem 8.

THEOREM 9. *If $\xi_1 = \xi_1(e), \xi_2 = \xi_2(e), \ldots$ are random variables, then the functions*

$$\overline{\xi}(e) = \begin{cases} \overline{\lim_{n \to \infty}} \xi_n(e) & \text{if } \overline{\lim_{n \to \infty}} \xi_n(e) < \infty, \\ 0 & \text{otherwise} \end{cases}$$

and

$$\underline{\xi}(e) = \begin{cases} \underline{\lim_{n \to \infty}} \xi_n(e) & \text{if } \underline{\lim_{n \to \infty}} \xi_n(e) > -\infty, \\ 0 & \text{otherwise} \end{cases}$$

are random variables.

We leave the proof to the reader.

§ 24. EXPECTATIONS OF RANDOM VARIABLES WHICH ASSUME A FINITE NUMBER OF VALUES

We shall now introduce one of the most important concepts in probability theory, namely that of expectation of a random variable.

Before giving a formal definition, we shall present the following intuitive reasoning. Suppose we are to throw a die and the owner of the die promises to pay us as many zlotys as there are points in the face that will come up in our throw. He requires, however, that we pay a fixed amount for the right of throwing a die. The question arises how much we should pay. In throwing the die we may obtain the results 1, 2, 3, 4, 5 or 6, all with equal probabilities 1/6. One can expect that, in a long series of throws, in about one sixth of all throws we shall obtain the result 1, in one sixth of all throws we shall obtain the result 2, and so on. If the series of throws consists of N throws, we shall have about $N/6$ results with 1, about $N/6$ results with 2, and so on. The total payment which we obtain from the owner of the die will be approximately

$$\frac{N}{6} + 2 \cdot \frac{N}{6} + 3 \cdot \frac{N}{6} + 4 \cdot \frac{N}{6} + 5 \cdot \frac{N}{6} + 6 \cdot \frac{N}{6};$$

hence for one throw we shall obtain on the average

$$\frac{1}{6} + 2 \cdot \frac{1}{6} + 3 \cdot \frac{1}{6} + 4 \cdot \frac{1}{6} + 5 \cdot \frac{1}{6} + 6 \cdot \frac{1}{6} = \frac{21}{6} = 3\frac{1}{2}.$$

Note that this result has been obtained by multiplying the values which can be thrown up with a die by their probabilities. Its intuitive meaning is that when we expect on the average $3\frac{1}{2}$ zł for one throw, we should be willing to pay that amount for the right of participating in the game.

V. RANDOM VARIABLES

Starting from the above intuitions, the following definition of expectation of a random variable which assumes only a finite number of values should appear natural:

DEFINITION 1. We say that the random variable $\xi = \xi(e)$ *assumes a finite number of values* if the space of elementary events can be partitioned into a finite number of disjoint events $A_1, A_2, \ldots, A_n$ in such a way that $\xi(e) = x_i$ for $e \in A_i$, where $x_1, \ldots, x_n$ are fixed numbers. If the numbers $x_1, \ldots, x_n$ are all different, we say that the random variable assumes the value x_i with the probability $p_i = P(A_i)$, $i = 1, \ldots, n$.

DEFINITION 2. If $\xi = \xi(e)$ is a random variable which assumes a finite number of values $x_1, \ldots, x_n$ on the events $A_1, \ldots, A_n$, then the *expectation* of this random variable is defined as

$$E\xi = \sum_{i=1}^{n} x_i P(A_i).$$

In order to ascertain that this definition is unambiguous, we must verify that $E\xi$ does not depend on the manner in which we partition the space of elementary events. Thus, if a random variable assumes values $x_1, \ldots, x_n$ on events $A_1, \ldots, A_n$ and also assumes values $y_1, \ldots, y_m$ on events $B_1, \ldots, B_m$, we must check that

$$\sum_{i=1}^{n} x_i P(A_i) = \sum_{j=1}^{m} y_j P(B_j).$$

Put

$$C_{ij} = A_i \cap B_j.$$

Since the events A_i are disjoint and their sum equals the space of all elementary events, and the same is true for events B_j, the events C_{ij} are also disjoint and their sum equals the space of all elementary events X. We easily note that the random variable is constant on the events C_{ij}; more precisely, the event C_{ij} can be non-empty only if $x_i = y_j$. Thus, let x_{ij} denote the value assumed by ξ on the event C_{ij}. If C_{ij} is non-empty, we have $x_{ij} = x_i = y_j$. If $C_{ij} = 0$, we may define x_{ij} in an arbitrary manner, say $x_{ij} = 0$. In view of the equalities

$$P(A_i) = \sum_{j=1}^{m} P(C_{ij}), \quad P(B_j) = \sum_{i=1}^{n} P(C_{ij})$$

and in view of the fact that $x_{ij} = x_i = y_j$ except for $C_{ij} = 0$, we have

$$\sum_{i=1}^{n} x_i P(A_i) = \sum_{i=1}^{n} x_i \sum_{j=1}^{m} P(C_{ij}) = \sum_{i=1}^{n} \sum_{j=1}^{m} x_{ij} P(C_{ij})$$

$$= \sum_{j=1}^{m} \sum_{i=1}^{n} x_{ij} P(C_{ij}) = \sum_{j=1}^{m} y_j \sum_{i=1}^{n} P(C_{ij}) = \sum_{j=1}^{m} y_j P(B_j),$$

which proves that Definition 2 is unambiguous.

We shall now prove some theorems on expectations, which are true in general, for all random variables with finite expectations. Thus, in our formulations we shall omit the assumption that the random variables assume only a finite number of values, though the proofs which we give will refer to such random variables.

THEOREM 1. *The expectation of the product of a random variable by a constant equals the product of that constant by the expectation of the random variable:*

$$Ea\xi = aE\xi.$$

PROOF. If the random variable ξ assumes values $x_1, \ldots, x_n$ on sets $A_1, \ldots, A_n$, then the random variable $a\xi$ assumes on the same sets the values $ax_1, \ldots, ax_n$. Thus

$$Ea\xi = ax_1 P(A_1) + \ldots + ax_n P(A_n)$$
$$= a(x_1 P(A_1) + \ldots + x_n P(A_n)) = aE\xi$$

which proves Theorem 1.

THEOREM 2. *The expectation of the sum of two random variables equals the sum of the expectations of those random variables:*

$$E(\xi + \eta) = E\xi + E\eta.$$

PROOF. Suppose that the random variable ξ assumes values $x_1, \ldots, x_n$ on sets $A_1, \ldots, A_n$ and that the random variable η assumes values $y_1, \ldots, y_m$ on sets $B_1, \ldots, B_m$. Then the random variable $\xi + \eta$ assumes values $x_i + y_j$ on sets C_{ij} where $C_{ij} = A_i \cap B_j$, $i = 1, 2, \ldots, n$, $j = 1, 2, \ldots, m$.

In view of the equations

$$P(A_i) = \sum_{j=1}^{m} P(C_{ij}), \quad i = 1, 2, \ldots, n,$$

$$P(B_j) = \sum_{i=1}^{n} P(C_{ij}), \quad j = 1, 2, \ldots, m,$$

we have

$$E(\xi + \eta) = \sum_{i=1}^{n} \sum_{j=1}^{m} (x_i + y_j) P(C_{ij}) = \sum_{i=1}^{n} \sum_{j=1}^{m} x_i P(C_{ij}) + \sum_{j=1}^{m} \sum_{i=1}^{n} y_j P(C_{ij})$$

$$= \sum_{i=1}^{n} x_i \sum_{j=1}^{m} P(C_{ij}) + \sum_{j=1}^{m} y_j \sum_{i=1}^{n} P(C_{ij})$$

$$= \sum_{i=1}^{n} x_i P(A_i) + \sum_{j=1}^{m} y_j P(B_j) = E\xi + E\eta,$$

which proves Theorem 2.

By induction we obtain from Theorems 1 and 2

COROLLARY 1. *The expectation of any linear combination of a finite number of random variables equals the linear combination of the expectations of those random variables*, i.e.,

$$E(a_1\xi_1+a_2\xi_2+ \ldots +a_n\xi_n) = a_1E\xi_1+a_2E\xi_2+ \ldots +a_nE\xi_n.$$

In particular, we have

COROLLARY 2. *If random variables* $\xi_1, \ldots, \xi_n$ *have identical expectations equal to* m, *i.e., if*

$$E\xi_1 = E\xi_2 = \ldots = E\xi_n = m,$$

then the expectation of the average of these random variables, is also equal to m, i.e.,

$$E\left(\frac{\xi_1+ \ldots +\xi_n}{n}\right) = m.$$

Next, we give some other theorems:

THEOREM 3. *The expectation of the absolute value of a random variable cannot be smaller than the absolute value of the expectation of that random variable*, i.e.,

$$E|\xi| \geq |E\xi|.$$

PROOF. In fact, if ξ assumes values $x_1, \ldots, x_n$ on events $A_1, \ldots, A_n$, then $|\xi|$ assumes on these events the values $|x_1|, \ldots, |x_n|$. We have therefore

$$-|x_i|P(A_i) \leq x_iP(A_i) \leq |x_i|P(A_i), \quad i = 1, 2, \ldots, n.$$

Adding these inequalities we obtain

$$-E|\xi| = -\sum_{i=1}^{n}|x_i|P(A_i) \leq E\xi = \sum_{i=1}^{n}x_iP(A_i) \leq \sum_{i=1}^{n}|x_i|P(A_i) = E|\xi|,$$

which proves Theorem 3.

THEOREM 4. *If for every* e *the random variable* ξ *satisfies the inequality*

$$\xi(e) < a,$$

then

$$E\xi < a.$$

PROOF. If ξ assumes values $x_1, \ldots, x_n$ on events $A_1, \ldots, A_n$, then for every i $(i = 1, 2, \ldots, n)$ we have

$$x_iP(A_i) \leq aP(A_i).$$

with the strict inequality if $P(A_i) > 0$. By addition, we get

$$E\xi = \sum_{i=1}^{n} x_i P(A_i) < \sum_{i=1}^{n} a P(A_i) = a \sum_{i=1}^{n} P(A_i) = a.$$

We shall now present some conclusions which follow from the last theorem.

COROLLARY 2. *If for every e the random variable ξ satisfies the inequality*

$$a < \xi(e),$$

then

$$a < E\xi.$$

COROLLARY 3. *If for every e the random variable ξ satisfies the inequality*

$$|\xi(e)| < \varepsilon,$$

then

$$|E\xi| \leqslant E|\xi| < \varepsilon.$$

THEOREM 5. *If the random variables ξ and η satisfy for every e the relation*

$$\xi(e) \geqslant \eta(e),$$

then

$$E\xi \geqslant E\eta.$$

We leave the proof to the reader.

THEOREM 6. *If the random variables ξ and η satisfy for every e the relation*

$$|\xi(e) - \eta(e)| < \varepsilon,$$

then

$$|E\xi - E\eta| < \varepsilon.$$

PROOF. Using Theorems 3 and 5 we may write

$$|E\xi - E\eta| = |E(\xi - \eta)| \leqslant E|\xi - \eta| < \varepsilon.$$

§ 25. EXPECTATION OF A RANDOM VARIABLE IN THE GENERAL CASE

We shall now define the expectation of a random variable in a general case. For this definition we shall need some considerations of an analytical character connected with passing to the limit.

As we already know, random variables are real valued functions defined on the space of all elementary events and measurable with respect to the field of events. Thus, let X be a fixed space of elementary events, let S be a field of events, and let P be the probability defined on S.

V. RANDOM VARIABLES

We shall show how one can define expectations for a large class of random variables.

We start from bounded random variables. Suppose that $\xi = \xi(e)$ is a bounded random variable, i.e. suppose that for some M we have the inequality

$$-M < \xi(e) < M$$

for every e. Let us partition the interval $(-M, M)$ into n parts by $n+1$ numbers a_i, $i = 0, 1, \ldots, n$, such that

(25.1) $\qquad -M = a_0 < a_1 < \ldots < a_n = M.$

Denote by A_i the set of elementary events e such that we have the inequality $a_{i-1} \leqslant \xi(e) < a_i$:

$$A_i = \{e: a_{i-1} \leqslant \xi(e) < a_i\}, \quad i = 1, 2, \ldots, n.$$

Clearly, the events A_i are disjoint and their sum equals the space of all elementary events.

Let us form the sums

$$l = \sum_{i=1}^{n} a_{i-1} P(A_i), \quad L = \sum_{i=1}^{n} a_i P(A_i).$$

We easily check that $-M \leqslant l \leqslant L \leqslant M$. The number l will be called the *lower integral sum*, and the number L will be called the *upper integral sum*, corresponding to partition (25.1).

Denote by λ the upper bound of the lower integral sums with respect to all the possible partitions of the interval $(-M, M)$ into a finite number of subintervals; similarly, let $\varLambda$ denote the lower bound of all the upper integral sums. We claim that $\lambda = \varLambda$. Accordingly, we introduce the following definition:

DEFINITION 1. If $\xi = \xi(e)$ is a bounded random variable, then the common value of the lower bound of the upper integral sums and the upper bound of the lower integral sums will be called the *expectation of the random variable* ξ and will be denoted by $E\xi$.

We shall now prove that $\lambda = \varLambda$. In any case, we have $\lambda \leqslant \varLambda$. This follows from the fact that every lower integral sum is smaller than or equal to every upper integral sum. Indeed, if

(25.2) $\qquad -M = a_0 < a_1 < \ldots < a_r = M$

and

(25.3) $\qquad -M = b_0 < b_1 < \ldots < b_s = M$

are two arbitrary partitions of interval $(-M, M)$, and if

$$-M = c_0 < c_1 < \ldots < c_n = M$$

is a partition resulting from connecting these two partitions, then we obviously have

$$\sum_{i=1}^{r} a_{i-1} P(\{e: a_{i-1} \leqslant \xi(e) < a_i\}) \leqslant \sum_{i=1}^{n} c_{i-1} P(\{e: c_{i-1} \leqslant \xi(e) < c_i\})$$

$$< \sum_{i=1}^{n} c_i P(\{e: c_{i-1} \leqslant \xi(e) < c_i\}) \leqslant \sum_{i=1}^{s} b_i P(\{e: b_{i-1} \leqslant \xi(e) < b_i\}).$$

These inequalities show that the lower integral sum corresponding to partition (25.2) is smaller than the upper integral sum corresponding to partition (25.3).

In view of the inequality $\lambda \leqslant \Lambda$ it suffices to show that for every $\varepsilon > 0$ there exist a lower integral sum and an upper integral sum such that their difference is smaller than ε. But the upper and lower integral sums corresponding to the partition of $(-M, M)$ into parts smaller than ε have this very property. Indeed, if in the partition

$$-M = a_0 < a_1 < \ldots < a_n = M$$

we have for every $i = 1, 2, \ldots, n$

$$a_i - a_{i-1} < \varepsilon,$$

then

$$\sum_{i=1}^{n} a_i P(\{e: a_{i-1} \leqslant \xi(e) < a_i\}) - \sum_{i=1}^{n} a_{i-1} P(\{e: a_{i-1} \leqslant \xi(e) < a_i\})$$

$$= \sum_{i=1}^{n} (a_i - a_{i-1}) P(\{e: a_{i-1} \leqslant \xi(e) < a_i\})$$

$$< \varepsilon \sum_{i=1}^{n} P(\{e: a_{i-1} \leqslant \xi(e) < a_i\}) = \varepsilon.$$

From the above it follows that we have

THEOREM 1. *Every bounded random variable has an expectation; this expectation equals to the limit of the lower and upper integral sums corresponding to the sequence of partitions of $(-M, M)$ such that $\max_{0 < i \leqslant n} (a_i - a_{i-1})$ decreases to zero as the number of partition increases.*

We leave to the reader the problem of verification whether Definition 1 is consistent with Definition 2 of § 24 for random variables with a finite number of values.

Definition 1 cannot be extended directly to unbounded random variables, since in partitions of $(-\infty, +\infty)$ into a finite number of parts one cannot define lower and upper integral sums in any natural way. A book by Mazurkiewicz* gives a systematic exposition of the theory of expectations based on lower and upper integral sums where the line $(-\infty, +\infty)$ is partitioned into a countable number of parts.

We shall extend the definition of expectation to the case of some unbounded random variables starting from bounded random variables. Suppose first that the random variable $\xi = \xi(e)$ is non-negative, i.e. that it satisfies for every e the inequality

$$\xi(e) \geqslant 0.$$

We now define the new, *truncated* random variables as follows:

(25.4) $$\xi_n(e) = \begin{cases} \xi(e) & \text{if } \xi(e) \leqslant n, \\ 0 & \text{otherwise.} \end{cases}$$

Obviously, the random variables $\xi_n = \xi_n(e)$ are bounded, hence they have expectations according to Definition 1. Since in our case we have for every e

$$\xi_n(e) \leqslant \xi_{n+1}(e)$$

in view of Theorem 5 of § 24, the sequence $E\xi_1, E\xi_2, \ldots$ is non-decreasing. Thus, this sequence either has a finite limit, or increases to infinity. If this limit is finite, we call it the expectation of the random variable ξ.

In a similar way we define the expectation of a non-positive random variable, i.e. a random variable such that

$$\xi(e) \leqslant 0$$

for every e. We define the auxiliary sequence of random variables by the equalities

(25.5) $$\xi_n(e) = \begin{cases} \xi(e) & \text{if } \xi(e) \geqslant -n, \\ 0 & \text{otherwise.} \end{cases}$$

We have now $\xi_{n+1}(e) \leqslant \xi_n(e)$ for every e, whence the sequence $E\xi_1, E\xi_2, \ldots$ is non-increasing. Thus, it either has a finite limit or decreases to $-\infty$. In the first case we call this limit the expectation of the random variable ξ.

We can put the above in the form of the following definition:

DEFINITION 2. If a random variable ξ is non-negative, and the sequence of expectations $E\xi_n$ of truncated random variables ξ_n defined by (25.4) has a finite

---

* S. Mazurkiewicz, *Podstawy rachunku prawdopodobieństwa* (*Foundations of Probability Theory*), Warszawa 1956.

limit, then this limit is called the *expectation of the random variable* ξ, or, by definition,
$$E\xi = \lim_{n\to\infty} E\xi_n.$$

The expectation of non-positive random variables is defined in an analogous way.

EXAMPLE. Let X be the interval $(0, 1) = \{e: 0 < e < 1\}$ and let P be the uniform distribution on $(0, 1)$. The random variable $\xi = \xi(e)$ defined by the equalities
$$\xi(e) = 2^n \quad \text{for} \quad \frac{1}{2^n} \leqslant e < \frac{1}{2^{n-1}},$$

$n = 1, 2, \ldots$ is a non-negative random variable which has no expectation.

In the general case, expectation is defined as follows: we represent the random variable ξ as the sum of non-negative and non-positive random variables, putting for every e
$$\xi(e) = \xi_+(e) + \xi_-(e),$$
where
$$\xi_+(e) = \begin{cases} \xi(e) & \text{if} \quad \xi(e) \geqslant 0, \\ 0 & \text{otherwise,} \end{cases}$$

$$\xi_-(e) = \begin{cases} \xi(e) & \text{if} \quad \xi(e) \leqslant 0, \\ 0 & \text{otherwise.} \end{cases}$$

The random variables $\xi_+(e)$ and $\xi_-(e)$ are called *positive* and *negative parts* of the random variable ξ. If both these random variables have finite expectations, we define the expectation of ξ as the sum

(25.6) $$E\xi = E\xi_+ + E\xi_-.$$

We shall formulate the above as follows:

DEFINITION 3. *If the positive part ξ_+ and the negative part, ξ_-, of a random variable ξ have finite expectations, then the sum (25.6) is called the expectation of the random variable ξ.*

We leave to the reader the problem of verifying that for random variables which assume only a finite number of values Definitions 1 and 3 are consistent.

THEOREM 2. *If a random variable ξ assumes countably many values $x_1, x_2, \ldots$ with probabilities $p_i = P(\{e: \xi(e) = x_i\})$, $i = 1, 2, \ldots$, then the expectation of* the random variable ξ exists if and only if the series

$$\sum_{i=1}^{\infty} x_i P(\{e: \xi(e) = x_i\})$$

converges absolutely; in this case the expectation equals the sum of this series. We leave the proof to the reader.

§ 26. EXPECTATION AND INTEGRATION. PROBABILITY DISTRIBUTION FUNCTION AND PROBABILITY DISTRIBUTION OF A RANDOM VARIABLE

In this section we shall discuss briefly the relations between the concept of expectation of a random variable and other concepts of probability theory and mathematical analysis.

Note first that the definitions of expectations which we gave in the last paragraph coincide with the abstract formulation of the definition of the Lebesgue integral adapted to the needs of probability theory. Our definition differs from the definition of the Lebesgue integral of a function defined, say, on the interval $(a, b) = \{x: a < x < b\}$ only at the point where in the integral sums for the Lebesgue integral the linear Lebesgue measure appears (of sets where the function integrated satisfies certain conditions), while we consider the probabilities of corresponding events, i.e. sets of elementary events on which the random variable satisfies certain conditions. Another difference lies in the fact that we have defined the expectation as the "integral" extended over the whole space X of elementary events. Apart from these points, the analogy with the Lebesgue integral is complete. That is why the expectation of the random variable $\xi = \xi(e)$ defined on the space of elementary events X, with the field of events S and probability measure P, will be denoted by means of the integral symbol as follows

$$E\xi = \int_X \xi(e) dP(e)$$

or simply

$$E\xi = \int_X \xi dP$$

and we shall treat these equalities as definition of expectation.

By replacing in the definitions of §§ 24 and 25 the whole space of elementary events X by an arbitrary event A we obtain the definitions of the integral of the random variable ξ with respect to the probability measure P extended over the event A. We shall denote such integrals by

$$\int_A \xi(e)\,dP(e)$$

or

$$\int_A \xi\,dP.$$

As an example, we shall paraphrase the definition of the expectation of a random variable with a finite number of values: if the event A can be split into a finite number of events $A_1, \ldots, A_n$ such that

$$\xi(e) = x_i \quad \text{for} \quad e \in A_i$$

then the integral of the random variable ξ over the event A is defined as the sum

$$\int_A \xi\,dP = \sum_{i=1}^n x_i P(A_i).$$

All the properties of expectation expressed in the theorems of § 24 extend to the integral of the random variable ξ over an arbitrary event A. Let us write them down in the form of the following formulas:

(26.1) $$\int_A a\xi\,dP = a\int_A \xi\,dP;$$

(26.2) $$\int_A (\xi+\eta)\,dP = \int_A \xi\,dP + \int_A \eta\,dP;$$

(26.3) $$\int_A |\xi|\,dP \geqslant \left|\int_A \xi\,dP\right|.$$

If $\xi(e) < a$ for $e \in A$ and $P(A) > 0$, then

(26.4) $$\int_A \xi(e)\,dP < aP(A).$$

If $\xi(e) \geqslant \eta(e)$ for $e \in A$, then

(26.5) $$\int_A \xi\,dP \geqslant \int_A \eta\,dP.$$

If $|\xi(e) - \eta(e)| < \varepsilon$ for $e \in A$ and $P(A) > 0$, then

(26.6) $$\left|\int_A \xi\,dP - \int_A \eta\,dP\right| < \varepsilon P(A).$$

Let us introduce some definitions.

V. RANDOM VARIABLES

DEFINITION 1. By the *probability distribution function*, or shortly, the *distribution function*, of a random variable ξ we shall mean a function $F_\xi(x)$ of a real variable x defined as
$$F_\xi(x) = P(\{e\colon \xi(e) < x\}).$$
We shall usually omit the index ξ, writing simply $F(x)$, unless this leads to misunderstanding.

We easily check that the probability distribution function of a random variable satisfies all the conditions given in § 10 for the probability distribution on the real line. The proof consists of a repetition of the proof of Theorem 2 of § 10, with the only difference that instead of the intervals $(a, b) = \{x\colon a < x < b\}$ we speak here of the events $\{e\colon a \leqslant \xi(e) < b\}$. According to § 12, the distribution function $F(x)$ determines uniquely a certain distribution on the real line. This remark justifies the following definition:

DEFINITION 2. The probability distribution on the real line determined by the probability distribution function $F_\xi(x)$ will be called the *probability distribution of the random variable* ξ.

Let us now investigate the relation between the expectation of a random variable ξ and its distribution function.

Suppose first that the random variable ξ is bounded, i.e. that for every $e \in X$ we have the inequality
$$-M < \xi(e) < M$$
where M is a fixed constant. Partitioning the interval $(-M, M)$ by the numbers a_i:
$$-M = a_0 < a_1 < \ldots < a_n = M$$
we defined in § 25 the lower and upper integral sums corresponding to these partitions:
$$l = \sum_{i=1}^n a_{i-1} P(\{e\colon a_{i-1} \leqslant \xi(e) < a_i\}),$$
$$L = \sum_{i=1}^n a_i P(\{e\colon a_{i-1} \leqslant \xi(e) < a_i\}).$$
We note, however, that
$$l = \sum_{i=1}^n a_{i-1}\bigl(F_\xi(a_i) - F_\xi(a_{i-1})\bigr),$$
$$L = \sum_{i=1}^n a_i\bigl(F_\xi(a_i) - F_\xi(a_{i-1})\bigr);$$

the latter are lower and upper integral sums used in defining the Stieltjes integral

$$\int_{-M}^{M} x \, dF_\xi(x).$$

It is also true that the extension of the definition of expectation to unbounded random variables coincides with the definition of the Stieltjes integral over the real line, where the latter is defined as the limit of integrals extended over finite intervals as follows:

$$\int_{-\infty}^{+\infty} x \, dF(x) = \lim_{\substack{a \to \infty \\ b \to \infty}} \int_a^b x \, dF(x).$$

It follows that the expectation of a random variable can be expressed by the Stieltjes integral with respect to the probability distribution function of that random variable. Namely, we have the following theorem:

THEOREM 1. *If the random variable ξ has a finite expectation then*

$$E\xi \stackrel{df}{=} \int_X \xi \, dP = \int_{-\infty}^{+\infty} x \, dF_\xi(x),$$

where both integrals converge in the case of random variables which have finite expectations, and both integrals diverge in the case of random variables which do not have expectation.

We omit the proof.

Finally, note that Definitions 1 and 2 can be extended to the case of systems of several random variables.

DEFINITION 3. If $\xi_1 = \xi_1(e), \ldots, \xi_n = \xi_n(e)$ are random variables, then the function on n real variables $x_1, \ldots, x_n$ defined as

$$F_{\xi_1, \ldots, \xi_n}(x_1, \ldots, x_n) = P(\{e: \xi_1(e) < x_1, \ldots, \xi_n(e) < x_n\})$$

will be called the *probability distribution function of the system $\xi_1, \ldots, \xi_n$ of random variables.*

We leave to the reader the problem of checking whether the probability distribution function of an arbitrary system of random variables satisfies all the properties of the probability distribution functions of the probability distributions in an n-dimensional Euclidean space, as given in § 13. Thus, according to § 13, this probability distribution function determines a certain n-dimensional probability distribution.

DEFINITION 4. The n-dimensional probability distribution determined by the probability distribution function of a system of n random variables will be called the *distribution of that system*.

Note that if points x of the real line are treated as elementary events, P is a probability distribution on the real line, and F is the probability distribution function of that distribution, then the random variable $\xi(x) = x$ has the distribution function F, and P is its probability distribution. Similarly, if points $x_1, \ldots, x_n$ on an n-dimensional Euclidean space play the role of elementary events, P is a probability distribution on that space, and $F(x_1, \ldots, x_n)$ is the probability distribution function of that distribution, then the random variables

$$\xi_1(x_1, \ldots, x_n) = x_1, \quad \ldots, \quad \xi_n(x_1, \ldots, x_n) = x_n$$

have $F(x_1, \ldots, x_n)$ as their distribution function, and P as their distribution; here marginal distributions are simply probability distributions of these random variables. This remark shows how one can construct a random variable, or a system of random variables with an arbitrarily given distribution. This remark shows also that from the theoretical point of view studying finite systems of random variables is equivalent to studying probability distributions in multidimensional Euclidean spaces. On the other hand, the concept of a random variable leads to a concise and intuitive formulation of many theorems in probability theory. It is true also that many theorems of probability theory were formulated in the terminology of random variables before it was discovered that random variables should be treated as numerical functions defined on the space of elementary events. This discovery, which constituted the cornerstone of the famous book by Kolmogorov*, led to the construction of the foundation of probability theory, and to including this theory in the domain of mathematics, as a branch of the metric theory of real functions.

§ 27. VARIANCE, MOMENTS, AND OTHER NUMERICAL CHARACTERISTICS OF THE DISTRIBUTION OF A RANDOM VARIABLE

In statistical applications, the concept of a random variable appears usually in situations similar to that described below.

We are given a population consisting of a large number of items, say bricks. We are interested in some characteristic of these items which can be expressed numerically, say resistance to crushing. Thus, to each item there corresponds a number, characterizing it with respect to the characteristic in question. However, we do not know those numbers. In order to learn them, we have to perform

---

* A. N. Kolmogorov, *Grundbegriffe der Wahrscheinlichkeitsrechnung*, Berlin 1933.

a test which may lead to the destruction of the item, say, by trying to crush the brick, and observing the pressure necessary to achieve this.

In spite of the fact that the numbers characterizing particular items are unknown, we may assume that for every x there is defined a fraction $F(x)$ of those items of our population for which the characteristic number is smaller than x. The fraction $F(x)$, treated as a function of the variable x, has all the properties of the probability distribution on the real line. We call it the *probability distribution of the characteristic in the population*.

To learn something about the probability distribution of the characteristic under investigation, we draw some items from the population and measure the value of the characteristic for the selected items. We say that we *take a random sample*.

Suppose that only one item is to be a sample. Then, the unknown value of the characteristic of the item which we are going to select is treated as a random variable with the probability distribution $F(x)$. When the item has already been selected and the value of its characteristic measured, we treat this value as the value $\xi(e)$ of the random variable ξ for some elementary event e. The elementary events in this case correspond to the items which can be selected from the population.

Similarly, if n items are to be included in the sample, then the values of the characteristics of the items before they are selected are treated as n random variables $\xi_1, \xi_2, \ldots, \xi_n$, each with the probability distribution $F(x)$. The values of the characteristic in the items already selected are treated as value $\xi_1(e), \xi_2(e), \ldots, \xi_n(e)$ of random variables $\xi_1, \xi_2, \ldots, \xi_n$ for some fixed elementary event e. Thus, here the elementary events play the role of all the possible systems of items which can be selected from the population.

To determine the probability distribution $F(x)$ we would have to measure the characteristic in all items of the population. Usually, this is impossible either because such a procedure would destroy the whole population (as in the case of the resistance of bricks to pressure), or, because such an investigation would be too costly. Thus, as a rule, the sample is small in comparison with the size of the whole population. We usually want to evaluate on its basis some numerical parameters which characterize to a certain extent the distribution $F(x)$ of the population rather than to determine the function $F(x)$ itself.

We shall now discuss such numerical characteristics. We shall relate them both to the distribution of the characteristic in the population and to the distribution function $F(x)$; primarily, however, we shall relate them to random variables. Thus, in some contexts we shall speak of the variance of the population characteristic, but most of all we shall speak of the variance of a random variable. We shall also speak sometimes of the variance of the distribution determined by the probability distribution function $F(x)$, or shortly, of the variance of the distribution $F(x)$.

Expectation. One such characteristic is already known to us—namely expectation, discussed in §§ 24–26. According to the remarks made at the beginning of § 24, expectation can be interpreted as a characteristic of position, since it indicates, roughly speaking, where the values assumed by a random variable are concentrated. If a random variable assumed only one value x with probability one, then the expectation of this random variable would be equal to x.

Other positional characteristics include

Quantiles. We shall first give the definition of quantiles for the case of continuous distribution functions. We shall define the *quantile of order p of a random variable* ξ with the continuous distribution function $F(x)$, where p is a fixed number from the interval $0 < p < 1$, as the number z_p which satisfies the equation

(27.1) $$F(z_p) = p.$$

In other words, the quantile of order p is defined as a number z such that

$$P(\{e: \xi(e) < z_p\}) = p.$$

Note that quantiles need not be unique. For instance, if a random variable ξ has a probability distribution with the density

$$f(x) = \begin{cases} 1/3 & \text{for} \quad 0 < x < 1 \text{ and } 2 < x < 4, \\ 0 & \text{otherwise} \end{cases}$$

then as the quantile of order 1/3 we may take any number from the interval $1 \leqslant z \leqslant 2$. The quantiles are, however, determined uniquely when the distribution function is strictly increasing.

The definition of the quantile given above does not apply in cases where the random variable ξ has a discrete distribution: in those cases numbers satisfying relation (27.1) may be non-existent. Thus, if a random variable assumes value 0 with probability 1/2 and value 1 with probability 1/2, there is no number x such that $F(x) = 1/3$: for $z \leqslant 0$ this function is equal to 0 and for $z > 0$ it is greater than 1/3. To be able to speak of quantiles for discrete random variables, we shall modify their definition correspondingly.

DEFINITION 1. By a *quantile of order p*, $0 < p < 1$, of a random variable ξ with probability distribution function $F(x)$ we shall mean any number z_p such that

(27.2) $$F(z_p) \leqslant p \leqslant F(z_p+0).$$

In other words: a quantile of the order p equals any number z_p such that
$$P(\{e: \xi(e) < z_p\}) \leqslant p \leqslant P(\{e: \xi(e) \leqslant z_p\}).$$
This means that the probability that the random variable ξ assumes value smaller than z_p does not exceed p, while the probability that the random variable ξ assumes value smaller than z_p or equal to z_p is at least equal to p, i.e. it is either equal to p or exceeds it.

A quantile of the order $1/2$ is called a *median*, while quantiles of orders $1/4$ and $3/4$ are called, respectively, *lower* and *upper quantiles*. The median of a random variable ξ will be denoted by $\text{Me}\,\xi$.

According to this definition, the quantile of order $1/3$ of a random variable assuming values 0 and 1 with equal probabilities $1/2$ is equal to 0.

Note that if the random variable ξ has a symmetric distribution, i.e. if there exists a number α such that for every x
$$P(\{e: \xi(e) < \alpha - x\}) = P(\{e: \alpha + x < \xi(e)\})$$
and if the finite expectation exists, then the median coincides with the expectation, i.e.
$$\text{Me}\,\xi = E\xi = \alpha.$$

As we see from the definitions of quantiles, they are defined for all random variables, with no exceptions.

At the beginning of § 24 we found that the expected number of points scored with a single throw of a die equals $3\frac{1}{2}$. It is clear, however, that this number cannot occur in any throw. Thus, the values will be dispersed around the number $3\frac{1}{2}$. We shall characterize this dispersion numerically. The measure of dispersion of a random variable around its expectation, or around any other positional characteristic, can be defined in many ways. The simplest way is to use the mean absolute deviation.

DEFINITION 2. The *mean absolute deviation* of a random variable ξ with respect to a number a is defined as the expectation $E|\xi - a|$ of the random variable $|\xi - a|$.

Most frequently, one considers the mean deviation with respect to the median or with respect to the expectation.

The question arises for what value of a the expectation $E|\xi - a|$ assumes for a given random variable ξ its minimal value. One can easily prove that the minimum is attained for $a = \text{Me}\,\xi$. This property results from the following

THEOREM 1. *For every a we have the inequality*
$$E|\xi - \text{Me}\,\xi| \leqslant E|\xi - a|.$$

V. RANDOM VARIABLES

This theorem shows that the median, as the location measure corresponds to the mean deviation as the measure of dispersion.

Nevertheless, the most important both from the theoretical and from the practical point of view is another dispersion measure, called the mean square deviation.

DEFINITION 3. The *mean square deviation* of a random variable ξ with respect to the number a is defined as the arithmetic square root of the expectation $E(\xi-a)^2$. The mean square deviation with respect to the expectation is called the *standard deviation* of the random variable ξ. The square of the standard deviation, i.e. the expectation $E(\xi-E\xi)^2$, is called the *variance* of the random variable ξ. The variance of the random variable ξ will be denoted by $D^2\xi$, and its standard deviation by $D\xi$.

It could be asked now for what value of a the expectation $E(\xi-a)^2$ attains its minimum. The answer is given by the following

THEOREM 2. *For every a we have the equality*

$$E(\xi-a)^2 = E(\xi-E\xi)^2 + (a-E\xi)^2.$$

PROOF.

$$E(\xi-a)^2 = E(\xi-E\xi+E\xi-a)^2$$
$$= E(\xi-E\xi)^2 + 2E(\xi-E\xi)(E\xi-a) + E(E\xi-a)^2.$$

Since $E\xi-a$ is a numerical constant, and $E(\xi-E\xi) = 0$, in the last expression the second term vanishes and the third equals $(E\xi-a)^2$, which completes the proof.

Theorem 2 shows that the expectation as a location measure, is distinguished with respect to the mean square deviation as the measure of dispersion. Here is another property of the variance:

THEOREM 3. *The variance of a product of random variable by a constant equals the product of the variance of that random variable by the square of the constant*:

$$D^2 a\xi = a^2 D^2 \xi,$$

whence

$$Da\xi = |a|D\xi.$$

PROOF. We have

$$D^2 a\xi = E(a\xi - Ea\xi)^2 = E\big(a(\xi-E\xi)\big)^2$$
$$= Ea^2(\xi-E\xi)^2 = a^2 E(\xi-E\xi)^2 = a^2 D^2 \xi,$$

which was to be proved.

We suggest that the reader investigate himself the behaviour of the remaining characteristics under the multiplication of a random variable by a constant.

Proceeding one step further along the line of the definitions of expectation and variance we arrive at the concept of moments of a random variable.

DEFINITION 4. The expectation $E(\xi-a)^\nu$, where $\nu = 0, 1, 2, \ldots$ is called the *νth moment* of a random variable ξ with respect to a. The expectation $E|\xi-a|^\nu$ is called the *νth absolute moment* of ξ with respect to a. Moments with respect to the number 0 are called simply *moments*, and moments with respect to the expectation $E\xi$ are called *central moments*. Moments of the random variable ξ will be denoted by α_ν or $\alpha_\nu(\xi)$, absolute moments will be denoted by β_ν or $\beta_\nu(\xi)$; finally, central moments will be denoted by μ_ν or $\mu_\nu(\xi)$.

Note that

(a) The expectation equals the first moment;

(b) The moment of order 0 always exists and equals 1;

(c) The variance equals the central moment of order two.

Here are some further properties of moments:

(d) A moment of the order ν exists if and only if there exists an absolute moment of order ν.

Indeed, a moment of order ν exists if and only if the expectations of the positive part η_+ and the negative part η_- of the random variable $\eta = \xi^\nu$ exist. We have, however,

$$|\xi(e)|^\nu = \eta_+(e) - \eta_-(e),$$

which implies property (d).

(e) If a moment of a certain order exists, then all moments of lower orders also exist.

This property follows from the fact that if for a certain e we have $|\xi(e)| > 1$, then

(27.3) $\qquad |\xi(e)|^\nu > |\xi(e)|^{\nu-1}, \quad \nu = 1, 2, \ldots$

and we can write

$$E|\xi|^\nu = \int_X |\xi|^\nu dP = \int_A |\xi|^\nu dP + \int_{X-A} |\xi|^\nu dP$$

where $A = \{e: |\xi(e)| \leq 1\}$. The integral extended over A exists for every ν, since we always have $|\xi|^\nu \leq 1$, and the existence of the integral $\int_{X-A} |\xi|^\nu dP$, inequality (27.3) and inequality (26.5) imply that the integral $\int_{X-A} |\xi|^{\nu-1} dP$ exists, which was to be proved.

V. RANDOM VARIABLES

(f) Central moments may be expressed in terms of ordinary moments with respect to zero, and conversely; here are the first few expressions:

$$\mu_1 = 0,$$
$$\mu_2 = \alpha_2 - \alpha_1^2,$$
$$\mu_3 = \alpha_3 - 3\alpha_2\alpha_1 + 2\alpha_1^3,$$
$$\mu_4 = \alpha_4 - 4\alpha_3\alpha_1 + 6\alpha_2\alpha_1^2 - 3\alpha_1^4,$$
$$\alpha_1 = \alpha_1,$$
$$\alpha_2 = \mu_2 + \alpha_1^2,$$
$$\alpha_3 = \mu_3 + 3\mu_2\alpha_1 + \alpha_1^3,$$
$$\alpha_4 = \mu_4 + 4\mu_3\alpha_1 - 6\mu_2\alpha_1^2 + \alpha_1^4.$$

Note that there is a certain assymetry in the above formulas: central moments are indeed expressed in terms of moments with respect to zero, but moments with respect to zero are expressed in terms of central moments starting with the second, and by the moment with respect to zero. This is explained by the definition of central moments, according to which $\mu_1 = \alpha_1 - \alpha_1 = 0$.

CHAPTER VI

Laws of Large Numbers

§ 28. INDEPENDENCE OF RANDOM VARIABLES. ADDITIVITY OF VARIANCE

At the end of § 26 we mentioned that—theoretically—the concept of a random variable is superfluous and could be omitted. For the reasons stated there, however, experiments or observations are most frequently described in terms of random variables. At the beginning of § 27 we showed how this is done in the case of a single experiment or observation. In §§ 18 and 19 we discussed the problem of constructing a probability distribution describing a series of independent experiments. We shall now formulate the same considerations in terms of random variables. Thus, we give a formal definition of independence of random variables and some theorems concerning this concept.

DEFINITION 1. *Random variables $\xi = \xi(e)$ and $\eta = \eta(e)$ are called independent if for any two Borel sets A and B on the real line the events $\{e: \xi(e) \in A\}$ and $\{e: \eta(e) \in B\}$ are independent,* i.e. if
$$P(\{e: \xi(e) \in A, \eta(e) \in B\}) = P(\{e: \xi(e) \in A\})P(\{e: \eta(e) \in B\}).$$

Note that the above definition can be formulated in an equivalent manner as follows: random variables ξ and η are *independent* if the joint probability distribution of the pair ξ and η equals the Cartesian product of probability distributions of these random variables (see Definition 4 of § 19). According to this remark, the theorem below is simply a modification of Theorem 4 of § 19:

THEOREM 1. *Random variables ξ and η are independent if and only if the probability distribution function of the pair of random variables ξ and η equals the product of the probability distribution functions of these random variables, i.e. for arbitrary x and y we have $F_{\xi,\eta}(x, y) = F_\xi(x)F_\eta(y)$.*

THEOREM 2. *Discrete random variables ξ and η assuming values $x_1, x_2, \ldots$ with probabilities $p_1, p_2, \ldots$ and $y_1, y_2, \ldots$ with probabilities $q_1, q_2, \ldots$ respectively are independent if and only if for every i, j*
$$P(\{e: \xi(e) = x_i, \eta(e) = y_j\}) = p_i q_j$$
(cf. Theorem 1 in § 19).

VI. LAWS OF LARGE NUMBERS

THEOREM 3. *If random variables $\xi = \xi(e)$ and $\eta = \eta(e)$ are independent, and $f(x)$ and $g(x)$ are arbitrary Borel functions, then the random variables $\zeta(e) = f(\xi(e))$ and $\theta(e) = g(\eta(e))$ are independent.*

PROOF. For any Borel sets A and B we have

$$\{e: \zeta(e) \in A\} = \{e: \xi(e) \in f^{-1}(A)\},$$
$$\{e: \theta(e) \in B\} = \{e: \eta(e) \in g^{-1}(B)\},$$

where $f^{-1}(A) = \{x: f(x) \in A\}$, $g^{-1}(B) = \{x: g(x) \in B\}$. From these equations and from the fact that for arbitrary Borel sets A and B the sets $f^{-1}(A)$ and $g^{-1}(B)$ are also Borel sets, we obtain the assertion of Theorem 3.

In particular, from the independence of random variables ξ and η it follows that the random variables $\xi - a$ and $\eta - b$ are also independent for any numerical constants a and b.

The following property of independent random variables is of particular importance:

THEOREM 4. *If random variables $\xi = \xi(e)$ and $\eta = \eta(e)$ are independent and have finite expectations, then the expectation of the product of these random variables equals the product of their expectations*

$$E\xi\eta = E\xi E\eta.$$

PROOF. The theorem is true in the general formulation presented; we shall, however, give its proof only for random variables assuming a finite number of values. Suppose then that the random variable ξ assumes values $x_1, \ldots, x_n$ with probabilities $p_1, \ldots, p_n$, and the random variable η assumes values $y_1, \ldots, y_m$ with probabilities $q_1, \ldots, q_m$. By definition

$$E\xi = \sum_{i=1}^{n} x_i p_i, \quad E\eta = \sum_{j=1}^{m} y_j q_j,$$

$$E\xi\eta = \sum_{i=1}^{n}\sum_{j=1}^{m} x_i y_j P(\{e: \xi(e) = x_i, \eta(e) = y_j\}).$$

Since the random variables ξ and η are independent, we have for every i and j

$$P(\{e: \xi(e) = x_i, \eta(e) = y_j\}) = p_i q_j.$$

Combining these formulas, we obtain

$$E\xi = \sum_{i=1}^{n}\sum_{j=1}^{m} x_i y_j p_i q_j = \sum_{i=1}^{n} x_i p_i \sum_{j=1}^{m} y_j q_j = \left(\sum_{i=1}^{n} x_i p_i\right)\left(\sum_{j=1}^{m} y_j q_j\right) = E\xi E\eta,$$

which was to be proved.

Theorem 4 implies the following important

THEOREM 5. *If independent random variables ξ and η have finite variances, then the variance of their sum $\xi+\eta$ equals the sum of their variances:*

$$D^2(\xi+\eta) = D^2\xi + D^2\eta.$$

PROOF. By the definition of variance and by the theorems on expectation which we have already proved, we have

$$\begin{aligned}D^2(\xi+\eta) &= E(\xi+\eta-E(\xi+\eta))^2 = E(\xi+\eta-(E\xi+E\eta))^2\\ &= E((\xi-E\xi)+(\eta-E\eta))^2\\ &= E((\xi-E\xi)^2+2(\xi-E\xi)(\eta-E\eta)+(\eta-E\eta)^2)\\ &= E(\xi-E\xi)^2+2E(\xi-E\xi)(\eta-E\eta)+E(\eta-E\eta)^2.\end{aligned}$$

The first and last terms of the last sum are the variances of the random variables ξ and η respectively. By Theorem 3, the independence of the random variables ξ and η implies the independence of the random variables $\xi-E\xi$ and $\eta-E\eta$, and $E(\xi-E\xi) = E(\eta-E\eta) = 0$; thus we obtain from Theorem 4

$$2E(\xi-E\xi)(\eta-E\eta) = 0,$$

which completes the proof.

Let us also add that using the arguments of § 19 we can construct independent random variables with any given distribution functions. Given two distribution functions $F_1(x)$ and $F_2(x)$ it suffices to take as elementary events the points (x, y) on the Euclidean plane, and take as P the probability distribution corresponding to the distribution function $F(x, y) = F_1(x)F_2(x)$. Then the random variables ξ and η, defined as $\xi(x, y) = x$, $\eta(x, y) = y$, will be independent and will have such distributions that $F_\xi(x) \equiv F_1(x)$, $F_\eta(x) \equiv F_2(x)$.

All that we have said so far about the independence of two random variables can easily be extended to a larger number of random variables. We give below the definition of independence:

DEFINITION 2. *Random variables $\xi_1 = \xi_1(e), \ldots, \xi_n = \xi_n(e)$ are called independent* if for any Borel sets $A_1, \ldots, A_n$ the events

$$Z_i = \{e\colon \xi_i(e) \in A_i\}, \quad i = 1, 2, \ldots, n$$

satisfy the relation

$$P(Z_1 \cap Z_2 \cap \ldots \cap Z_n) = P(Z_1)P(Z_2) \ldots P(Z_n).$$

Clearly, the last relation holds when the events $Z_1, \ldots, Z_n$ are independent, but this relation alone does not establish the independence of these events (cf. Definition 2 in § 17). However, one can easily deduce that if the random

VI. LAWS OF LARGE NUMBERS 125

variables $\xi_1, \ldots, \xi_n$ are independent, then the events $Z_1, \ldots, Z_n$ defined above are always independent *en bloc*. It suffices to note that by substituting for some of the sets A_i the whole real line we obtain relations required in the definition of independence of system of events, namely the multiplicative conditions for subsystems of $Z_1, \ldots, Z_n$. Thus, we can say that random variables $\xi_1, \ldots, \xi_n$ are independent if and only if for any Borel sets $A_1, \ldots, A_n$ the events $Z_i = \{e \colon \xi_i(e) \in A_i\}$, $i = 1, 2, \ldots, n$, are independent *en bloc*.

Equivalently, Definition 2 can be formulated as follows (see Definition 9 of § 19): random variables $\xi_1, \ldots, \xi_n$ are independent if and only if the joint probability distribution of the whole system equals the Cartesian product of probability distributions of the random variables $\xi_1, \ldots, \xi_n$.

Below we give some consequences of Definition 2.

THEOREM 6. *If random variables $\xi_1, \ldots, \xi_n$ are independent, and $\xi_{i_1}, \ldots, \xi_{i_k}$ are some of the variables from among $\xi_1, \ldots, \xi_n$, then the random variables $\xi_{i_1}, \ldots, \xi_{i_k}$ are also independent. In other words, independence of a system of random variables implies the independence of every subsystems of it.*

PROOF. Put the whole real line instead of A_i for all indices i from outside the set $\{i_1, \ldots, i_k\}$. Then Definition 2 reduces to the definition of the independence of random variables $\xi_{i_1}, \ldots, \xi_{i_k}$.

THEOREM 7. *If random variables $\xi_1, \ldots, \xi_n$ are independent and $f(x_1, \ldots, x_m)$, $m < n$, is a Borel function of m real variables, $\eta(e) = f(\xi_1(e), \ldots, \xi_m(e))$, then the random variables $\eta, \xi_{m+1}, \ldots, \xi_n$ are independent.*

Sketch of the proof. For any Borel set A we have
$$\{e \colon \eta(e) \in A\} = \{e \colon (\xi_1(e), \ldots, \xi_m(e)) \in f^{-1}(A)\}$$
where
$$f^{-1}(A) = \{(x_1, \ldots, x_m) \colon f(x_1, \ldots, x_m) \in A\}.$$
Thus, it suffices to show that, for any m-dimensional Borel set A and Borel sets $A_{m+1}, \ldots, A_n$ on the real line, the events
$$Z_m = \{e \colon (\xi_1(e), \ldots, \xi_m(e)) \in A\},$$
$$Z_i = \{e \colon \xi_i(e) \in A_i\}, \quad i = m+1, \ldots, n,$$
are independent. We easily prove that the class K of m-dimensional sets A for which the events $Z_m, Z_{m+1}, \ldots, Z_n$ defined above are independent for all Borel sets $A_{m+1}, \ldots, A_n$ has the following properties:

(a) Class K contains all sets of the form
$$\{(x_1, \ldots, x_m) \colon x_1 < a_1, \ldots, x_m < a_m\};$$

(b) Class K is complementative;

(c) Class K contains the sums of every sequence of pairwise disjoint elements of this class;

By a rather involved argument, one can deduce from (b) and (c) that

(c') Class K contains the sums of every sequence of its elements.

Thus, (b) and (c') imply that K is a countably additive field of sets; by (a) it contains all m-dimensional Borel sets.

By simple induction, using Theorem 7 we obtain the following generalizations of Theorems 4 and 5:

THEOREM 8. *If random variables $\xi_1, \ldots, \xi_n$ have finite expectations and are independent, then*

$$E\xi_1 \xi_2 \ldots \xi_n = E\xi_1 E\xi_2 \ldots E\xi_n.$$

THEOREM 9. *If random variables $\xi_1, \ldots, \xi_n$ have finite variances and are independent, then*

$$D^2(\xi_1 + \ldots + \xi_n) = D^2\xi_1 + \ldots + D^2\xi_n.$$

Theorems 5 and 9 state the significant fact of the additivity of variance for independent random variables. This property distinguishes variance from other characteristics of dispersion. As we shall see in the next sections, this property implies another important fact: popularly speaking, given a large number of independent observations one can reconstruct the distribution of the characteristic in the population and its numerical parameters. In theory, this fact is contained in the so-called laws of large numbers, to be discussed in the present chapter.

Let us also note the following direct consequence of Theorem 9 and Theorem 3 of § 27:

COROLLARY. *If random variables $\xi_1, \ldots, \xi_n$ are independent and have identical variances equal to σ^2, then*

$$D^2\left(\frac{\xi_1 + \ldots + \xi_n}{n}\right) = \frac{\sigma^2}{n}.$$

In conclusion, let us add that the concept of independence of random variables can be extended to sequences of random variables:

DEFINITION 3. We say that random variables $\xi_1 = \xi_1(e), \xi_2 = \xi_2(e), \ldots$ form a *sequence of independent random variables* if for every n the random variables $\xi_1, \ldots, \xi_n$ are independent.

VI. LAWS OF LARGE NUMBERS

The existence of a sequence of independent random variables with any distributions given in advance is guaranteed by Theorem 9 of § 19. It suffices to consider, for any sequence of probability distribution functions $F_1(x), F_2(x), \ldots$ the Cartesian product of probability distributions on real lines, these distributions corresponding to the distribution functions under consideration, and to put for $i = 1, 2, \ldots$

$$\xi_i(x_1, x_2, \ldots) = x_i,$$

where $(x_1, x_2, \ldots)$ is a point of the above-mentioned Cartesian product of real lines. We easily note that the random variables defined by the above equality constitute a sequence of independent random variables with the probability distribution functions $F_1(x), F_2(x), \ldots$

§ 29. ČEBYŠEV INEQUALITY

In § 27 we introduced the concept of standard deviation and its square—variance—as measures of the dispersion of a random variable around its expectation. Variance as a measure of dispersion has various advantages. Thus, it is equal to zero if and only if the random variable assumes with probability one only one value. It increases when the values of the random variable are multiplied by a constant exceeding one in absolute value, and decreases if we multiply the values of the random variable by a constant smaller than one in absolute value (see Theorem 3 of § 27). These properties are consistent with the requirements which any measure of dispersion should satisfy. Indeed, under multiplication of values of a random variable by a constant exceeding one in absolute value its values spread out.

When we want to evaluate a certain characteristic of the distribution of a population on the basis of a sample (say, the expectation) other problems arise. Suppose that for some practical reasons connected with the way in which we shall use the knowledge of the population mean, we need to evaluate it with an error smaller than $\varepsilon > 0$. Then the question naturally arises, what is the probability that, under a given manner of estimation, the value obtained will differ from the true population mean by more than ε. The smaller is this probability the better is the manner of estimation (from the point of view of this criterion).

Indeed, there exists a general relation between this probability and the variance as a measure of dispersion. This relation plays a fundamental role in probability theory and serves as a foundation for the so-called laws of large numbers, mentioned at the end of § 28. It is contained in the famous Čebyšev inequality*.

---
* P. L. Čebyšev, 1821–1894.

This inequality shows that the smaller is the variance, the smaller is (for a fixed $\varepsilon > 0$) the probability that the value of the random variable differs from its expectation by more than ε. Here is the inequality:

THEOREM (Čebyšev inequality). *If a random variable $\xi = \xi(e)$ has a finite variance $\sigma^2 = D^2\xi$, then for every $\varepsilon > 0$*

$$P(\{e:\ |\xi(e)-E\xi| \geqslant \varepsilon\}) \leqslant \frac{\sigma^2}{\varepsilon^2}.$$

PROOF. From the definition we have

$$\sigma^2 = D^2\xi = E(\xi-E\xi)^2 = \int_X (\xi(e)-E\xi)^2 dP.$$

Let us denote by A the event that the value of the random variable differs from its expectation by at least ε:

$$A = \{e:\ |\xi(e)-E\xi| \geqslant \varepsilon\}.$$

For $e \in A$ we have the inequality

$$(\xi(e)-E\xi)^2 \geqslant \varepsilon^2.$$

We can therefore write

$$\sigma^2 = \int_X (\xi(e)-E\xi)^2 dP \geqslant \int_A (\xi(e)-E\xi)^2 dP \geqslant \int_A \varepsilon^2 dP = \varepsilon^2 P(A).$$

Dividing both sides of the inequality $\sigma^2 \geqslant \varepsilon^2 P(A)$ by ε^2, we obtain

$$P(A) \leqslant \frac{\sigma^2}{\varepsilon^2}$$

which is equivalent to the assertion of our theorem.

We leave to the reader the proofs of the following analogues of the last theorem:

(a) *If the random variable ξ has a finite expectation, then for every $\varepsilon > 0$*

$$P(\{e:\ |\xi(e)| \geqslant \varepsilon\}) \leqslant \frac{E|\xi|}{\varepsilon};$$

(b) *If for a natural v there exists a finite moment of the order $2v$, then for every $\varepsilon > 0$*

$$P(\{e:\ |\xi(e)| \geqslant \varepsilon\}) \leqslant \frac{E\xi^{2v}}{\varepsilon^{2v}}.$$

§ 30. EMPIRICAL DISTRIBUTION FUNCTION AND SAMPLE CHARACTERISTICS

Before we proceed further with the formulations and proofs of the laws of large numbers, we shall extend a little the considerations of § 27. This will allow us to grasp the practical sense of the laws of large numbers.

VI. LAWS OF LARGE NUMBERS

When we want to infer something about the distribution of population characteristics on the basis of a sample, we usually treat the sample as a miniature of the population, and we base our procedures on the hope that the sample characteristics computed in the same manner as the population characteristics should be approximately equal to the latter. Thus, if $x_1, \ldots, x_n$ are the values observed in the sample, we define the *sample distribution function*, or: the *empirical distribution function*, of the characteristic as the function $F_n(x)$ equal, for a fixed x, to the fraction of items in the sample with the value of the characteristic smaller than x. Formally, this definition can be expressed as follows:

$$F_n(x) = \frac{1}{n} \text{Card } \{i: x_i < x\},$$

where Card A denotes the number of elements in the set A. The numerical characteristics, such as expectation, or mean, variance, moments, quantiles, etc, computed for the empirical distribution function $F_n(x)$ will be called *empirical*, or *sample* parameters. Thus, the *sample mean*, or *sample expectation*, is defined as the number

$$\bar{x} = \frac{x_1 + \ldots + x_n}{n};$$

the *sample variance* will be defined as

$$s^2 = \frac{(x_1 - \bar{x})^2 + \ldots + (x_n - \bar{x})^2}{n}$$

and so on. However, the empirical distribution function depends on the results of sampling, and consequently all sample characteristics depend on the sample. As already mentioned, the results of sampling are described in our theory in terms of a system of random variables $\xi_1 = \xi_1(e), \ldots, \xi_n = \xi_n(e)$, which are independent and have the same distribution function, equal to the distribution function $F(x)$ of the population characteristic. The result of sampling of a given number of items corresponds in our theory to the values of our system of random variables for a fixed elementary event:

$$\xi_1(e) = x_1, \quad \ldots, \quad \xi_n(e) = x_n.$$

Thus, the empirical distribution function, and all the sample characteristics become functions of the elementary event, and hence random variables. Clearly, as regards the empirical distribution function, the random variable here is the value of this distribution function at any specified point x; thus, we have here a whole family of random variables "numbered" by real numbers. Consequently the empirical distribution function is a random variable whose values are real functions having the form of special distribution functions.

As has been said, we hope that the larger is the sample, the better miniature of the population it constitutes. We neglect here the fact that the population may become exhausted or that its distribution may change in the process of sampling. Thus, one should think either of sampling with replacement, or—instead of sampling from a finite population—on a random experiment which can be repeated under unchanged conditions an arbitrary number of times. We expect that, as the sample size increases, the sample characteristics—as random variables—will become more and more concentrated around the corresponding population characteristics. The precise formulation of the last property and the proof that it does in fact hold constitute—speaking popularly—the content of the laws of large numbers. In our theory, adding new random variables to the system corresponds to the concept of increasing the size of the sample. If we want to speak of the behaviour of sample characteristics when the size of the sample increases unboundedly, we must speak of a sequence of independent random variables with the same distribution function. Therefore, the object of our considerations will be a sequence of random variables $\xi_1, \ldots, \xi_n, \ldots$ with the same distribution function $F(x)$. Next, the sample characteristics, as functions whose arguments will be random variables of our system, will also constitute a sequence of random variables. Our object will be to study the properties of such sequences of random variables.

§ 31. WEAK LAW OF LARGE NUMBERS. CONVERGENCE IN PROBABILITY AND CONVERGENCE WITH PROBABILITY ONE

We shall now prove the so-called weak law of large numbers.

THEOREM 1 (Weak law of large numbers). *If $\xi_1 = \xi_1(e), \xi_2 = \xi_2(e), \ldots$ is a sequence of independent random variables with the same distribution with finite expectation m and variance σ^2, and $s_n = s_n(e)$ denote the average of the first n variables, i.e.*

$$s_n(e) = \frac{1}{n}(\xi_1(e) + \ldots + \xi_n(e)),$$

then for every $\varepsilon > 0$

$$\lim_{n \to \infty} P(\{e: |s_n(e) - m| \geqslant \varepsilon\}) = 0.$$

It can easily be seen that using the concept of limit, we can formulate assertion of our theorem equivalently as follows:

For every $\varepsilon > 0$ and $\eta > 0$ there exists an n_0 such that for $n > n_0$

$$P(\{e: |s_n(e) - m| \geqslant \varepsilon\}) \leqslant \eta.$$

This theorem asserts, roughly speaking, that by choosing a sufficiently large n we may ensure the estimation of the expectation m with an arbitrarily given error $\varepsilon > 0$ and with an arbitrarily given probability of exceeding this error.

PROOF. By the Corollary in § 29 we have
$$D^2 s_n = D^2\left(\frac{\xi_1 + \ldots + \xi_n}{n}\right) = \frac{\sigma^2}{n}.$$
Moreover, by Corollary 2 of § 24 we get
$$E s_n = E\left(\frac{\xi_1 + \ldots + \xi_n}{n}\right) = m.$$
Thus, using the Čebyšev inequality (Theorem in § 29) we can write
$$P(\{e\colon |s_n(e) - m| \geqslant \varepsilon\}) \leqslant \frac{\sigma^2}{n\varepsilon^2},$$
which immediately implies the assertion of our theorem.

The first variant of this theorem was proved in 1713 by Jacob Bernoulli. His assertion was: if m_n denotes the number of successes in n independent trials with the same probability of success equal to p, then the probability that the difference $\left|\frac{m_n}{n} - p\right|$ will exceed a given arbitrary positive number ε tends to zero as n increases to infinity. Let us note that this is a special case of our theorem, since m_n can be represented as a sum of n independent random variables $\xi_1, \ldots, \xi_n$, each assuming value 1 with probability p and value 0 with probability $q = 1-p$.

Let us come back to the assertion of our theorem. It can be formulated roughly as follows: when n increases to infinity, the random variables $s_n = s_n(e)$ differ less and less from the random variable $s = s(e) = m$, i.e. from a random variable which assumes the same value m for every elementary event e. This approximation is made more precise in the following definition:

DEFINITION 1. We say that the random variables $\xi_1 = \xi_1(e)$, $\xi_2 = \xi_2(e)$, ... *converge in probability* (or: *converge stochastically*) to the random variable $\xi = \xi(e)$ if for every $\varepsilon > 0$
$$\lim_{n \to \infty} P(\{e\colon |\xi_n(e) - \xi(e)| \geqslant \varepsilon\}) = 0.$$

Using this definition, we may formulate the assertion of our theorem as follows: the averages $s_n(e)$ are stochastically convergent to the random variable $s(e) \equiv m$.

EXAMPLE 1. Let us take as the space X of elementary events the unit interval $X = \{e\colon 0 < e < 1\}$, and consider the uniform probability distribution on this interval. Then the random variables $\xi_i(e)$, $i = 1, 2, \ldots$, defined by the equalities

$$\xi_i(e) = \begin{cases} 1 & \text{for } j/n \leqslant e < (j+1)/n, \\ 0 & \text{otherwise}, \end{cases}$$

where n is a positive integer such that

$$1+2+ \ldots +n \leqslant i < 1+2+ \ldots +n+(n+1)$$

and $j = i-(1+2+ \ldots +n)$, are convergent in probability to the random variable ξ defined by the equality

$$\xi(e) = 0 \quad \text{for} \quad 0 < e < 1.$$

We invite the reader to draw the graphs of the first few random variables of this sequence.

This example shows that random variables $\xi_1, \xi_2, \ldots$ can converge in probability to a random variable in spite of the fact that the sequence $\xi_1(e), \xi_2(e), \ldots$ is not convergent for any elementary event e. Indeed, in our example for every e the sequence $\xi_1(e), \xi_2(e), \ldots$ contains infinitely many zeros and infinitely many ones, whence it is not convergent. Thus, Theorem 1, which asserts that for large n the probability of a big difference between the average s_n and the expectation m becomes small does not exclude the possibility that as n increases, big differences will appear from time to time, perhaps not frequently. This is the reason why the theorem is called the weak law of large numbers.

The question arises to what extent the convergence in probability determines the limit. In other words, if random variables $\xi_1, \xi_2, \ldots$ converge in probability to a random variable ξ, can they converge also to another random variable η, different from ξ? The answer is given in the following two theorems, which the reader can easily prove for himself.

First, if the random variables $\xi_1, \xi_2, \ldots$ converge in probability to the random variable $\xi = \xi(e)$, and the random variable $\eta = \eta(e)$ differs from the random variable ξ on an event of probability zero, i.e. if

$$P(\{e: \xi(e) \neq \eta(e)\}) = 0,$$

then the random variables $\xi_1, \xi_2, \ldots$ converge in probability also to the random variable η. Second, if

$$P(\{e: \xi(e) \neq \eta(e)\}) > 0$$

then the random variable $\eta(e)$ is not the limit in probability of the sequence $\xi_1, \xi_2, \ldots$ if it converges in probability to $\xi(e)$. The content of both these theorems can be expressed by saying that the limit in probability is unique up to the values on events with probability zero, or shortly, unique almost surely.

VI. LAWS OF LARGE NUMBERS

The definition below gives another type of convergence of sequences of random variables:

DEFINITION 2. We say that random variables $\xi_1 = \xi_1(e)$, $\xi_2 = \xi_2(e)$, ... *converge to a random variable* $\xi = \xi(e)$ *with probability one*, or: *almost surely*, if there exists an event A such that $P(A) = 1$ and for every $e \in A$ the numerical sequence $\xi_1(e), \xi_2(e), ...$ converges to $\xi(e)$.

We can easily see that in the case of almost sure convergence the limit is determined uniquely up to events of probability zero.

We shall now investigate the mutual relation of these two types of convergence. We have already seen, that sequences convergent in probability need not be convergent with probability one. We shall prove that the second convergence is stronger in the sense that it implies the first. This explains their mutual relation.

THEOREM 2. *If random variables* $\xi_1, \xi_2, ...$ *are convergent to a random variable* ξ *with probability one, then they converge to this random variable in probability*.

PROOF. We assume that

(31.1) $$P(\{e: \lim_{n \to \infty} \xi_n(e) = \xi(e)\}) = 1.$$

We have to deduce that for every $\varepsilon > 0$

(31.2) $$\lim_{n \to \infty} P(\{e: |\xi_n(e) - \xi(e)| \geqslant \varepsilon\}) = 0.$$

This implication will become almost obvious if we express both of the above conditions in an alternative "epsilon" form.

Thus, for a fixed $\varepsilon > 0$ let A_r be the set of elementary events e such that the random variable ξ_r differs from the random variable ξ by less than ε. Let B_n denote the set of elementary events e such that starting from $r = n$ all random variables ξ_r differ from the random variable ξ by less than ε. In other words, let

$$A_r = \{e: |\xi_r(e) - \xi(e)| < \varepsilon\},$$

$$B_n = \bigcap_{r=n}^{\infty} A_r = \bigcap_{r=n}^{\infty} \{e: |\xi_r(e) - \xi(e)| < \varepsilon\}.$$

We claim that

(a) Condition (31.1) holds if and only if for every $\varepsilon > 0$ and $\eta > 0$ there exists an n such that

$$P(B_n) > 1 - \eta;$$

(b) Condition (31.2) holds if and only if for every $\varepsilon > 0$ and $\eta > 0$ there

exists an n such that for $r \geq n$
$$P(A_r) > 1-\eta.$$

From the fact that $B_n \subset A_k$ for $k \geq n$ it follows that if $P(B_n) > 1-\eta$, then $P(A_k) > 1-\eta$ for $k \geq n$, which proves our theorem. Thus, we have to prove only (a) and (b). Since (b) is quite obvious, it remains to prove (a).

Thus, let A denote the set of e such that
$$\lim_{n \to \infty} \xi_n(e) = \xi(e).$$

Let C_ε denote the set of e for which there exists an n such that for $r \geq n$ we have $|\xi_r(e) - \xi(e)| < \varepsilon$.

We shall show (a) in two steps. We prove first that the condition in (a) is necessary and sufficient in order that $P(C_\varepsilon) = 1$ for every $\varepsilon > 0$, and then we show that the equality $P(C_\varepsilon) = 1$ for every $\varepsilon > 0$ is equivalent to the equation $P(A) = 1$. These two properties will obviously imply (a).

To prove the first equivalence note that
$$C_\varepsilon = \bigcup_{n=1}^{\infty} B_n$$
and for every n
$$B_n \subset B_{n+1}.$$

Thus, C_ε is a sum of an increasing sequence of events. By Theorem 5 of § 6 we have
$$P(C_\varepsilon) = \lim_{n \to \infty} P(B_n),$$
which proves the first equivalence.

Now, for every $\varepsilon > 0$ we have the inclusion
$$A \subset C_\varepsilon.$$

Thus, if $P(A) = 1$, it follows that $P(C_\varepsilon) = 1$ for every $\varepsilon > 0$. To prove that the converse is also true, i.e. that if $P(C_\varepsilon) = 1$ for every $\varepsilon > 0$ then also $P(A) = 1$, note the following: relation
$$\lim_{n \to \infty} \xi_n(e) = \xi(e)$$
for a certain elementary event e holds if and only if, for every positive integer k there exists an n such that for all $r \geq n$
$$|\xi_r(e) - \xi(e)| < 1/k.$$

Thus, we can write
$$A = \bigcap_{k=1}^{\infty} C_{1/k},$$

hence A can be represented as an intersection of a sequence of events of probability one. It follows that $P(A) = 1$, since we have

$$P(A') = P\left(\left(\bigcap_{k=1}^{\infty} C_{1/k}\right)'\right) = P\left(\bigcup_{k=1}^{\infty} C'_{1/k}\right) \leqslant \sum_{k=1}^{\infty} P(C'_{1/k}) = 0;$$

we have used here the de Morgan laws and the fact that the complement of an event has probability zero if and only if that event has probability one.

The proof of Theorem 2 is complete.

§ 32. STRONG LAW OF LARGE NUMBERS

We shall now prove that under the assumptions of Theorem 1 of § 31 the averages s_n converge to m not only in probability but also with probability one. More precisely, we have

THEOREM (Strong law of large numbers). *If* $\xi_1 = \xi_1(e)$, $\xi_2 = \xi_2(e), \ldots$ *is a sequence of independent random variables with the same probability distribution with expectation m and variance σ^2, and if $s_n = s_n(e)$ is the average of first n of these variables*:

$$s_n(e) = \frac{1}{n}(\xi_1(e) + \ldots + \xi_n(e)),$$

then

$$P(\{e: \lim_{n \to \infty} s_n(e) = m\}) = 1.$$

PROOF.* By the weak law of large numbers, the averages s_n are convergent in probability to the random variable $s(e) \equiv m$. Thus, our theorem will be proved if we show that the set A of elementary events e for which the sequence $s_1(e)$, $s_2(e), \ldots$ converges has probability one. Indeed, denoting by $S(e)$ the random variable equal to $\lim_{n\to\infty} s_n(e)$ for $e \in A$ and equal to 0 for $e \in A'$ we shall be able to claim that the random variables $s_1(e), s_2(e), \ldots$ are convergent to $S(e)$ with probability one. By Theorem 2 of § 31 these random variables will also be convergent to $S(e)$ in probability. Thus, the averages $s_1(e), s_2(e), \ldots$ will converge in probability to $s(e) \equiv m$ and to $S(e)$. It follows that $S(e)$ differs from $s(e)$ at most on a set of probability zero. Thus, the averages $s_1(e), s_2(e), \ldots$ are convergent to $s(e)$ with probability one.

We thus have to show that

$$P(A) = 1 \quad \text{where} \quad A = \{e: \lim_{n \to \infty} s_n(e) \text{ exists}\}.$$

---

* We present the proof taken from B. de Finetti's paper, *La legge dei grandi numeri nel caso dei numeri aleatoari equivalenti*, Atti R. Acc. Naz. Lincei 18 (1933), pp. 203–207.

By definition, the limit $\lim_{n\to\infty} s_n(e)$ exists if and only if the sequence $s_n(e)$ is fundamental, i.e. if for every $\varepsilon > 0$ there exists an n such that for $r \geqslant n$ and an arbitrary q
$$|s_r(e) - s_{r+q}(e)| < \varepsilon.$$
Since instead of n we can write always a larger number in the above condition, we can require in addition that n be of the form $n = k^2$.

We show first that for every $\varepsilon > 0$ the set C_ε of points e for which there exists a k such that for $r \geqslant k^2$ and any q
$$|s_r(e) - s_{r+q}(e)| < \varepsilon$$
has probability one. From this property we shall deduce that $P(A) = 1$.

Let us fix $\varepsilon > 0$, and denote for fixed positive integers k and c
$$T_c = \bigcap_{i=k^2+1}^{(k+c)^2} \{e: |s_{k^2}(e) - s_i(e)| < \tfrac{1}{2}\varepsilon\}.$$
Note that if $e \in T_c$, then for arbitrary i and j such that $k^2 < i < j \leqslant (k+c)^2$ we have
$$|s_i(e) - s_j(e)| < \varepsilon$$
which follows from the simultaneous inequalities
$$|s_{k^2}(e) - s_i(e)| < \tfrac{1}{2}\varepsilon \quad \text{and} \quad |s_{k^2}(e) - s_j(e)| < \tfrac{1}{2}\varepsilon.$$
Now let
$$E = \bigcap_{j=k}^{k+c-1} \{e: |s_{j^2}(e) - s_{(k+c)^2}(e)| < \tfrac{1}{6}\varepsilon\},$$
$$A_j = \bigcap_{i=1}^{2j} \{e: |s_{j^2}(e) - s_{j^2+i}(e)| < \tfrac{1}{6}\varepsilon\},$$
where $j = k, k+1, \ldots, k+c-1$ and let
$$S = E \cap A_k \cap A_{k+1} \cap \ldots \cap A_{k+c-1}.$$
We shall prove that
$$S \subset T_c.$$
Indeed, if $e \in S$, then for every i such that $k^2 < i \leqslant (k+c)^2$ we have
$$|s_{k^2}(e) - s_i(e)| = |s_{k^2}(e) - s_{(k+c)^2}(e) + s_{(k+c)^2}(e) - s_{j^2}(e) + s_{j^2}(e) - s_i(e)|$$
$$\leqslant |s_{k^2}(e) - s_{(k+c)^2}(e)| + |s_{(k+c)^2}(e) - s_{j^2}(e)| + |s_{j^2}(e) - s_i(e)|$$
$$< 3 \cdot \tfrac{1}{6}\varepsilon = \tfrac{\varepsilon}{2}.$$

In these inequalities j is a positive integer such that $j^2 \leqslant i < (j+1)^2$.

VI. LAWS OF LARGE NUMBERS

Our first aim is to show that for a sufficiently large k the set T_c is an event with an arbitrarily large probability, irrespective of the value of c. By the inclusion $S \subset T_c$ it suffices to show that this property holds for the event S. Instead, we prove that the complement S' of the event S has an arbitrarily small probability irrespective of the value of c, provided only that k is sufficiently large. The previous assertion will follow immediately. By de Morgan's laws we have

$$S' = E' \cup A'_k \cup A'_{k+1} \cup \ldots \cup A'_{k+c-1}$$

and

$$P(S') \leqslant P(E') + P(A'_k) + P(A'_{k+1}) + \ldots + P(A'_{k+c-1}).$$

Thus, our assertion will be proved if we show that the events E', A'_k, $A'_{k+1}, \ldots, A'_{k+c-1}$ have sufficiently small probabilities. We shall use here the Čebyšev inequality applied to the random variables $s_n - s_{n+q}$.

Clearly, the expectation of the random variable $s_n - s_{n+q}$ is equal to 0, since $Es_n = m$ for every n. Let us compute the variance of this random variable. We have

$$s_n - s_{n+q} = \frac{1}{n}(\xi_1 + \ldots + \xi_n) - \frac{1}{n+q}(\xi_1 + \ldots + \xi_{n+q})$$

$$= \left(\frac{1}{n} - \frac{1}{n+q}\right)(\xi_1 + \ldots + \xi_n) - \frac{1}{n+q}(\xi_{n+1} + \ldots + \xi_{n+q})$$

$$= \frac{q}{n(n+q)} \sum_{i=1}^{n} \xi_i - \frac{1}{n+q} \sum_{i=n+1}^{n+q} \xi_i.$$

Since the random variables $\xi_1, \ldots, \xi_{n+q}$ are independent, we obtain

$$D^2(s_n - s_{n+q}) = D^2\left(\frac{q}{n(n+q)} \sum_{i=1}^{n} \xi_i - \frac{1}{n+q} \sum_{i=n+1}^{n+q} \xi_i\right)$$

$$= \frac{q^2}{n^2(n+q)^2} D^2\left(\sum_{i=1}^{n} \xi_i\right) + \frac{1}{(n+q)^2} D^2\left(\sum_{i=n+1}^{n+q} \xi_i\right)$$

$$= \frac{q^2}{n^2(n+q)^2} \sum_{i=1}^{n} D^2\xi_i + \frac{1}{(n+q)^2} \sum_{i=n+1}^{n+q} D^2\xi_i$$

$$= \frac{q^2}{n^2(n+q)^2} n\sigma^2 + \frac{1}{(n+q)^2} q\sigma^2$$

$$= \sigma^2\left(\frac{q^2}{n(n+q)^2} + \frac{q}{(n+q)^2}\right) = \sigma^2 \frac{q^2 + nq}{n(n+q)^2}$$

$$= \sigma^2 \frac{q(n+q)}{n(n+q)^2} = \sigma^2 \frac{q}{n(n+q)}.$$

Finally,
$$E(s_n-s_{n+q}) = 0, \quad D^2(s_n-s_{n+q}) = \sigma^2 \frac{q}{n(n+q)}.$$

Now, using the Čebyšev inequality we may estimate the probabilities of the events E', A'_k, A'_{k+1}, ..., A'_{k+c-1}. Let us start from the event E'. We have
$$E' = \bigcup_{j=1}^{k+c-1} \{e: \ |s_{j^2}(e)-s_{(k+c)^2}(e)| \geqslant \tfrac{1}{6}\varepsilon\}.$$

By the Čebyšev inequality, for every j such that $k \leqslant j < (k+c)$ we have
$$P(\{e: \ |s_{j^2}(e)-s_{(k+c)^2}(e)| \geqslant \tfrac{1}{6}\varepsilon\}) \leqslant \frac{D^2(s_{j^2}-s_{(k+c)^2})}{(\tfrac{1}{6}\varepsilon)^2}$$
$$= \frac{\sigma^2}{(\tfrac{1}{6}\varepsilon)^2} \cdot \frac{(k+c)^2-j^2}{j^2(k+c)^2} < \frac{\sigma^2}{(\tfrac{1}{6}\varepsilon)^2} \cdot \frac{1}{j^2}.$$

It follows that
$$P(E') \leqslant \sum_{j=k}^{k+c-1} P(\{e: \ |s_{j^2}(e)-s_{(k+c)^2}(e)| \geqslant \tfrac{1}{6}\varepsilon\})$$
$$< \frac{\sigma^2}{(\tfrac{1}{6}\varepsilon)^2} \sum_{j=k}^{k+c-1} \frac{1}{j^2} < \frac{\sigma^2}{(\tfrac{1}{6}\varepsilon)^2} R(k),$$

where $R(k)$ denotes the remainder
$$R(k) = \frac{1}{k^2} + \frac{1}{(k+1)^2} + \cdots$$

of the convergent series $1+(\tfrac{1}{2})^2+(\tfrac{1}{3})^2+\cdots$

Next, we have
$$A'_j = \bigcup_{i=1}^{2j} \{e: \ |s_{j^2}(e)-s_{j^2+i}(e)| \geqslant \tfrac{1}{6}\varepsilon\}.$$

By the Čebyšev inequality, for every i such that $1 \leqslant i \leqslant 2j$ we have
$$P(\{e: \ |s_{j^2}(e)-s_{j^2+i}(e)| \geqslant \tfrac{1}{6}\varepsilon\}) \leqslant \frac{D^2(s_{j^2}-s_{j^2+i})}{(\tfrac{1}{6}\varepsilon)^2} = \frac{\sigma^2}{(\tfrac{1}{6}\varepsilon)^2} \cdot \frac{i}{j^2(j^2+i)}.$$

Thus
$$P(A'_j) \leqslant \sum_{i=1}^{2j} P(\{e: \ |s_{j^2}(e)-s_{j^2+i}(e)| \geqslant \tfrac{1}{6}\varepsilon\}) \leqslant \frac{\sigma^2}{(\tfrac{1}{6}\varepsilon)^2} \sum_{i=1}^{2j} \frac{i}{j^2(j^2+i)}$$
$$< \frac{\sigma^2}{(\tfrac{1}{6}\varepsilon)^2} \sum_{i=1}^{2j} \frac{2j}{j^4} = \frac{\sigma^2}{(\tfrac{1}{6}\varepsilon)^2} 2j \frac{2j}{j^4} = \frac{4\sigma^2}{(\tfrac{1}{6}\varepsilon)^2} \cdot \frac{1}{j^2}.$$

VI. LAWS OF LARGE NUMBERS

Finally,

$$P(S') \leqslant P(E') + \sum_{j=k}^{k+c-1} P(A'_j) \leqslant \frac{\sigma^2}{(\frac{1}{6}\varepsilon)^2} R(k) + \frac{4\sigma^2}{(\frac{1}{6}\varepsilon)^2} \sum_{j=k}^{k+c-1} \frac{1}{j^2} < \frac{5\sigma^2}{(\frac{1}{6}\varepsilon)^2} R(k),$$

and since $P(T'_c) \leqslant P(S')$, we obtain for every c

$$P(T'_c) \leqslant \frac{5\sigma^2}{(\frac{1}{6}\varepsilon)^2} R(k).$$

However, we have

$$T'_c = \bigcup_{i=k^2+1}^{(k+c)^2} \{e: |s_{k^2}(e) - s_i(e)| \geqslant \tfrac{1}{2}\varepsilon\}$$

which implies that

$$T'_c \subset T'_{c+1}.$$

Thus, the sets $T'_1, T'_2, \ldots$ form an increasing sequence, and their sum

$$B'_k = \bigcup_{i=1}^{\infty} T'_i$$

satisfies the relation

$$P(B'_k) = \lim_{c \to \infty} P(T'_c).$$

From the above relation, combined with the inequality proved before, we obtain

$$P(B'_k) \leqslant \frac{5\sigma^2}{(\frac{1}{6}\varepsilon)^2} R(k).$$

Thus, for the set $B_k = (B'_k)'$ we have

$$P(B_k) \geqslant 1 - \frac{5\sigma^2}{(\frac{1}{6}\varepsilon)^2} R(k).$$

The set B_k equals the set of those elementary events e for which, starting from $i = k^2$, all the differences $s_{k^2}(e) - s_i(e)$ are smaller than $\varepsilon/2$ in absolute value:

$$B_k = \bigcap_{i=k^2+1}^{\infty} \{e: |s_{k^2}(e) - s_i(e)| < \tfrac{1}{2}\varepsilon\}.$$

The union C_ε of all sets B_k equals to the set of those elementary events e for which, starting from a certain k, for all $i \geqslant k^2$ we have

$$|s_{k^2}(e) - s_i(e)| < \tfrac{1}{2}\varepsilon,$$

i.e., it is the set of those points e for which there exists a k such that for $r \geqslant k^2$ and all q we have

$$|s_r(e) - s_{r+q}(e)| < \varepsilon.$$

For every k we have
$$B_k \subset C_\varepsilon,$$
hence for every k
$$P(C_\varepsilon) \geq P(B_k) \geq 1 - \frac{5\sigma^2}{(\frac{1}{6}\varepsilon)^2} R(k).$$
Since $R(k)$, as the remainder of a convergent series, tends to zero as k increases to infinity, it follows from the last inequality that
$$P(C_\varepsilon) = 1.$$
We shall now use the fact that for a sequence $s_1(e), s_2(e), \ldots$ to be fundamental for a certain e it is sufficient that for every positive integer r there exist a k such that for $j \geq k^2$ and all positive integer q
$$|s_j(e) - s_{j+q}(e)| < 1/r.$$
Thus, the set A can be represented as the intersection of countably many sets $C_{1/r}$:
$$A = \bigcap_{r=1}^{\infty} C_{1/r}.$$
As we have shown, all sets $C_{1/r}$ are events with probability one, hence A, as their intersection, is also an event with probability one. The relation
$$P(A) = 1$$
is therefore proved, and the proof of Theorem 1 is complete.

Some remarks are here in order. Note first that in the proof we used the assumption of independence of the random variables $\xi_1, \xi_2, \ldots$ only in computing the variances of $s_n - s_{n+q}$, where we used the fact that the variance of the sum of our random variables equals the sum of their variances. This property is a consequence of the fact that for the independent random variables ξ and η
$$E(\xi - E\xi)(\eta - E\eta) = 0.$$
When this relation holds, we say that the random variables ξ and η are *uncorrelated*, or *orthogonal*. Thus, we have only used the fact that our random variables, being independent, are also uncorrelated. Uncorrelated random variables, however, need not be independent, and in Theorem 1 we could have replaced the assumption of independence of the random variables by the assumption of their uncorrelatedness, thus obtaining a more general theorem.

Secondly the assumption that our random variables have identical distributions was not fully exploited, either. We only used the fact that they have identical expectations and variances.

VI. LAWS OF LARGE NUMBERS

Using a different method of proof, where the assumptions of independence and identical distributions are fully utilized, Kolmogorov proved* that the averages s_n in Theorem 1 tend with probability one to the common expectation of the random variables $\xi_1, \xi_2, \ldots$ provided only that a finite expectation exists (while the variance may not exist).

Finally, let us add that the assumption that our random variables are independent and have the same distribution is natural when we consider the sampling of elements from the same population, or when we consider independent repetitions of a random experiment performed in the same conditions. However, when each item is drawn from a different population, or when the conditions of experiment change from one repetition to another, our assumptions cease to be adequate. One could give more examples of such situations. The simplest example is provided by drawing without replacement from a finite population—in this case the results of previous drawings influence the probability distributions of subsequent results of drawings. This example, however, is of little use for us at the moment, since here the number of trials is bounded from above by the size of the population. Another example, where we have no such boundedness, is provided by sampling texts of telegrams sent from a given post office, in order to test the frequencies of various letters. In this case it is clear that successive letters are not completely random—among other reasons, simply because some combinations of letters cannot appear. In any observations performed at successive moments of time one can expect that the conditions of observations will somehow depend on time. The problem how to include these deviations, either from independence, or from identity of distributions, into a probabilistic scheme has been extensively studied. The theory of stochastic processes has grown out mainly from the need of adapting the probabilistic description to those phenomena which are not adequately described by the scheme of independent and identically distributed random variables. Here we want to mention briefly the generalizations of Theorem 1 which stem from these ideas. One direction of generalizations consists, roughly speaking, in retaining the assumption of identical distributions but weakening the assumptions on moments (one example of such generalizations is provided by the above mentioned theorem of Kolmogorov). An extension of this direction lies in weakening the assumption of independence and replacing it by some more general analogues of identical distributions, expressed in terms of joint probability distributions of systems of random variables. This direction culminates in the so-called ergodic theory, in which one proves the convergence of averages under very

---

* See for instance B. V. Gnedenko, *Lectures in Probability Theory*, Moscow–Leningrad 1950, Chapter 6.

general conditions; the assumptions of Theorem 1 constitute a very special case of these conditions.

The second direction consists in retaining fairly strong assumptions on moments, in particular the assumption of the existence of variance, and weakening the assumptions of independence and identity of distributions. One assumes only the equality of the first two moments and the uncorrelatedness of the random variables. Further generalizations are provided by the correlation theory of stochastic processes.

§ 33. ON THE CONVERGENCE OF SAMPLE CHARACTERISTICS

In this section we shall put together the conclusions which may be drawn from the laws of large numbers and which show, generally, that sample characteristics, such as sample mean, sample variance, moments, quantiles, probability distributions, tend—as the sample size increases—to the corresponding characteristics of the population. Let us start with moments.

THEOREM 1. *If $\xi_1 = \xi_1(e)$, $\xi_2 = \xi_2(e)$, ... is a sequence of independent random variables with identical distributions and if there exists a finite moment of the order 2ν, i.e. $\alpha_2 = E\xi_1^{2\nu} < \infty$, then the sample moments of the order ν defined as*

$$a_\nu^{(n)}(e) = \frac{1}{n}\left(\xi_1^\nu(e) + \ldots + \xi_n^\nu(e)\right)$$

converge with probability one to $\alpha_\nu = E\xi_1^\nu$.

PROOF. This theorem follows directly from Theorem 1 of § 32. It suffices to note that under our assumptions the random variables $\eta_1 = \xi_1^\nu$, $\eta_2 = \xi_2^\nu$, ... satisfy the assumptions of this theorem and $E\eta_1 = E\xi_1^\nu = \alpha_\nu$.

This conclusion can easily be extended to arbitrary continuous functions of moments. We have

THEOREM 2. *If $\xi_1 = \xi_1(e)$, $\xi_2 = \xi_2(e)$, ... are independent random variables with identical distributions, and if there exists a finite moment of the order 2ν, i.e. $\alpha_{2\nu} = E\xi_1^{2\nu} < \infty$, and $f(x_1, \ldots, x_m)$ is a function defined in an m-dimensional Euclidean space, continuous at the point $x_1 = \alpha_{i_1}, x_2 = \alpha_{i_2}, \ldots, x_m = \alpha_{i_m}$, where α_{i_k} is the moment of order i_k of our random variables, i.e. $\alpha_{i_k} = E\xi_1^{i_k}$, with $i_k \leqslant \nu$ for $k = 1, 2, \ldots, m$, then the random variables*

$$\eta_n(e) = f\left(a_{i_1}^{(n)}(e), a_{i_2}^{(n)}(e), \ldots, a_{i_m}^{(n)}(e)\right)$$

where

$$a_{i_k}^{(n)}(e) = \frac{1}{n}\left(\xi_1^{i_k}(e) + \ldots + \xi_n^{i_k}(e)\right)$$

converge with probability one to the constant $f(\alpha_{i_1}, \ldots, \alpha_{i_m})$.

VI. LAWS OF LARGE NUMBERS

PROOF. The continuity of $f(x_1, \ldots, x_m)$ at the point $x_1 = \alpha_{i_1}, \ldots, x_m = \alpha_{i_m}$ means that if $(x_1^{(n)}, \ldots, x_m^{(n)})$, $n = 1, 2, \ldots$, is a sequence of points in R_m convergent to the point $(\alpha_{i_1}, \ldots, \alpha_{i_m})$, i.e. if

$$\lim_{n \to \infty} x_k^{(n)} = \alpha_{i_k}, \quad k = 1, 2, \ldots, m,$$

then
$$\lim_{n \to \infty} f(x_1^{(n)}, \ldots, x_m^{(n)}) = f(\alpha_{i_1}, \ldots, \alpha_{i_m}).$$

By our assumptions, for every $k = 1, 2, \ldots, m$ there exists a set A_k of elementary events, with probability equal to 1, such that for $e \in A_k$

$$\lim_{n \to \infty} a_{i_k}^{(n)}(e) = \alpha_{i_k}.$$

Since $P(A_k) = 1$ for $k = 1, \ldots, m$, we have also $P(A) = 1$, where

$$A = A_1 \cap A_2 \cap \ldots \cap A_m.$$

However, for $e \in A$, the points $(a_{i_1}^{(n)}(e), \ldots, a_{i_m}^{(n)}(e))$ constitute a sequence of points in R_m convergent to $(\alpha_{i_1}, \ldots, \alpha_{i_m})$. Thus, the assertion of our theorem follows.

In particular, it follows from this theorem that the sample variance and other central sample moments converge with probability one to the corresponding moments of the population, since these moments are continuous functions of ordinary sample moments.

Below we give a theorem which could be called the strong law of large numbers for the frequency of successes in independent trials.

THEOREM 3. *If* $\xi_1 = \xi_1(e)$, $\xi_2 = \xi_2(e), \ldots$ *are independent random variables, such that for every* i

$$P(\{e: \xi_i(e) = 1\}) = p,$$
$$P(\{e: \xi_i(e) = 0\}) = 1 - p = q,$$

then the averages

$$s_n(e) = \frac{1}{n}(\xi_1(e) + \ldots + \xi_n(e))$$

converge with probability one to p.

This theorem is an obvious consequence of the theorem of § 32. It can also be formulated as follows: given a sequence of independent trials with the probability of success equal to p, if m_n denotes the number of successes in the first n trials, then with probability one we have the relation

$$\lim_{n \to \infty} \frac{m_n}{n} = p.$$

The reader should compare this theorem with Bernoulli's theorem in § 31.

Another important consequence of the above theorem is the following:

Theorem 3a. *If $\xi_1 = \xi_1(e)$, $\xi_2 = \xi_2(e)$, ... is a sequence of independent random variables with the same probability distribution function $F(x)$ and $F_n(x)$ is the empirical probability distribution function, i.e.*

$$F_n(x) = F_n(x; e) = \frac{1}{n} \operatorname{Card}\{i: \xi_i(e) < x, \ i = 1, 2, ..., n\},$$

then for every fixed x the random variables $F_n(x; e)$ converge with probability one to $F(x)$, i.e. for every x

$$P(\{e: \lim_{n \to \infty} F_n(x; e) = F(x)\}) = 1$$

(here and in the sequel $\operatorname{Card} A$ denotes the number of elements of the set A).

Proof. It suffices to note that

$$F_n(x; e) = \frac{1}{n}\left(\eta_1(e) + ... + \eta_n(e)\right)$$

where the random variables η_i are defined as

$$\eta_i(e) = \begin{cases} 1 & \text{if } \xi_i(e) < x, \\ 0 & \text{otherwise.} \end{cases}$$

The random variables η_i satisfy the assumptions of Theorem 3 with $p = F(x)$, which proves Theorem 3a.

The above theorem states that empirical distribution functions converge with probability one for every x separately. The question arises whether we can strenghten this theorem so as to obtain the simultaneous convergence of empirical distribution functions for every value of x; this is not guaranteed by Theorem 3a. It turns out, however, that one can prove much more: the famous theorem of Glivenko asserts that we have such simultaneous convergence uniformly in x. Note that this theorem, which we shall presently formulate and prove, can be treated as a theorem on "random variables" whose values are not numbers but probability distribution functions.

Theorem 4 (Glivenko). *If $\xi_1 = \xi_1(e)$, $\xi_2 = \xi_2(e)$, ... is a sequence of independent random variables with the common probability distribution function $F(x)$, and if $F_n(x)$ denotes the empirical distribution function, i.e.*

$$F_n(x) = F_n(x; e) = \frac{1}{n} \operatorname{Card}\{i: \xi_i(e) < x, i = 1, 2, ..., n\},$$

then the empirical distribution functions $F_n(x)$ converge uniformly to the distribution $F(x)$ with probability one.

VI. LAWS OF LARGE NUMBERS

PROOF. We have to show that the set A of elementary events for which the functions $F_n(x; e)$ converge uniformly to $F(x)$ as $n \to \infty$ is an event with probability one. From the definition of A it follows that if $e \in A$, then for every $\varepsilon > 0$ there exists an n such that for $r \geq n$ and all x we have

$$|F_r(x; e) - F_n(x)| < \varepsilon.$$

Note, however, that $e \in A$ if the functions $F_n(x; e)$ are such that for every k there exists an n such that for $r \geq n$ and all x we have

$$|F_r(x; e) - F(x)| < 1/k.$$

It follows that if C_ε denotes the set of points e for which the functions $F_n(x; e)$ have the property that there exists an r such that for $n \geq r$ and all x

$$|F_n(x; e) - F(x)| < \varepsilon,$$

then the set A can be represented as a product of countably many sets $C_{1/k}$, i.e.

$$A = \bigcap_{k=1}^{\infty} C_{1/k}.$$

To prove the theorem we shall show that for every $\varepsilon > 0$ the set C_ε has probability one; thus, A as a product of a sequence of events with probability one will also have probability one.

Let us fix $\varepsilon > 0$. We want to show that C_ε is an event with probability one. Suppose first that the probability distribution function $F(x)$ is continuous. Then one can find a finite system of numbers $x_1 < x_2 < \ldots < x_N$ such that

(33.1)
$$F(x_1) < \varepsilon/2,$$
$$F(x_{i+1}) - F(x_i) < \varepsilon/2, \quad i = 1, 2, \ldots, N-1,$$
$$F(x_N) > 1 - \varepsilon/2.$$

By Theorem 3a, for every $i = 1, 2, \ldots, N$ there exists an event B_i with probability one such that

$$\lim_{n \to \infty} F_n(x_i; e) = F(x_i).$$

Thus, the event

$$D = B_1 \cap B_2 \cap \ldots \cap B_N$$

also has probability one. We shall show that

$$D \subset C_\varepsilon,$$

which implies that

$$P(C_\varepsilon) = 1.$$

Indeed, if $e \in D$, then for every i we have $\lim_{n\to\infty} F_n(x_i; e) = F(x_i)$. Thus, we can find an r such that for $n \geqslant r$ and all $i = 1, 2, \ldots, N$ we have

$$|F_n(x_i; e) - F(x_i)| < \varepsilon/2.$$

We shall show that for every x

(33.2) $\qquad |F_n(x; e) - F(x)| < \varepsilon,$

which implies that $e \in C_\varepsilon$. To prove (33.2) denote for convenience $x_0 = -\infty$ and $x_{N+1} = +\infty$. Then, instead of (33.1) we can write

(33.3) $\qquad |F(x_{i+1}) - F(x_i)| < \varepsilon/2 \quad \text{for} \quad i = 0, 1, \ldots, N.$

Let us fix x. We can find an i such that $x_i \leqslant x \leqslant x_{i+1}$. We have

$$F(x_i) \leqslant F(x) \leqslant F(x_{i+1})$$

and

$$F_n(x_i) \leqslant F_n(x) \leqslant F_n(x_{i+1}).$$

The above inequalities and (33.3) imply (33.2). Indeed, on the one hand we have

$$F_n(x) - F(x) \leqslant F_n(x_{i+1}) - F(x_i) \leqslant F_n(x_{i+1}) - \left(F(x_{i+1}) - \tfrac{1}{2}\varepsilon\right)$$
$$\leqslant \tfrac{1}{2}\varepsilon + F_n(x_{i+1}) - F(x_{i+1})$$
$$\leqslant \tfrac{1}{2}\varepsilon + |F_n(x_{i+1}) - F(x_{i+1})| < \tfrac{1}{2}\varepsilon + \tfrac{1}{2}\varepsilon = \varepsilon.$$

On the other hand, we have

$$F(x) - F_n(x) \leqslant F(x_{i+1}) - F_n(x_i) \leqslant \left(\tfrac{1}{2}\varepsilon + F(x_i)\right) - F_n(x_i)$$
$$\leqslant \tfrac{1}{2}\varepsilon + |F(x_i) - F_n(x_i)| < \tfrac{1}{2}\varepsilon + \tfrac{1}{2}\varepsilon = \varepsilon,$$

which proves relation (33.2).

Thus, we have proved our theorem for continuous distribution functions. In the case where the distribution function has jumps the above proof is not valid; it can, however, easily be modified. One should take into account all values which are assumed with positive probability and use the fact that the average frequency of occurrence of these values in the sequence $\xi_1(e), \xi_2(e), \ldots$ converges with probability one to the probabilities of these values. We shall not consider the details of the modification of this proof.

To complete this section we shall prove a theorem on the convergence of sample quantiles. However, let us make some remarks first.

As we know from § 27, the quantile of order p, $0 < p < 1$, of a random variable with the probability distribution function $F(x)$ has been defined as any number z_p which satisfies the relation

(33.4) $\qquad F(z_p) \leqslant p \leqslant F(z_p + 0).$

VI. LAWS OF LARGE NUMBERS

The quantile need not be unique if the distribution function has an interval of constancy in which it assumes the value p. However, in view of the left continuity of the distribution function there always exists a largest quantile of order p. The empirical distribution function

(33.5) $$F_n(x) = \frac{1}{n} \operatorname{Card}\{i: x_i < x, 1 \leqslant i \leqslant n\},$$

corresponding to an n-element sample $x_1, \ldots, x_n$, as a distribution function assigning the same weights of $1/n$ to all points of the sample necessarily has intervals of constancy. Thus, in order to have a unique definition of the quantile for an empirical distribution function, we must consider the largest quantile of order p.

Let $x_1^* < \ldots < x_n^*$ be elements of the sample $x_1, \ldots, x_n$ ordered according to their magnitude. Since the empirical distribution function $F_n(x)$ assumes the value k/n in the interval $x_k^* < x \leqslant x_{k+1}^*$, the conditions for the quantile z_p of order p given in definition (33.4) may be written as follows:

$$z_p = x_{k+1}^* \quad \text{if} \quad k/n < p < (k+1)/n, \quad k = 0, 1, \ldots, n-1,$$
$$x_k^* \leqslant z_p \leqslant x_{k+1}^* \quad \text{if} \quad p = k/n, \quad k = 1, 2, \ldots, n-1.$$

Thus, we see that the largest quantile of the order p for the empirical distribution function $F_n(x)$ is given by the relation

$$z_p = x_{[np]+1}^*,$$

where $[x]$ is the greatest integer not exceeding x.

We can now formulate the theorem on the convergence of sample quantiles:

THEOREM 5. *Let $\xi_1 = \xi_1(e), \xi_2 = \xi_2(e), \ldots$ be a sequence of independent random variables with the same distribution function $F(x)$. Let z_p be the quantile of order p of the distribution function $F(x)$, where p is a number from the interval $0 < p < 1$ and let*

$$F_n(x) = \frac{1}{n} \operatorname{Card}\{i: \xi_i(e) < x, i = 1, 2, \ldots, n\}$$

be the empirical distribution function. Further, let $\zeta_p^{(n)}$ be the (largest) quantile of order p of $F_n(x)$. If the distribution function $F(x)$ is increasing at the point $x = z_p$, then

$$P\left(\{e: \lim_{n \to \infty} \zeta_p^{(n)} = z_p\}\right) = 1.$$

PROOF. Let us fix $\varepsilon > 0$. Since the distribution function $F(x)$ is increasing at the point $x = z_p$, we have

$$F(z_p - \varepsilon) < F(z_p) \leqslant p \leqslant F(z_p + 0) < F(z_p + \varepsilon).$$

Denote by η the smaller of the numbers $p-F(z_p-\varepsilon)$ and $F(z_p+\varepsilon)-p$. By Theorem 3a the empirical distribution functions $F_n(x)$ converge to the distribution function $F(x)$ at the points $x = z_p-\varepsilon$ and $x = z_p+\varepsilon$. Thus, with probability one, for sufficiently large n we shall have

$$|F_n(z_p-\varepsilon)-F(z_p-\varepsilon)| < \eta/2,$$

$$|F_n(z_p+\varepsilon)-F(z_p+\varepsilon)| < \eta/2,$$

whence

$$F_n(z_p-\varepsilon) < p < F_n(z_p+\varepsilon).$$

Thus, we shall have

$$z_p-\varepsilon < \zeta_p^{(n)} < z_p+\varepsilon.$$

We have thus proved that the event C_ε, consisting of elementary events e such that, starting from some index n, all the sample quantiles $\zeta_p^{(n)}$ of order p lie in the interval $z_p-\varepsilon < z < z_p+\varepsilon$, has probability one. Thus, the event $C_1 \cap C_{1/2} \cap \ldots$, consisting of those elementary events for which $\lim_{n\to\infty} \zeta_p^{(n)} = z_p$, also has probability one, which completes the proof.

§ 34. COMPLEMENTS AND EXAMPLES

In this section we present some direct calculations of expectations and variances of distributions which we have already encountered, and we formulate some general remarks connected with these calculations.

EXAMPLE 1. Compute the expectation of a random variable ξ with the binomial distribution $b(k; n, p)$.

SOLUTION. According to Definition 2 of § 24 we have

$$E\xi = \sum_{k=0}^{n} k b(k; n, p) = \sum_{k=1}^{n} k \binom{n}{k} p^k (1-p)^{n-k}.$$

When passing to the second term we omitted the term corresponding to $k = 0$. In the last sum, we may put the factor np before the summation sign, and obtain (on changing the numbering of the terms and using the fact that $\frac{k}{n}\binom{n}{k} = \binom{n-1}{k-1}$):

$$E\xi = np \sum_{k=1}^{n} \frac{k}{n}\binom{n}{k} p^{k-1}(1-p)^{n-k} = np \sum_{k=1}^{n} \binom{n-1}{k-1} p^{k-1}(1-p)^{(n-1)-(k-1)}$$

VI. LAWS OF LARGE NUMBERS

$$= np \sum_{k=0}^{n-1} \binom{n-1}{k} p^k (1-p)^{(n-1)-k} = np \sum_{k=0}^{n-1} b(k; n-1, p) = np.$$

Thus, the required expectation equals np.

EXAMPLE 2. Compute the expectation of a random variable ξ with the Poisson distribution $p(k; \lambda)$.

SOLUTION. According to Theorem 2 of § 21 and the definition of the Poisson distribution, we have

$$E\xi = \sum_{k=1}^{\infty} k \frac{\lambda^k}{k!} e^{-\lambda} = \sum_{k=1}^{\infty} \frac{\lambda^k}{(k-1)!} e^{-\lambda} = \lambda \sum_{k=1}^{\infty} \frac{\lambda^{k-1}}{(k-1)!} e^{-\lambda} = \lambda \sum_{k=0}^{\infty} \frac{\lambda^k}{k!} e^{-\lambda} = \lambda.$$

The second equality has been obtained by omitting the term with $k = 0$, the third—by putting the common factor λ before the summation sign, and the fourth—by changing the numbering of the terms. The sum appearing in the penultimate expression equals one as the sum of all the probabilities occurring in the Poisson distribution. Thus, the required expectation equals λ.

The following corollary to Theorem 1 of § 26 is useful in calculating expectations:

COROLLARY 1. *If a random variable ξ with finite expectation has a continuous distribution function with density $f(x)$, then its expectation equals*

$$\int_{-\infty}^{+\infty} xf(x)\, dx.$$

EXAMPLE 3. Compute the expectation of a random variable with normal distribution with the density

(34.1) $$f(x) = \frac{1}{\sigma\sqrt{2\pi}} e^{-(x-m)^2/2\sigma^2}.$$

SOLUTION. We have, by Corollary 1

$$E\xi = \int_{-\infty}^{+\infty} \frac{1}{\sigma\sqrt{2\pi}} xe^{-(x-m)^2/2\sigma^2}\, dx.$$

If we subtract from and add to the right-hand side the expression

$$m \int_{-\infty}^{+\infty} \frac{1}{\sigma\sqrt{2\pi}} e^{-(x-m)^2/2\sigma^2}\, dx = m$$

(this equality follows from the fact that the integrand is a probability density), we obtain

$$E\xi = \int_{-\infty}^{+\infty} \frac{1}{\sigma\sqrt{2\pi}} (x-m) e^{-(x-m)^2/2\sigma^2} dx + m = \int_{-\infty}^{+\infty} \frac{1}{\sigma\sqrt{2\pi}} u e^{-u^2/2\sigma^2} du + m.$$

The second equality has been obtained by means of the substitution $x-m = u$. The integral in the last expression vanishes. Thus, the required expectation equals m, which explains the meaning of this parameter.

When computing moments of random variables one can use the following identities, obtained in the same way as Theorem 1 in § 26, or Corollary 1 above.

COROLLARY 2. *If the random variable ξ has a continuous probability distribution with density $f(x)$, then its moments $\alpha_\nu(\xi)$ and $\beta_\nu(\xi)$, if they exist, can be obtained from the formulas*

$$\alpha_\nu(\xi) = \int_{-\infty}^{+\infty} x^\nu f(x) dx, \quad \beta_\nu(\xi) = \int_{-\infty}^{+\infty} |x|^\nu f(x) dx.$$

Similarly, the central moments are equal to integrals

$$\mu_\nu(\xi) = \int_{-\infty}^{+\infty} (x-m)^\nu f(x) dx \quad \text{where} \quad m = E\xi.$$

EXAMPLE 4. Calculate the variance of a random variable ξ which has a normal distribution given by formula (34.1).

SOLUTION. By Corollary 2 we have

$$D^2\xi = E(\xi - E\xi)^2 = \int_{-\infty}^{+\infty} (x-m)^2 \frac{1}{\sigma\sqrt{2\pi}} e^{-(x-m)^2/2\sigma^2} dx.$$

Substituting $u = (x-m)/\sigma$ we obtain

$$D^2\xi = \sigma^2 \int_{-\infty}^{+\infty} \frac{1}{\sqrt{2\pi}} u^2 e^{-u^2/2} du = \sigma^2,$$

since the integral in the above expressions equals 1. Thus, the required variance equals σ^2, which explains the meaning of the second parameter of normal distribution.

The fact that

(34.2) $$I = \int_{-\infty}^{+\infty} \frac{1}{\sqrt{2\pi}} u^2 e^{-u^2/2} du = 1$$

VI. LAWS OF LARGE NUMBERS

can be verified as follows. The use of equality (22.12) and the fact that the double integral of a product of two functions each of them depending of one variable only equals to the product of single integrals gives us

$$I = \int_{-\infty}^{+\infty} \frac{1}{\sqrt{2\pi}} u^2 e^{-u^2/2} du \cdot \int_{-\infty}^{+\infty} \frac{1}{\sqrt{2\pi}} e^{-v^2/2} dv$$

$$= \frac{1}{2\pi} \int_{-\infty}^{+\infty}\int_{-\infty}^{+\infty} u^2 e^{-(u^2+v^2)/2} du\, dv = \frac{1}{4\pi} \int_{-\infty}^{+\infty}\int_{-\infty}^{+\infty} (u^2+v^2) e^{-(u^2+v^2)/2} du\, dv.$$

We now pass to polar coordinates. Applying the substitution $x = r\cos\varphi$ and $y = r\sin\varphi$, we change the double integral into the product of single integrals. After simple transformations we get

$$I = \frac{1}{4\pi} \int_0^{2\pi}\int_0^{\infty} r^3 e^{-r^2/2} dr\, d\varphi = \frac{1}{2} \int_0^{\infty} r^3 e^{-r^2/2} dr.$$

Substituting $t = r^2/2$ and integrating twice by parts we obtain

$$I = \int_0^{\infty} te^{-t} dt = [-te^{-t}]_0^{\infty} + \int_0^{\infty} e^{-t} dt = [-e^{-t}]_0^{\infty} = 1.$$

EXAMPLE 5. Find the variance of a random variable ξ with a binomial distribution $b(k; n, p)$.

SOLUTION. According to the formulas given in § 27, point (f), the variance (or: second central moment) can be expressed by the first two moments with respect to zero as follows:

$$D^2\xi = E\xi^2 - (E\xi)^2.$$

The expectation we already know from Example 1. To compute the expectation of the square, we first compute the expectation of the product $\xi(\xi-1)$, since this is easier in our case. We have

$$E\xi(\xi-1) = E\xi^2 - E\xi = \sum_{k=0}^{n} k(k-1) b(k; n, p)$$

$$= \sum_{k=2}^{n} k(k-1) \binom{n}{k} p^k (1-p)^{n-k}$$

$$= \sum_{k=0}^{n-2} (k+2)(k+1) \binom{n}{k+2} p^{k+2} (1-p)^{n-k-2}.$$

Using the equality $(k+2)(k+1)\binom{n}{k+2} = n(n-1)\binom{n-2}{k}$, we get, after putting the appropriate expressions before the summation sign,

$$E\xi(\xi-1) = n(n-1)p^2 \sum_{k=0}^{n-2} \binom{n-2}{k} p^k(1-p)^{(n-2)-k}$$

$$= n(n-1)p^2 \sum_{k=0}^{n-2} b(k;n-2,p) = n(n-1)p^2.$$

It follows that
$$E\xi^2 = n(n-1)p^2 + np,$$
whence
$$D^2\xi = E\xi^2 - (E\xi)^2 = n(n-1)p^2 + np - (np)^2 = np(1-p).$$

Thus, the variance of a random variable with a binomial distribution $b(k;n,p)$ equals $np(1-p)$.

In our case, it has proved easier to compute the expectation of $\xi(\xi-1)$ than the expectation of the square; the former expectation equals the so-called *second factorial moment*. Generally, the *k*th *factorial moment* $\alpha_{[k]}(\xi) = \alpha_{[k]}$ of a random variable ξ, $k = 1, 2, \ldots$ is defined as the expectation of the product $\xi(\xi-1)(\xi-2) \ldots (\xi-k+1)$. In a similar way, we define the *central factorial moment of order k* as the expectation of the product $(\xi-E\xi)(\xi-E\xi-1) \ldots (\xi-E\xi-k+1)$. The reader can find the relations between factorial and ordinary moments.

EXAMPLE 6. Find the variance of a random variable which has a Poisson distribution with the parameter λ.

SOLUTION. In this case it is also more convenient to use the second factorial moment. We have

$$E\xi(\xi-1) = \sum_{k=0}^{\infty} k(k-1)\frac{\lambda^k}{k!}e^{-\lambda} = \sum_{k=2}^{\infty} k(k-1)\frac{\lambda^k}{k!}e^{-\lambda}$$

$$= \sum_{k=0}^{\infty} (k+2)(k+1)\frac{\lambda^{k+2}}{(k+2)!}e^{-\lambda} = \lambda^2 \sum_{k=0}^{\infty} \frac{\lambda^k}{k!}e^{-\lambda} = \lambda^2.$$

Using this equality and the result of Example 2, we obtain
$$D^2\xi = E\xi^2 - (E\xi)^2 = E\xi(\xi-1) + E\xi - (E\xi)^2 = \lambda^2 + \lambda - \lambda^2 = \lambda.$$

Thus, the variance of a random variable with the Poisson distribution is equal to λ, whence it is equal to the expectation of this random variable.

Problems

Find all factorial moments for binomial and Poisson distributions.

CHAPTER VII

Central Limit Theorems of Probability Theory

§ 35. GENERATING FUNCTIONS

When in § 20 we introduced the binomial distribution, we considered the number of successes in n Bernoulli trials. The number of successes in n Bernoulli trials may be treated as the sum of n independent random variables defined in such a way that the ith variable assumes value 1 in the case of success in the ith trial and 0 otherwise. Thus, the binomial distribution coincides with the distribution of a sum of n independent random variables with identical probability distributions.

In §§ 21 and 22 we proved a limit theorem asserting that the binomial distributions, under suitable manipulations with parameters n and p, and suitable norming, tend to Poisson or to normal distributions. As we have noted, the Bernoulli distribution with parameters n and p coincides with the distribution of a sum of n independent random variables, each of them assuming value 1 with probability p and value 0 with probability $1-p$. The question arises, whether the same convergence of probability distributions extends also to the case where the summands have distributions other than the above zero-one distributions.

It turns out that the answer is positive; in particular, convergence to a normal distribution is a very general phenomenon. The sum of a large number of independent terms has a distribution close to normal practically independently of the probability distributions of the summands. However, when we want to verify this, we are at once faced with great difficulty in calculating the probability distribution of sums of random variables. In the cases considered in §§ 21 and 22 we succeeded mainly because the probability distributions of the sums of the random variables turned out to be sufficiently simple and the convergence in question could be verified directly.

When looking for a general answer to the above problems, it is convenient to use a certain indirect method: roughly speaking, we look for functions which could be assigned to the probability distributions in question in a one-to-one way, such that to the probability distribution of a sum of two independent random variables corresponds the product of functions assigned to the terms. Thus, a rather involved operation of calculating the probability distribution of a sum of two independent random variables is replaced by the much simpler operation of multiplication of functions of a real variable. There are several such methods, differing in simplicity and in the range of probability distributions to which they can be applied. We shall begin with a discussion of the method of generating functions, applicable in the case of random variables which assume non-negative integer values.

Let us first consider a simple example.

EXAMPLE 1. Random variables ξ and η are independent and assume values 0, 1, 2 with probabilities p_0, p_1, p_2 and q_0, q_1, q_2. Find the probability distribution of the sum $\zeta = \xi + \eta$.

In this case the solution is simple. For instance, the sum ζ assumes value 2 if and only if ξ assumes value 0 and η assumes value 2, or ξ assumes value 1 and η assumes value 1, or ξ assumes value 2 and η assumes value 0. Since these events are disjoint, and ξ and η are independent, we obtain for the probability that $\zeta = 2$ the expression $p_0 q_2 + p_1 q_1 + p_2 q_0$.

In general, if ξ and η are independent random variables which assume non-negative integer values, and p_i and q_i are probabilities with which these random variables assume value i $(i = 0, 1, 2, \ldots)$, then the probability that the sum $\zeta = \xi + \eta$ assumes value k equals

$$(35.1) \qquad P(\{e : \zeta = k\}) = \sum_{i=0}^{k} p_i q_{k-i}.$$

As we see, the calculation is simple, but rather inconvenient for studying limits.

Now let us consider, for the random variables ξ and η of Example 1, the polynomials
$$w_\xi(t) = p_0 + p_1 t + p_2 t^2$$
and
$$w_\eta(t) = q_0 + q_1 t + q_2 t^2$$

and compare the coefficients at the successive powers of the variable t in the product $w_\zeta(t) = w_\xi(t) w_\eta(t)$ with the corresponding probabilities for the sum ζ. We have
$$w_\zeta(t) = w_\xi(t) w_\eta(t) = p_0 q_0 + (p_0 q_1 + q_0 p_1) t$$
$$+ (p_0 q_2 + p_1 q_1 + p_2 q_0) t^2 + (p_1 q_2 + p_2 q_1) t^3 + p_2 q_2 t^4.$$

The coefficient at the kth power of t equals the probability that the random variable ζ assumes value k. This example shows how the problem of finding the probability distribution of the sum of two independent random variables

VII. CENTRAL LIMIT THEOREMS

can be reduced to that of finding the product of two polynomials. What we have done in our example may be repeated with regard to arbitrary integer valued random variables. Instead of polynomials, in the case of random variables which assume infinitely many integer values with a positive probability we shall have analytic functions of a variable t; these analytic functions, expanded into power series with respect to t, have the probabilities of the particular values k as coefficients at the corresponding powers of t. We shall formulate the results of the above considerations in the form of a definition and a theorem.

DEFINITION 1. *If ξ is a random variable assuming non-negative integer values k with probabilities p_k, then the function of the real variable t defined by the equality*

$$w_\xi(t) = p_0 + p_1 t + p_2 t^2 + \ldots = \sum_{k=0}^{\infty} p_k t^k$$

is called the generating function of the random variable *ξ, or shortly, the* generating function.

Since the non-negative coefficients p_k add up to 1, the generating function is well defined at least in the closed interval $-1 \leqslant t \leqslant 1$. According to this definition, we have a one-to-one correspondence between analytic functions of a certain form, and the probability distributions of non-negative integer valued random variables. This correspondence has the property that to the sum of two independent random variables corresponds the product of those analytic functions.

THEOREM 1. *If ξ and η are independent random variables assuming non-negative integer values, then the generating function of their sum $\zeta = \xi + \eta$ equals the product of the generating functions of the terms*:

$$w_{\xi+\eta}(t) = w_\xi(t) w_\eta(t).$$

Generating functions have one more property which makes them useful in studying limit theorems: the convergence of generating functions is equivalent to the convergence of the corresponding probability distributions. We state this property in the form of the following theorem:

THEOREM 2. *Suppose that for every $n = 1, 2, \ldots$*

(35.2) $$p_{0,n}, p_{1,n}, p_{2,n}, \ldots$$

is the probability distribution of a certain non-negative integer valued random variable. Then the relation

(35.3) $$\lim_{n \to \infty} p_{k,n} = p_k$$

for every $k = 0, 1, 2, \ldots$ holds if and only if for every t from the interval $0 \leqslant t \leqslant 1$ we have

(35.4) $$\lim_{n \to \infty} w_n(t) = w(t),$$

where

$$w_n(t) = p_{0,n} + p_{1,n}t + p_{2,n}t^2 + \ldots,$$
$$w(t) = p_0 + p_1 t + p_2 t^2 + \ldots$$

Note that relation (35.3) implies automatically that $0 \leqslant p_k \leqslant 1$ and $\sum_{k=0}^{\infty} p_k \leqslant 1$. Thus, the limit function $w(t)$ is defined at least in the interval $-1 \leqslant t \leqslant 1$. However, the sequence $p_0, p_1, p_2, \ldots$ is not necessarily a probability distribution. For example, if in the sequence $p_{0,n}, p_{1,n}, p_{2,n}, \ldots$ the first n terms vanish, the limit sequence consists of zeros only. This theorem, however, allows us to infer from the convergence of a sequence of generating functions to a generating function that the corresponding sequence of distributions converges to the limit distribution.

It is worth while to present the proof of this theorem. Its main idea is the same as the idea of the proof of an analogous theorem for characteristic functions, which will be presented later.

PROOF. Suppose first that relations (35.3) hold. For a fixed t from the interval $0 < t < 1$ and a fixed $\varepsilon > 0$ we can find an integer r such that $t^r/(1-t) < \varepsilon$. Obviously, for every n we have

$$\sum_{k=r+1}^{\infty} p_{k,n} t^k < \sum_{k=r+1}^{\infty} t^k = \frac{t^{r+1}}{1-t} < \varepsilon$$

and

$$\sum_{k=r+1}^{\infty} p_k t^k < \sum_{k=r+1}^{\infty} t^k = \frac{t^{r+1}}{1-t} < \varepsilon.$$

We infer that

$$|w_n(t) - w(t)| < \sum_{k=0}^{r} |p_{k,n} - p_k| t^k + 2\varepsilon.$$

Indeed, the right-hand side has only a finite number of terms tending to zero. Thus, $|w_n(t) - w(t)|$ is arbitrarily small for a sufficiently large n; this proves that relation (35.3) implies (35.4).

Suppose now that (35.4) holds. Then one can find an increasing sequence of positive integers $n_1, n_2, \ldots$ which determines a subsequence of distributions (35.2) with the property that for every k we shall have $\lim_{j \to \infty} p_{k,n_j} = p'_k$. This can

VII. CENTRAL LIMIT THEOREMS

be achieved by choosing first a subsequence for which the convergence in question holds for $k = 0$, we then select from this sequence a subsequence for which the convergence holds for $k = 1$, and then select from it a subsequence for which the convergence holds for $k = 2$ and so on. Then, by the well known diagonal method one can form a sequence from such a sequence of sequences by choosing as its kth term the kth term in the kth of our sequences. If relation (35.3) were not satisfied, we could find another sequence of positive integers $n_1' < n_2' < \ldots$ such that for every k we would have $\lim_{j \to \infty} p_{k,n_j'} = p_k''$, with the limit perhaps different. But then, for every t from the interval $0 < t < 1$ we would have on the one hand

$$\lim_{j \to \infty} w_{n_j}(t) = \sum_{k=0}^{\infty} p_k' t^k,$$

and on the other hand

$$\lim_{j \to \infty} w_{n_j'}(t) = \sum_{k=0}^{\infty} p_k'' t^k,$$

which contradicts equality (35.4). It follows that relation (35.4) implies relation (35.3), which was to be shown.

As an example of the application of generating functions to the investigation of limit distributions we present here a new proof of Theorem 1 of § 21.

PROOF OF THEOREM 1 OF § 21. By definition, the generating function of a random variable which assumes value 1 with probability p and value 0 with probability q is $q+pt$. Thus, $(q+pt)^n = (1-p(1-t))^n$ is the generating function of the binomial distribution $b(k; n, p)$, equal to the distribution of the sum of n independent terms with the above zero-one distribution. Let us now find the generating function of the Poisson distribution $p(k; \lambda)$. Using the definition, we obtain after some simple transformations:

$$\sum_{k=0}^{\infty} p(k; \lambda) t^k = \sum_{k=0}^{\infty} \frac{\lambda^k}{k!} e^{-\lambda} t^k = e^{-\lambda} \sum_{k=0}^{\infty} \frac{(\lambda t)^k}{k!} = e^{-\lambda} e^{\lambda t} = e^{-\lambda(1-t)}.$$

We shall show that if p_n, $0 < p_n < 1$ changes with n so that $\lim_{n \to \infty} np_n = \lambda$, then for every t, $0 < t < 1$, we have

(35.5) $$\lim_{n \to \infty} (1-p_n(1-t))^n = e^{-\lambda(1-t)}.$$

It suffices to show that for every t from the interval $0 \leqslant t < 1$ we have

(35.6) $$\lim_{n \to \infty} n \log(1-p_n(1-t)) = -\lambda(1-t).$$

This relation can easily be verified by using the expansion

$$\log(1+x) = x + O(x^2)$$

valid in the neighbourhood of the point $x = 0$ (we write $f(t) = O(t)$ if in some interval $|t| < \varepsilon$ we have $|f(t)| < K|t|$ for a certain K). Let us fix t. Substituting $x = p_n(1-t)$ and using the fact that $p_n = O(1/n)$ (since $np_n \to \lambda$), we obtain

$$\log(1 - p_n(1-t)) = -p_n(1-t) + O(1/n^2),$$

hence

$$n\log(1 - p_n(1-t)) = -np_n(1-t) + O(1/n).$$

Relation (35.6), whence also relation (35.5) follows. Thus, by Theorem 2, we obtain the assertion of Theorem 1 of § 21.

Problems

1. Prove, by means of differentiating the power series appearing in the definition of the generating function, that the expectation of a non-negative integer-valued random variable equals the derivative of its generating function at the point $t = 1$; generally, prove that the nth derivative of the generating function at the point $t = 1$ equals the nth factorial moment of the random variable, provided this moment exists.

2. For an infinite sequence of Bernoulli trials with probability p of success, find the probability distribution of the waiting time for the first success and the generating function of this waiting time. The *waiting time for the first success* is defined as the random variable which assumes value k ($k = 0, 1, ...$) if the first success occurs in the $(1+k)$th trial. This distribution depends on the parameter p and is called the *geometric distribution*.

3. Find the expectation and variance of the geometric distribution.

4. Using generating functions, find, for the infinite sequence of Bernoulli trials with the probability of success p the *probability distribution of the waiting time for the rth success*. The latter random variable is defined as equal to k ($k = 0, 1, ...$) if the rth success occurred in $(r+k)$th trial. This is the so-called *Pascal distribution*; it depends on two parameters, r and p.

Hint. The waiting time for the rth success may be treated as the sum of r independent random variables with a probability distribution such as that for the waiting time for the first success.

5. Prove that if in the Pascal distribution the parameters p and r change in such a way that $r \to \infty$ and $r(1-p) \to \lambda$ then in the limit we obtain the Poisson distribution $p(k; \lambda)$.

§ 36. CHARACTERISTIC FUNCTIONS

Generating functions, considered in the preceding paragraph, constitute a very convenient tool for studying probability distributions of sums of independent random variables and finding limit distributions. Unfortunately, their application is restricted only to non-negative integer valued random variables, while even simple multiplication by a number may lead beyond this class

VII. CENTRAL LIMIT THEOREMS

of random variables. It turns out, however, that an operation assigning functions of a real variable to probability distributions under which addition of independent random variables corresponds to multiplication of these functions may be extended to all probability distributions. In this manner we obtain a convenient analytic tool for studying limit theorems in their full generality.

Note first that the generating function of a non-negative integer valued random variable ξ, which we defined formally as a power series of the variable t, can be defined equivalently as the expectation of the random variable t^ξ, i.e. $w_\xi(t) = Et^\xi$. If we realize this, then the theorem asserting that the generating function of the sum of two independent random variables equals to the product of the generating functions of the terms of the sum becomes an immediate consequence of the theorem asserting that the expectation of the product of two independent random variables equals the product of their expectations.

An analytical tool similar to that of generating functions, but convenient in its application to a wider class of distributions, can be obtained if instead of the function t^x we take the function e^{tx}. This leads to the so-called moment generating function: the *moment generating function* of a random variable ξ with the probability distribution $F_\xi(x)$ is defined as the function $W_\xi(t)$ of the real variable t:

$$(36.1) \qquad W_\xi(t) = Ee^{t\xi} = \int e^{tx} dF_\xi(x).$$

Moment generating functions have properties analogous to those of generating functions, and their use has proved convenient in studying convergence to the normal distribution. Their name is explained by the fact that in view of the expansion

$$e^{t\xi} = 1 + t\xi + \frac{(t\xi)^2}{2!} + \frac{(t\xi)^3}{3!} + \cdots$$

we can represent the moment generating function (by taking expectations on both sides) as the series:

$$(36.2) \qquad W_\xi(t) = 1 + \alpha_1(\xi)t + \frac{\alpha_2(\xi)t^2}{2!} + \frac{\alpha_3(\xi)t^3}{3!} + \cdots$$

Here the coefficients at different powers of t are proportional to the successive moments of the random variable ξ. This expansion allows us in some cases to prove convergence to a certain limit distribution by showing that all the moments are convergent to the corresponding moments of the limit distribution.

Still, moment generating functions are not yet defined for all random variables. This is connected with the fact that the random variable $e^{t\xi}$ is unbounded from above when the random variable ξ is unbounded. Thus, the expectation of the random variable $e^{t\xi}$ may be non-existent. To obtain the tool which we need we must replace the function e^{tx} by a function bounded for all t and x. We take the complex function $e^{itx} = \cos tx + i \sin tx$, where i denotes the complex unit. This leads us to the consideration of complex random variables. Such random variables may always be treated as pairs of real random variables, consisting of a real and an imaginary part, and apart from the proper treatment of the imaginary unit in the addition of random variables, in their multiplication, and in other operations, we can treat complex random variables in the same way as we treat two dimensional random variables. Thus, the concepts of probability distribution, moments, and the like, can be directly extended to complex case without essential changes. The expectation of a complex random variable equals a complex number whose real part is the expectation of the real part of the random variable, and whose imaginary part equals the expectation of the imaginary part of the random variable. Various theorems, in particular the theorem asserting that the expectation of the absolute value of a random variable is not smaller than the absolute value of the expectation of that random variable and the theorem asserting that if the absolute value of a random variable has finite expectation, so does the random variable in question, remain valid in the complex case.

Thus, we arrive at the most important concept, namely that of the characteristic function.

DEFINITION 1. A function of the real variable t defined by the equality

$$\varphi_\xi(t) = E e^{it\xi} = \int e^{itx} dF_\xi(x)$$

will be called the *characteristic function of the random variable* ξ with the probability distribution function $F_\xi(x)$.

All random variables with the same distribution have the same characteristic functions. Thus, we shall also speak of the characteristic functions of probability distributions, or characteristic functions corresponding to the distribution function $F(x)$, or (in view of Theorem 4) of the distribution function $F(x)$ corresponding to the characteristic function $\varphi(t)$.

Since for any t and x we have

(36.3) $$|e^{itx}| = 1,$$

the characteristic functions are defined for every random variable on the whole real line.

Below we give some analytic properties of characteristic functions, and their connection with moments.

VII. CENTRAL LIMIT THEOREMS

THEOREM 1. *Every characteristic function $\varphi(t)$ of a probability distribution has the following properties*:
(a) $\varphi(0) = 1$,
(b) $|\varphi(t)| \leq 1$ *for every* t,
(c) $\varphi(t) = \overline{\varphi(-t)}$ *where the bar denotes the conjugate complex number*,
(d) $\varphi(t)$ *is uniformly continuous on the whole real line*.

PROOF. Property (a) follows directly from the definition. In view of (36.3), the application of the inequality $|E\xi| \leq E|\xi|$ to the random variable $e^{it\xi}$ leads to inequality (b). Equality (c) follows from the fact that e^{itx} is conjugate to e^{-itx}. We have

$$\varphi(t+h) - \varphi(t) = Ee^{i(t+h)\xi} - Ee^{it\xi} = Ee^{it\xi}(e^{ih\xi} - 1).$$

Thus, from the equality

$$|e^{it\xi}(e^{ih\xi} - 1)| = |e^{ih\xi} - 1|$$

we obtain

$$|\varphi(t+h) - \varphi(t)| \leq E|e^{ih\xi} - 1|.$$

The right-hand side is independent of t and tends to zero with h. This implies that the characteristic function is uniformly continuous.

THEOREM 2. *If the νth moment of a random variable ξ with the characteristic function $\varphi(t)$ exists, then $\varphi(t)$ is differentiable ν times, the νth derivative is a continuous function and*

(36.4) $$\alpha_\nu(\xi) = \frac{1}{i^\nu}\left[\frac{d^\nu \varphi(t)}{dt^\nu}\right]_{t=0}.$$

PROOF. By definition

$$\varphi(t) = \int_{-\infty}^{+\infty} e^{itx} dF(x).$$

Differentiating formally ν times under the integral sign we get

$$\frac{d^\nu \varphi(t)}{dt^\nu} = \int_{-\infty}^{+\infty} (ix)^\nu e^{itx} dF(x).$$

We know from analysis that the sufficient condition for differentiability under the integral sign and for the equality of the integral of the derivative and the derivative of the integral is that the partial derivative of the integrand, treated as a function of the variable under integration, be bounded in absolute value by a function integrable with respect to the variable of integration. In view of the equality

$$|(ix)^\nu e^{itx}| = |x|^\nu$$

this condition is satisfied in our case, provided that the νth moment exists.

Moreover, as the function under the integral sign in the formal expression of the νth derivative is continuous with respect to t, the same is true for the νth derivative. This completes the proof of Theorem 2.

The most important properties of characteristic functions (from the point of view of limit theorems) are given in Theorems 3, 4 and 5.

THEOREM 3. *If ξ and η are two independent random variables, then their sum has the characteristic function*

$$\varphi_{\xi+\eta}(t) = \varphi_\xi(t)\varphi_\eta(t).$$

PROOF. In view of the independence of the random variables ξ and η the random variables $e^{it\xi}$ and $e^{it\eta}$ are also independent. We have therefore

$$\varphi_{\xi+\eta}(t) = Ee^{it(\xi+\eta)} = Ee^{it\xi}e^{it\eta} = Ee^{it\xi}Ee^{it\eta} = \varphi_\xi(t)\varphi_\eta(t),$$

which was to be proved.

By definition, to every probability distribution corresponds exactly one characteristic function. We do not know, however, whether or not the same characteristic function may correspond to several distributions. We shall prove that the correspondence between probability distributions and characteristic functions is one-to-one.

THEOREM 4 (Conversion formula). *If $\varphi(t)$ is the characteristic function of the distribution function $F(x)$, and x' and x'' are points of continuity of the function $F(x)$, then*

$$(36.5) \qquad F(x'') - F(x') = \frac{1}{2\pi} \int_{-\infty}^{+\infty} \frac{e^{-itx'} - e^{-itx''}}{it} \varphi(t) dt,$$

whence the characteristic function determines uniquely the probability distribution function. If, in addition, the absolute value of the characteristic function is integrable over the whole line, then there exists a continuous density $f(x)$ given by the formula

$$(36.6) \qquad f(x) = \frac{1}{2\pi} \int_{-\infty}^{+\infty} e^{-itx} \varphi(t) dt.$$

PROOF. Let $x' < x''$. Denote

$$I_c = \frac{1}{2\pi} \int_{-c}^{c} \frac{\exp(-itx') - \exp(-itx'')}{it} \varphi(t) dt$$

(here and in the sequel we write for convenience $\exp(x)$ instead of e^x).

VII. CENTRAL LIMIT THEOREMS

Expressing $\varphi(t)$ by $F(x)$ and changing the order of integration, we obtain

$$I_c = \frac{1}{2\pi} \int_{-c}^{c} \frac{\exp(-itx') - \exp(-itx'')}{it} \int_{-\infty}^{\infty} e^{itx} dF(x) dt$$

$$= \frac{1}{2\pi} \int_{-\infty}^{+\infty} \int_{-c}^{c} \frac{\exp[it(x-x')] - \exp[it(x-x'')]}{it} dt \, dF(x).$$

We have, however

$$\int_{-c}^{c} \frac{\exp[it(x-x')] - \exp[it(x-x'')]}{it} dt$$

$$= \int_{-c}^{c} \frac{\cos t(x-x') - \cos t(x-x'') + i[\sin t(x-x') - \sin t(x-x'')]}{it} dt$$

$$= -i \int_{-c}^{c} \frac{\cos t(x-x') - \cos t(x-x'')}{t} dt + \int_{-c}^{c} \frac{\sin t(x-x') - \sin t(x-x'')}{t} dt$$

$$= i \int_{-c}^{c} \frac{2 \sin t \left(x - \frac{x'+x''}{2}\right) \sin t \frac{x''-x'}{2}}{t} dt + \int_{-c}^{c} \frac{\sin t(x-x') - \sin t(x-x'')}{t} dt$$

$$= \int_{-c}^{c} \frac{\sin t(x-x') - \sin t(x-x'')}{t} dt = 2 \int_{0}^{c} \left(\frac{\sin t(x-x')}{t} - \frac{\sin t(x-x'')}{t}\right) dt$$

since the first of these integrals vanishes as an integral of an odd function extended over an interval symmetric about the origin. Thus,

$$I_c = \frac{1}{\pi} \int_{-\infty}^{\infty} \int_{0}^{c} \left(\frac{\sin t(x-x')}{t} - \frac{\sin t(x-x'')}{t}\right) dt \, dF(x).$$

From analysis we know that for arbitrary a and c

$$\left| \frac{1}{\pi} \int_{0}^{c} \frac{\sin at}{t} dt \right| = \left| \frac{1}{\pi} \int_{0}^{ac} \frac{\sin u}{u} du \right| < 1$$

and

(36.7) $$\lim_{c \to \infty} \frac{1}{\pi} \int_{0}^{c} \frac{\sin at}{t} dt = \begin{cases} \frac{1}{2} & \text{if } a > 0, \\ -\frac{1}{2} & \text{if } a < 0, \end{cases}$$

where the convergence is uniform in every interval of the form $a > \delta > 0$ and $a < -\delta < 0$.

Let us choose δ so small that $x'+\delta < x''-\delta$ and write I_c as a sum of five integrals

$$I_c = \int_{-\infty}^{x'-\delta} + \int_{x'-\delta}^{x'+\delta} + \int_{x'+\delta}^{x''-\delta} + \int_{x''-\delta}^{x''+\delta} + \int_{x''+\delta}^{\infty} \psi(c, x; x', x'') dF(x),$$

where

$$\psi(c, x; x', x'') = \int_0^c \left(\frac{\sin t(x-x')}{t} - \frac{\sin t(x-x'')}{t} \right) dt.$$

It follows from (36.7) that

$$\lim_{c \to \infty} \psi(c, x; x', x'') = 0 \quad \text{for} \quad x < x'-\delta \text{ and } x''+\delta < x$$

and

$$\lim_{c \to \infty} \psi(c, x; x', x'') = 1 \quad \text{for} \quad x'+\delta < x < x''-\delta,$$

where the convergence is uniform with respect to x. In the intervals $x'-\delta < x < x'+\delta$ and $x''-\delta < x < x''+\delta$ we shall have at any rate

$$|\psi(x, c; x', x'')| \leqslant 2.$$

Thus, for all (not too large) $\delta > 0$ we have

$$\lim_{c \to \infty} I_c = F(x''-\delta) - F(x'+\delta) + R(\delta, x', x'')$$

where

$$|R(\delta, x', x'')| \leqslant 2(F(x'+\delta) - F(x'-\delta) + F(x''+\delta) - F(x''-\delta)).$$

Using the fact that x' and x'' are points of continuity of the distribution function $F(x)$, we obtain, by passing to the limit with $\delta \to 0$, that $\lim_{c \to \infty} I_c = F(x'') - F(x')$, which proves the first part of the theorem.

Now, if in (36.5) we pass to the limit with x' tending to $-\infty$ over the set of continuity points of the distribution function $F(x)$ (which is permissible), we obtain the relation

$$F(x) = \frac{1}{2\pi} \lim_{y \to -\infty} \int_{-\infty}^{+\infty} \frac{\exp(-ity) - \exp(-itx)}{it} \varphi(t) dt,$$

which, in view of the continuity on the left of the distribution function, shows that the characteristic function determines uniquely the distribution function.

If the characteristic function is absolutely integrable over the real line, then the derivative of the integrand in (36.5) with respect to x'' satisfies the inequality

$$|e^{-itx} \varphi(t)| \leqslant |\varphi(t)|$$

VII. CENTRAL LIMIT THEOREMS

(where we write x instead of x''). Thus, differentiation under the integral sign is permissible, and relation (36.6) holds. The uniform continuity of $f(x)$ can be proved in the same way as the uniform continuity of the characteristic function in Theorem 1.

We have already mentioned that characteristic functions were introduced mainly for studying limit distributions. We have also formulated propositions like "distributions tend to the normal distribution". Before we state the theorem on the relation between convergence of distributions and convergence of characteristic functions, we must specify the type of convergence which we shall be investigating. We shall speak of convergence of probability distributions in the sense of the so-called fundamental convergence of probability distribution functions to the probability distribution function of the limit distribution.

DEFINITION 2. We say that probability distribution functions $F_1(x), F_2(x), \ldots$ *converge fundamentally* to a non-decreasing function $G(x)$, if for every point x which is a point of continuity of the limit function $G(x)$ we have

$$\lim_{n \to \infty} F_n(x) = G(x).$$

If the probability distribution functions $F_1(x), F_2(x), \ldots$ converge fundamentally to a distribution function $F(x)$ which determines the probability distribution $P(A)$, then we say that the sequence of probability distributions $P_n(A)$, $n = 1, 2, \ldots$ corresponding to distribution functions $F_n(x)$ are *weakly convergent* to the distribution $P(A)$.

Some remarks are here in order.

The limit function $G(x)$ does not have to be a probability distribution function: while for every x we must have $0 \leqslant G(x) \leqslant 1$ none of the relations $G(-\infty) = 0$ and $G(+\infty) = 1$ is obligatory. The reader will easily give examples.

From the weak convergence of distributions $P_n(A)$ to the distribution $P(A)$ it follows that if $u < v$ are points of continuity of distribution function of the limit distribution, and $A = \{x: u \leqslant x < v\}$, then $\lim_{n \to \infty} P_n(A) = P(A)$. In general, however, such convergence does not hold. Thus, for the distribution functions

$$F_n(x) = \begin{cases} 0 & \text{for } x \leqslant 1 - 1/n, \\ 1 & \text{otherwise} \end{cases}$$

we have fundamental convergence to the distribution function

$$F(x) = \begin{cases} 0 & \text{for } x \leqslant 1, \\ 1 & \text{otherwise}, \end{cases}$$

but for $A = \{x: -1 \leqslant x < 1\}$ we have $\lim_{n \to \infty} P_n(A) = 1$ and $P(A) = 0$.

Weak convergence has nothing to do with convergence of moments. A sequence of distributions without expectation may converge weakly to a distribution which has all the moments, and vice versa, a sequence of distributions which have all the moments may converge weakly to a distribution without expectation. The reader will easily construct suitable examples.

Let us note that we have the following lemma:

LEMMA 1. *For the fundamental convergence of distribution functions $F_1(x)$, $F_2(x), \ldots$ to a non-decreasing function $G(x)$ it is sufficient that the convergence*
$$\lim_{n \to \infty} F_n(x) = G(x)$$
hold for all x from a certain everywhere dense set D (say, for rational points x).

In fact, let x be an arbitrary point and let x' and x'' be two points of D such that $x' < x < x''$. Then we have
$$F_n(x') \leqslant F_n(x) \leqslant F_n(x'').$$
Therefore
$$\underline{\lim_{n \to \infty}} F_n(x') \leqslant \underline{\lim_{n \to \infty}} F_n(x) \leqslant \overline{\lim_{n \to \infty}} F_n(x) \leqslant \overline{\lim_{n \to \infty}} F_n(x'').$$
Since by assumption $\lim_{n \to \infty} F_n(x') = G(x')$ and $\lim_{n \to \infty} F_n(x'') = G(x'')$, we may write
$$G(x') \leqslant \underline{\lim_{n \to \infty}} F_n(x) \leqslant \overline{\lim_{n \to \infty}} F_n(x) \leqslant G(x'').$$
The middle terms of these inequalities do not depend on x' and x'', and we have therefore
$$G(x-0) \leqslant \underline{\lim_{n \to \infty}} F_n(x) \leqslant \overline{\lim_{n \to \infty}} F_n(x) \leqslant G(x+0).$$
If the function G is continuous at the point x, we have
$$G(x-0) = G(x) = G(x+0),$$
and from the last inequalities it follows that $\lim_{n \to \infty} F_n(x) = G(x)$.

Now we present the theorem which gives the connection between the convergence of probability distributions and the convergence of corresponding characteristic functions.

THEOREM 5. *Let $F_1(x), F_2(x), \ldots$ be a sequence of probability distribution functions, and let $\varphi_1(t), \varphi_2(t), \ldots$ be the sequence of the corresponding characteristic functions. Then, the sequence $F_1(x), F_2(x), \ldots$ is fundamentally convergent to a distribution function $F(x)$ if and only if for every t we have $\lim_{n \to \infty} \varphi_n(t) = \varphi(t)$, where $\varphi(t)$ is a continuous function; in this case $\varphi(t)$ is the characteristic function of the limit distribution with the distribution function $F(x)$.*

VII. CENTRAL LIMIT THEOREMS

We shall base the proof on two lemmas; the second of them concerns Stieltjes integrals.

LEMMA 2 (The first theorem of Helly). *Every sequence*

(36.8) $$F_1(x), F_2(x), \ldots$$

of distribution functions contains at least one subsequence

$$F_{n_1}(x), F_{n_2}(x), \ldots$$

convergent fundamentally to a certain non-decreasing function $G(x)$.

LEMMA 3 (The generalized second theorem of Helly). *Let $f(x)$ be a bounded and continuous function defined on the whole real line $-\infty < x < +\infty$ and let the sequence of distribution functions*

$$F_1(x), F_2(x), \ldots$$

converge fundamentally to a distribution function $G(x)$. Then

(36.9) $$\lim_{n \to \infty} \int_{-\infty}^{+\infty} f(x) dF_n(x) = \int_{-\infty}^{+\infty} f(x) dG(x).$$

PROOF OF LEMMA 2. Let D be a countable and everywhere dense set of points $x_1', x_2', \ldots$ Let us consider the values of sequence (36.8) at the point x_1':

$$F_1(x_1'), F_2(x_1'), \ldots$$

Since this set is bounded (as all these numbers are between zero and one), it must contain at least one subsequence convergent, say, to $G(x_1')$, also contained between zero and one. In other words, sequence (36.8) contains a subsequence

(S_1) $F_{11}(x), F_{12}(x), \ldots, F_{1k}(x), \ldots$

for which $\lim_{k \to \infty} F_{1k}(x_1') = G(x_1')$. Now, let us take the set of all values of sequence (S_1) at the point x_2'. Again, this set is bounded, hence it contains at least one subsequence convergent to a limit, say $G(x_2')$. Thus, we have

(S_2) $F_{21}(x), F_{22}(x), \ldots, F_{2k}(x), \ldots$

for which $\lim_{k \to \infty} F_{2k}(x_2') = G(x_2')$.

Continuing in this manner we may select sequences out of formerly selected sequences for every point x_n': we select for the point x_n' the sequence

(S_n) $F_{n1}(x), F_{n2}(x), \ldots, F_{nk}(x), \ldots$

such that $\lim_{k \to \infty} F_{nk}(x_n') = G(x_n')$.

The diagonal sequence

(S) $$F_{11}(x), F_{22}(x), \ldots, F_{kk}(x), \ldots$$

is convergent at every point of the set D. Indeed, this sequence is selected out of sequence (S_1), whence $\lim_{k \to \infty} F_{kk}(x_1') = G(x_1')$. Starting from the second term, this sequence is selected from the second subsequence (S_2), hence $\lim_{k \to \infty} F_{kk}(x_2') = G(x_2')$. Starting from the nth term, this sequence is selected from the sequence (S_n), whence $\lim_{k \to \infty} F_{kk}(x_n') = G(x_n')$.

Thus, sequence (S) is a subsequence of (36.8) convergent at every point of D to a certain function $G(x)$, at present defined only on D. This function is non-decreasing and bounded, since the functions $F_n(x)$ are non-decreasing and bounded. Thus, its definition can be extended to the whole real line in such a way that the above properties are preserved. It suffices to define $G(x)$ at every point x from outside D by the relation

$$G(x) = \lim_{x' \to x} G(x')$$

where x' is in D and converges on the left to x. The sequence (S) converges to the function $G(x)$ defined above on an everywhere dense set, hence by Lemma 1, it converges fundamentally. Thus, Lemma 2 is proved.

PROOF OF LEMMA 3. We prove first that for arbitrary a and b which are continuity points of the distribution $G(x)$ we have

(36.10) $$\lim_{n \to \infty} \int_a^b f(x) \, dF_n(x) = \int_a^b f(x) \, dG(x).$$

From the continuity of the function $f(x)$ in the interval $a \leqslant x \leqslant b$ it follows that for every $\varepsilon > 0$ there exists a partition of this interval, say $a = x_0 < x_1 < \ldots < x_N = b$, such that in every interval $x_{i-1} \leqslant x \leqslant x_i$ we have the inequality $|f(x) - f(x_i)| < \varepsilon$. Using this we may introduce an auxiliary function $f_\varepsilon(x)$ which assumes only a finite number of values, defined by the equalities $f_\varepsilon(x_0) = f(x_1)$, $f_\varepsilon(x) = f(x_i)$ for $x_{i-1} < x \leqslant x_i$. For this function, we have in the whole interval $a \leqslant x \leqslant b$ the inequality

(36.11) $$|f(x) - f_\varepsilon(x)| < \varepsilon.$$

We can choose points $x_0, x_1, \ldots, x_N$ of our partition so that they all are continuity points of the function $G(x)$. Then, by the fundamental convergence of the distributions $F_1(x), F_2(x), \ldots$ to the distribution function $G(x)$, for all sufficiently large n we shall have at all points of the partition

(36.12) $$|G(x_i) - F_n(x_i)| < \varepsilon/MN,$$

where M denotes the maximum of the absolute value of $f(x)$ in the interval $a \leqslant x \leqslant b$.

We have the following estimates

$$\left| \int_a^b f(x)\,dG(x) - \int_a^b f(x)\,dF_n(x) \right|$$

$$\leqslant \left| \int_a^b f(x)\,dG(x) - \int_a^b f_\varepsilon(x)\,dG(x) \right| + \left| \int_a^b f_\varepsilon(x)\,dG(x) - \int_a^b f_\varepsilon(x)\,dF_n(x) \right|$$

$$+ \left| \int_a^b f_\varepsilon(x)\,dF_n(x) - \int_a^b f(x)\,dF_n(x) \right|.$$

We easily note that the first term of the right-hand side of this inequality does not exceed $\varepsilon|G(b)-G(a)|$, while the third does not exceed $\varepsilon|F_n(b)-F_n(a)|$. This follows from (36.11). The second term of this inequality equals

$$\left| \sum_{i=1}^{N} f(x_i)\,[G(x_i)-G(x_{i-1})] - \sum_{i=1}^{N} f(x_i)\,[F_n(x_i)-F_n(x_{i-1})] \right|$$

$$= \left| \sum_{i=1}^{N} f(x_i)\,[G(x_i)-F_n(x_i)] - \sum_{i=1}^{N} f(x_i)\,[G(x_{i-1})-F_n(x_{i-1})] \right|,$$

hence, by (36.12), it does not exceed 2ε for sufficiently large n. Since the distributions $F_n(x)$ and $G(x)$ are uniformly bounded, the sum

$$\varepsilon[G(b)-G(a)] + \varepsilon[F_n(b)-F_n(a)] + 2\varepsilon$$

will be arbitrarily small for a sufficiently small ε. Thus, for a sufficiently large n the integrals $\int_a^b f(x)\,dG(x)$ and $\int_a^b f(x)\,dF_n(x)$ differ arbitrarily little, which implies relation (36.10).

Suppose now that $a < 0$ and $b > 0$ are continuity points of the distribution function $G(x)$ and let

$$I_1 = \left| \int_{-\infty}^{a} f(x)\,dG(x) - \int_{-\infty}^{a} f(x)\,dF_n(x) \right|,$$

$$I_2 = \left| \int_a^b f(x)\,dG(x) - \int_a^b f(x)\,dF_n(x) \right|,$$

$$I_3 = \left| \int_b^{\infty} f(x)\,dG(x) - \int_b^{\infty} f(x)\,dF_n(x) \right|.$$

Then of course we have

$$\left| \int_{-\infty}^{+\infty} f(x)\,dG(x) - \int_{-\infty}^{+\infty} f(x)\,dF_n(x) \right| \leqslant I_1 + I_2 + I_3.$$

To complete the proof of the lemma it suffices to show that the sum on the right-hand side is arbitrarily small for sufficiently large n.

Denote by M the upper bound of the absolute value of the function $f(x)$ over the whole real line. We then have

$$I_1 \leqslant M[G(a) + F_n(a)], \quad I_3 \leqslant M[1 - G(b) + 1 - F_n(b)].$$

We see that by choosing a sufficiently small a and a sufficiently large b we may make $G(a)$ and $1 - G(b)$ arbitrarily small. If, in addition, the points a and b are continuity points of $G(x)$, then by the fundamental convergence of $F_n(x)$ to $G(x)$, also $F_n(a)$ and $1 - F_n(b)$ will be sufficiently small for sufficiently large n. Thus, by choosing appropriate a and b we may make the sum $I_1 + I_3$ smaller than an arbitrarily chosen number $\varepsilon > 0$ for all sufficiently large n.

For fixed a and b, however, we have relation (36.10), hence for sufficiently large n, integral I_2 will also be smaller than ε. Thus, for an arbitrary $\varepsilon > 0$ one can find an n_0 such that for all $n \geqslant n_0$ the sum $I_1 + I_2 + I_3$ will be smaller than 2ε, which proves Lemma 3.

PROOF OF THEOREM 5. N e c e s s i t y. Suppose that the distribution functions $F_1(x), F_2(x), \ldots$ converge fundamentally to the distribution function $F(x)$, and let $\varphi_1(t), \varphi_2(t), \ldots$ and $\varphi(t)$ be the corresponding characteristic functions. Thus, we have

$$\varphi_n(t) = \int_{-\infty}^{+\infty} e^{itx}\,dF_n(x), \quad \varphi(t) = \int_{-\infty}^{+\infty} e^{itx}\,dF(x).$$

Since $e^{itx} = \cos tx + i\sin tx$, and the functions $\cos tx$ and $\sin tx$ are continuous and bounded over the whole real line for every t, we may apply Lemma 3, obtaining for every t

$$\lim_{n \to \infty} \varphi_n(t) = \varphi(t).$$

The function $\varphi(t)$, as a characteristic function, is continuous for $t = 0$, which proves the necessity.

S u f f i c i e n c y. We have to prove that if the characteristic functions $\varphi_1(t)$, $\varphi_2(t), \ldots$ converge for every t to a function $\varphi(t)$ which is continuous at $t = 0$, then $\varphi(t)$ is a characteristic function and the corresponding distribution function $F(x)$ equal to the limit to which the distributions $F_1(x), F_2(x), \ldots$ converge fundamentally.

VII. CENTRAL LIMIT THEOREMS

Our reasoning will be as follows: by Lemma 2 we may select a subsequence

$$F_{n_1}(x), F_{n_2}(x), \ldots$$

out of the sequence $F_1(x), F_2(x), \ldots$, this subsequence converging fundamentally to a certain non-decreasing function $F(x)$. The values of the function $F(x)$ are obviously non-negative and bounded by 1, and at points of discontinuity one can modify the function $F(x)$ without disturbing the fundamental convergence, so that $F(x)$ will be continuous on the left. We shall prove that $F(x)$ satisfies also the condition characterizing the distribution functions, namely $\lim_{x \to -\infty} F(x) = 0$ and $\lim_{x \to +\infty} F(x) = 1$. We shall show that the assumption

(36.13) $$\lim_{x \to +\infty} F(x) - \lim_{x \to -\infty} F(x) = \delta < 1$$

is inconsistent with the assumption that $\varphi(t)$ is continuous at $t = 0$. Indeed, for $t = 0$ all characteristic functions $\varphi_1(t), \varphi_2(t), \ldots$ have value 1, whence also $\varphi(0) = 1$. Since the function $\varphi(t)$ is continuous by assumption, we can find positive numbers τ and ε such that

(36.14) $$\frac{1}{2\pi} \left| \int_{-\tau}^{\tau} \varphi(t) dt \right| > 1 - \frac{\varepsilon}{2} > \delta + \frac{\varepsilon}{2}.$$

If relation (36.13) were true, we could find $X > 4/\tau\varepsilon$ and K such that for $k > K$

$$\delta_k = F_{n_k}(X) - F_{n_k}(-X) < \delta + \tfrac{1}{4}\varepsilon.$$

Since $\varphi_{n_k}(t)$ is the characteristic function corresponding to the distribution function $F_{n_k}(x)$, we have

$$\int_{-\tau}^{\tau} \varphi_{n_k}(t) dt = \int_{-\tau}^{\tau} \left\{ \int_{-\infty}^{+\infty} e^{itx} dF_{n_k}(x) \right\} dt = \int_{-\infty}^{+\infty} \left\{ \int_{-\tau}^{\tau} e^{itx} dt \right\} dF_{n_k}(x),$$

and since $|e^{itx}| = 1$, we may write

$$\left| \int_{-\tau}^{\tau} e^{itx} dt \right| \leqslant 2\tau.$$

From the equality

$$\int_{-\tau}^{\tau} e^{itx} dt = \frac{e^{i\tau x} - e^{-i\tau x}}{ix} = \frac{2}{x} \sin \tau x$$

and from the relation $|\sin \tau x| \leqslant 1$ we obtain for $|x| > X$ the estimate

$$\left| \int_{-\tau}^{\tau} e^{itx} dt \right| < \frac{2}{X}.$$

Using the first estimate for $|x| \leqslant X$ and the second for $|x| > X$ we have

$$\left| \int_{-\tau}^{\tau} \varphi_{n_k}(t)dt \right| \leqslant \left| \int_{|x| \leqslant X} \left(\int_{-\tau}^{\tau} e^{itx} dt \right) dF_{n_k}(x) \right| + \left| \int_{|x| > X} \left(\int_{-\tau}^{\tau} e^{itx} dt \right) dF_{n_k}(x) \right|$$

$$< \left| \int_{|x| \leqslant X} 2\tau \, dF_{n_k}(x) \right| + \left| \int_{|x| > X} \frac{2}{x} dF_{n_k}(x) \right| < 2\tau \delta_k + \frac{2}{X},$$

hence

$$\frac{1}{2\tau} \left| \int_{-\tau}^{\tau} \varphi_{n_k}(t) dt \right| < \delta_k + \frac{1}{X\tau} < \delta + \frac{\varepsilon}{2}.$$

The functions $\varphi_{n_k}(t)$ are uniformly bounded and converge to $\varphi(t)$, hence the same inequality would be preserved in the limit:

$$\frac{1}{2\tau} \left| \int_{-\tau}^{\tau} \varphi(t) dt \right| \leqslant \delta + \frac{\varepsilon}{2},$$

which contradicts (36.14). Thus, relation (36.13) is impossible. We have thus shown that there exists a subsequence $F_{n_1}(x)$, $F_{n_2}(x)$, ... convergent fundamentally to a certain distribution function $F(x)$.

By the first part of our theorem we may claim that the characteristic functions $\varphi_{n_1}(t)$, $\varphi_{n_2}(t)$, ... converge to a characteristic function which corresponds to the limit distribution function $F(x)$, and since by assumption they are convergent to $\varphi(t)$, we see that $\varphi(t)$ is the characteristic function of the distribution $F(x)$.

To complete the proof of Theorem 5 it suffices to show that the whole sequence $F_1(x), F_2(x), \ldots$ converges fundamentally to the distribution function $F(x)$. Suppose that this property is not satisfied, and that the sequence $F_1(x), F_2(x), \ldots$ does not converge fundamentally to $F(x)$. Then there exist a point c at which $F(x)$ is continuous and a subsequence $F_{n_1'}(x), F_{n_2'}(x), \ldots$ for which $\lim_{k \to \infty} F_{n_k'}(c)$ $= F^*(c) \neq F(c)$. Reasoning as before, we could select from this subsequence another subsequence, converging fundamentally to a distribution function $F^*(x)$ different from $F(x)$ at least at the point c. As before, the function $\varphi(t)$ would prove to be the characteristic function corresponding to the distribution function $F^*(x)$. In this way we would arrive at a contradiction; thus, Theorem 5 is proved.

§ 37. CHARACTERISTIC FUNCTION OF SOME IMPORTANT DISTRIBUTIONS. ADDITIVITY OF THESE DISTRIBUTIONS

In the preceding section we defined characteristic functions and we proved some basic theorems, which make characteristic function a convenient and universal tool for studying limit distributions. In this section we shall derive

VII. CENTRAL LIMIT THEOREMS

a few formulas useful in computing characteristic functions, find the characteristic functions of a few distributions and prove a certain conservative property of those distributions.

THEOREM 1. *The characteristic functions of random variables ξ and $a\xi+b$ are related by the formula*

(37.1) $$\varphi_{a\xi+b}(t) = e^{itb}\varphi_\xi(at).$$

PROOF. Relation (37.1) follows immediately from the equalities

$$\varphi_{a\xi+b}(t) \stackrel{df}{=} Ee^{it(a\xi+b)} = Ee^{iat\xi+itb} = e^{itb}Ee^{iat\xi} = e^{itb}\varphi_\xi(at).$$

THEOREM 2. *If a random variable ξ has a Bernoulli distribution, i.e., if*

$$P(\{e: \xi(e) = k\}) = b(k; n, p) = \binom{n}{k} p^k (1-p)^{n-k}$$

for $k = 0, 1, \ldots, n$, then its characteristic function is given by

(37.2) $$\varphi_\xi(t) = (q+pe^{it})^n,$$

where $q = 1-p$.

PROOF. Relation (37.2) follows immediately from the Newton binomial formula. Indeed, we have

$$\varphi_\xi(t) = Ee^{it\xi} = \sum_{k=0}^{n} e^{itk} \binom{n}{k} p^k(1-p)^{n-k} = \sum_{k=0}^{n} \binom{n}{k} (pe^{it})^k q^{n-k} = (pe^{it}+q)^n.$$

Since $q+pe^{it}$ is the characteristic function of the random variable which assumes value 1 with probability p and value 0 with probability $q = 1-p$, it follows from (37.2) (independently of the considerations of §§ 19–21), that the binomial distribution can be treated as the distribution of the sum of n independent random variables with identical distributions.

COROLLARY 1. *If the random variable ξ and η are independent, ξ has the binomial distribution $b(k; n_1, p)$ and η has the binomial distribution $b(k; n_2, p)$, then their sum $\xi+\eta$ has the binomial distribution $b(k; n_1+n_2, p)$.*

PROOF. Indeed, we have for the characteristic functions

$$\varphi_{\xi+\eta}(t) = \varphi_\xi(t)\varphi_\eta(t) = (q+pe^{it})^{n_1}(q+pe^{it})^{n_2} = (q+pe^{it})^{n_1+n_2}.$$

THEOREM 3. *If the random variable ξ has a Poisson distribution, i.e., if*

$$P(\{e: \xi(e) = k\}) = p(k; \lambda) = \frac{\lambda^k}{k!} e^{-\lambda},$$

then its characteristic function is

(37.3) $$\varphi_\xi(t) = e^{\lambda(e^{it}-1)}.$$

PROOF. We have

$$\varphi_\xi(t) = Ee^{it\xi} = \sum_{k=0}^{\infty} e^{itk} \frac{\lambda^k}{k!} e^{-\lambda}$$

$$= e^{-\lambda} \sum_{k=0}^{\infty} \frac{1}{k!} (\lambda e^{it})^k = e^{-\lambda} e^{\lambda e^{it}} = e^{\lambda(e^{it}-1)}.$$

COROLLARY 2. *If the random variables ξ and η are independent, the first has the Poisson distribution $p(k; \lambda_1)$ and the second has the Poisson distribution $p(k; \lambda_2)$, then their sum has the Poisson distribution $p(k; \lambda_1+\lambda_2)$.*

This corollary states that summation of independent random variables with Poisson distributions always leads to random variables with Poisson distributions.

PROOF. For the characteristic function we have

$$\varphi_{\xi+\eta}(t) = \varphi_\xi(t)\varphi_\eta(t) = \exp\{\lambda_1(e^{it}-1)\}\exp\{\lambda_2(e^{it}-1)\}$$

$$= \exp\{(\lambda_1+\lambda_2)(e^{it}-1)\}.$$

THEOREM 4. *If a random variable ξ has a normal distribution with the probability density*

$$f_\xi(x) = \frac{1}{\sigma\sqrt{2\pi}} \exp\left\{-\frac{(x-m)^2}{2\sigma^2}\right\}$$

(i.e., with expectation m and variance σ^2), then its characteristic function is given by the formula

(37.4) $$\varphi_\xi(t) = \exp(imt - \tfrac{1}{2}\sigma^2 t^2).$$

PROOF. We have

$$\varphi_\xi(t) = Ee^{it\xi} = \int_{-\infty}^{+\infty} e^{itx} f_\xi(x)\, dx = \frac{1}{\sigma\sqrt{2\pi}} \int_{-\infty}^{+\infty} e^{itx} \exp\left\{-\frac{(x-m)^2}{2\sigma^2}\right\} dx.$$

Substituting $u = (x-m)/\sigma$, we obtain

$$\varphi_\xi(t) = \frac{1}{\sqrt{2\pi}} \int_{-\infty}^{+\infty} \exp[it(\sigma u + m)] e^{-u^2/2}\, du = \frac{1}{\sqrt{2\pi}} e^{itm} \int_{-\infty}^{+\infty} \exp(it\sigma u - \tfrac{1}{2}u^2)\, du.$$

Transforming the exponent of the integrand according to the formula

$$it\sigma u - \tfrac{1}{2}u^2 = -\tfrac{1}{2}(u^2 - 2it\sigma u) = -\tfrac{1}{2}((u-it\sigma)^2 + t^2\sigma^2),$$

VII. CENTRAL LIMIT THEOREMS

we get
$$\varphi_\xi(t) = \exp(itm - \tfrac{1}{2}\sigma^2 t^2) \frac{1}{\sqrt{2\pi}} \int_{-\infty}^{+\infty} \exp\{-\tfrac{1}{2}(u - it\sigma)^2\} du.$$

In § 22 we proved that
$$\int_{-\infty}^{+\infty} \frac{1}{\sqrt{2\pi}} e^{-u^2/2} du = 1$$

(see formula 22.12). For the function $\dfrac{1}{\sqrt{2\pi}} e^{-z^2/2}$ of the complex variable z one can show that the integral of this function extended over any line parallel to the real axis has the same value as the integral over the real axis given by the last formula. Thus
$$\frac{1}{\sqrt{2\pi}} \int_{-\infty}^{+\infty} \exp\{-\tfrac{1}{2}(u - it\sigma)^2\} du = 1,$$
which completes the proof of (37.4).

COROLLARY 3. *If the random variables ξ and η are independent, ξ has normal distribution with expectation m_1 and variance σ_1^2, and η has normal distribution with expectation m_2 and variance σ_2^2, then their sum $\xi + \eta$ also has normal distribution with expectation $m_1 + m_2$ and variance $\sigma_1^2 + \sigma_2^2$.*

PROOF. For the characteristic function we have
$$\varphi_{\xi+\eta}(t) = \varphi_\xi(t) \varphi_\eta(t) = \exp(itm_1 - \tfrac{1}{2}\sigma_1^2 t^2) \exp(itm_2 - \tfrac{1}{2}\sigma_2^2 t^2)$$
$$= \exp\{it(m_1 + m_2) - \tfrac{1}{2}(\sigma_1^2 + \sigma_2^2) t^2\},$$
which proves Corollary 3.

In analysis we are familiar with the function $\Gamma(p)$ defined by the equality

(37.5) $$\Gamma(p) = \int_0^\infty x^{p-1} e^{-x} dx.$$

The integrand is non-negative, and if we define a new function by the equation

(37.6) $$f(x) = \begin{cases} 0 & \text{for } x < 0, \\ \dfrac{1}{\Gamma(p)} x^{p-1} e^{-x} & \text{for } x > 0, \end{cases}$$

this function will have all the properties of probability density. In general, substituting $x = au$ in (37.5), we obtain

(37.7) $$\Gamma(p) = \int_0^\infty a^p x^{p-1} e^{-ax} dx,$$

which leads to the following definition of probability density, depending on two parameters $a > 0$ and $p > 0$:

(37.8) $$\gamma(x; a, p) = \begin{cases} 0 & \text{for } x < 0. \\ \dfrac{a^p}{\Gamma(p)} x^{p-1} e^{-ax} & \text{for } x > 0, \end{cases}$$

DEFINITION 1. *The probability distribution defined by formula* (37.8) *is called the* gamma distribution.

We shall find its characteristic function.

THEOREM 5. *If a random variable ξ has a gamma distribution with density $\gamma(x; a, p)$ defined by* (37.8), *then its characteristic function is given by*

(37.9) $$\varphi_\xi(t) = \frac{1}{\left(1 - \dfrac{it}{a}\right)^p}.$$

PROOF. By definition, we have

$$\varphi_\xi(t) = \int_{-\infty}^{+\infty} e^{itx} \frac{a^p}{\Gamma(p)} x^{p-1} e^{-ax} dx = \frac{a^p}{\Gamma(p)} \int_0^{+\infty} x^{p-1} \exp[(it-a)x] dx.$$

The above integral can be treated as the integral of the function $z^{p-1} e^{(it-a)z}$ of a complex variable extended over the positive half of the real axis. It can be proved that the integral of this function has the same value for all rays leading from the origin and contained in the half-plane $\operatorname{Re} z > 0$. For us it will be convenient to replace it by extended along the ray $z = \dfrac{1}{a-it} u$, where $u > 0$ is the real parameter. We then have

$$\int_0^\infty x^{p-1} e^{(it-a)x} dx = \int_0^\infty \frac{1}{(a-it)^{p-1}} u^{p-1} e^{-u} \frac{1}{a-it} du$$

$$= \frac{1}{(a-it)^p} \int_0^\infty u^{p-1} e^{-u} du = \frac{\Gamma(p)}{(a-it)^p}.$$

Finally,

$$\varphi_\xi(t) = \frac{a^p}{\Gamma(p)} \cdot \frac{\Gamma(p)}{(a-it)^p} = \frac{1}{\left(1 - \dfrac{it}{a}\right)^p}.$$

COROLLARY 4. *If the random variables ξ and η are independent, ξ has a gamma distribution with the density $\gamma(x; a, p_1)$ and η has a gamma distribution with the*

VII. CENTRAL LIMIT THEOREMS

density $\gamma(x; a, p_2)$, then their sum $\xi+\eta$ has a gamma distribution with the density $\gamma(x; a, p_1+p_2)$.

PROOF. We have the following relations for the characteristic function:

$$\varphi_{\xi+\eta}(t) = \varphi_\xi(t)\varphi_\eta(t) = \frac{1}{\left(1-\dfrac{it}{a}\right)^{p_1}} \cdot \frac{1}{\left(1-\dfrac{it}{a}\right)^{p_2}} = \frac{1}{\left(1-\dfrac{it}{a}\right)^{p_1+p_2}}.$$

Since

(37.10) $$\int_{-\infty}^{+\infty} \tfrac{1}{2} e^{-|x|} dx = 1,$$

the function

(37.11) $$f(x) = \tfrac{1}{2} e^{-|x|}, \quad -\infty < x < \infty$$

is a probability density. The probability distribution with this density is called the *Laplace distribution*. We can easily see that if the random variable ξ has a Laplace distribution with density (37.11), then the random variable $\eta = \lambda\xi+\mu$ has a probability distribution with the density

(37.12) $$l(x; \mu, \lambda) = \frac{1}{2\lambda} \exp\left(-\frac{|x-\mu|}{2\lambda}\right),$$

where $\lambda > 0$ and μ are real parameters. The distribution with density (37.12) will also be called the *Laplace distribution*.

From analysis we know that

(37.13) $$\int_{-\infty}^{+\infty} \frac{1}{\pi} \cdot \frac{1}{1+x^2} dx = \left[\frac{1}{\pi} \arctan x\right]_{-\infty}^{\infty} = 1,$$

hence the function

(37.14) $$f(x) = \frac{1}{\pi} \cdot \frac{1}{1+x^2}, \quad -\infty < x < \infty$$

is a probability density. If the random variable ξ has probability density (37.14), then the random variable $\eta = \lambda\xi+\mu$ has the density

(37.15) $$c(x; \mu, \lambda) = \frac{1}{\pi} \cdot \frac{\lambda}{\lambda^2+(x-\mu)^2}, \quad -\infty < x < \infty$$

where $\lambda > 0$ and μ are real parameters. The probability distribution with density (37.15) is called the *Cauchy distribution*.

THEOREM 6. *The characteristic function of the Laplace distribution with density* (37.12) *is given by the formula*

(37.16) $$\varphi(t) = e^{it\mu}\frac{1}{1+(\lambda t)^2},$$

and the characteristic function of the Cauchy distribution with density (37.15) *is given by the formula*

(37.17) $$\varphi(t) = e^{it\mu-\lambda|t|}.$$

PROOF. Suppose that the random variable ξ has a Laplace distribution with density (37.11). Then its characteristic function is

$$\varphi_\xi(t) = \int_{-\infty}^{+\infty} e^{itx}\tfrac{1}{2}e^{-|x|}dx = \int_{-\infty}^{+\infty}(\cos tx + i\sin tx)\tfrac{1}{2}e^{-|x|}dx.$$

Using the fact that the function $\sin tx$ is odd, and the function $\cos tx$ is even, we obtain

$$\varphi_\xi(t) = \int_{-\infty}^{+\infty}\tfrac{1}{2}(\cos tx)e^{-|x|}dx = \int_0^{+\infty}(\cos tx)e^{-x}dx.$$

Integrating by parts we finally get

(37.18) $$\varphi_\xi(t) = \left[\frac{(t\sin tx - \cos tx)e^{-x}}{1+t^2}\right]_0^\infty = \frac{1}{1+t^2}.$$

Applying formula (37.1), we obtain (37.16), which completes the proof of the first part of Theorem 6.

Since the characteristic function given by (37.18) is integrable over the whole real axis, by Theorem 4 of § 36 we have the conversion formula

(37.19) $$\frac{1}{2}e^{-|x|} = \frac{1}{2\pi}\int_{-\infty}^{+\infty}e^{-itx}\frac{1}{1+t^2}dt.$$

In view of the fact that the function $\sin tx$ is odd, we have

$$\int_{-\infty}^{+\infty}e^{-itx}\frac{1}{1+t^2}dt = \int_{-\infty}^{+\infty}e^{itx}\frac{1}{1+t^2}dt.$$

Using this equality and (37.19) and interchanging the role of the variables t and x, we can write

(37.20) $$e^{-|t|} = \int_{-\infty}^{+\infty}e^{itx}\cdot\frac{1}{\pi}\cdot\frac{1}{1+x^2}dx.$$

This formula shows that $e^{-|t|}$ is a characteristic function corresponding to the Cauchy distribution given by (37.14). Formula (37.17) follows from (37.20) by applying (37.1), which completes the proof of Theorem 6.

Exercises

1. Prove that if a random variable ξ has a distribution function $F(x)$ and a density $f(x)$, then a random variable $\eta = a\xi+b$ has the distribution function $F\left(\dfrac{x-b}{a}\right)$ and density $\dfrac{1}{a}f\left(\dfrac{x-b}{a}\right)$ if $a>0$, and it has the distribution function $1-F\left(\dfrac{x-b}{a}\right)$ and density $-\dfrac{1}{a}f\left(\dfrac{x-b}{a}\right)$ if $a<0$.

2. Prove that if a random variable ξ has a normal distribution, then a random variable $a\xi+b$ ($a\neq 0$) also has a normal distribution.

3. Prove that if a random variable ξ has a gamma distribution, then a random variable $k\xi$ where $k>0$ also has a gamma distribution.

4. A gamma distribution with parameter $p = 1$ is called the *exponential distribution*. Differentiating the characteristic function at $t = 0$, calculate the moments of this distribution.

§ 38. CENTRAL LIMIT THEOREM OF PROBABILITY THEORY

In this section we shall prove the theorem asserting that distributions of sums of independent random variables tend to the normal distribution. The classical result concerning this convergence was already discussed in § 22, where we considered the binomial distribution, treating it as the distribution of a sum of independent random variables with zero-one distribution. Owing to the special properties of the binomial distribution we could give in § 22 a direct proof of convergence to the normal distribution. In this section we want to use the tool of characteristic functions to prove much more general results. We shall be interested in the general behaviour of averages under independent repetitions of the same random experiment. For the probabilistic description we shall use independent and identically distributed random variables. This is our starting point.

As we already know from Chapter VI on the laws of large numbers, in such a situation the averages converge in probability to the expectation, as n increases; that is to say, their distribution concentrates more and more around the expectation and therefore tends to a degenerate limit distribution which assigns probability one to the expectation. Thus, in order to bring out the similarity of the distribution of averages to the normal distribution, we shall have to extend the former in a suitable manner, i.e., to consider instead of averages, some random variables proportional to them. This is expressed by the following theorem:

THEOREM 1 (Lindeberg–Lévy). *If random variables $\xi_1, \xi_2, \ldots$ are independent and have the same distribution with finite expectation m and finite variance $\sigma^2 \neq 0$, then the probability distributions of the random variables*

$$\zeta_n = \frac{\xi_1 + \xi_2 + \ldots + \xi_n - nm}{\sigma \sqrt{n}}$$

converge weakly to the normal distribution with expectation 0 and variance 1, i.e., for every x we have the relation

(38.1) $$\lim_{n \to \infty} F_{\zeta_n}(x) = \lim_{n \to \infty} P(\{e: \zeta_n(e) < x\}) = \frac{1}{\sqrt{2\pi}} \int_{-\infty}^{x} e^{-t^2/2} dt.$$

PROOF. In view of Theorem 5 of § 36 it suffices to show that the characteristic functions $\varphi_{\zeta_n}(t)$ of the random variables ζ_n are, for every t, convergent to the characteristic function of the normal distribution with expectation 0 and variance 1. From Theorem 4 of § 37 we know that this characteristic function is equal to $\varphi(t) = e^{-t^2/2}$. Thus, we shall prove that

(38.2) $$\lim_{n \to \infty} \varphi_{\zeta_n}(t) = e^{-t^2/2}$$

for every t. By assumption, the random variables $\xi_1, \xi_2, \ldots$ have the same distribution with expectation m. Thus, the random variables $\eta_1 = \xi_1 - m$, $\eta_2 = \xi_2 - m, \ldots$ will have the same probability distribution with expectation 0, and we can write

(38.3) $$\zeta_n = \frac{1}{\sigma \sqrt{n}} (\eta_1 + \ldots + \eta_n).$$

By Theorem 2 of § 36, the characteristic function $\varphi_{\eta_1}(t)$ of the random variable η_1 has a continuous second derivative. Moreover, the relations

$$\alpha_1(\eta_1) = 0 = \frac{1}{i} \left[\frac{d\varphi_{\eta_1}(t)}{dt} \right]_{t=0},$$

$$\alpha_2(\eta_1) = \sigma^2 = \frac{1}{i^2} \left[\frac{d^2\varphi_{\eta_1}(t)}{dt^2} \right]_{t=0},$$

obtained from formula (36.4) allow us to write the following expansion, valid in the neighbourhood of the point $t = 0$:

(38.4) $$\varphi_{\eta_1}(t) = 1 - \tfrac{1}{2}\sigma^2 t^2 + o(t^2).$$

VII. CENTRAL LIMIT THEOREMS

As usual, $o(t^2)$ means that the remainder of this expansion, when divided by t^2, tends to 0 as $t \to 0$.

From analysis we know the Maclaurin formula:

$$f(x) = f(0) + \frac{f'(0)}{1!}x + \ldots + \frac{f^{(n-1)}(0)}{(n-1)!}x^{n-1} + \frac{f^{(n)}(\theta x)}{n!}x^n \quad (0 \leqslant \theta \leqslant 1),$$

valid under the assumption that function $f(x)$ has n derivatives defined in the closed interval $0 \leqslant t \leqslant x$ or in the interval $x \leqslant t \leqslant 0$ if $x < 0$. If the nth derivative is continuous the same formula can be written as follows:

$$f(x) = f(0) + \frac{f'(0)}{1!}x + \ldots + \frac{f^{(n)}(0)}{n!}x^n + R_n$$

where

$$R_n = \frac{f^{(n)}(\theta x) - f^{(n)}(0)}{n!}x^n.$$

Since $f^{(n)}(\theta x) - f^{(n)}(0)$ tends to zero as $x \to 0$ we have

$$R_n = o(x^n).$$

Expansion (38.4) is of the above type.

Next, the sum $\eta_1 + \ldots + \eta_n$ has the characteristic function

$$\varphi_{\eta_1 + \ldots + \eta_n}(t) = [\varphi_{\eta_1}(t)]^n,$$

and in view of relation (38.3) and Theorem 1 of § 37 we get

$$\varphi_{\zeta_n}(t) = \varphi_{\eta_1 + \ldots + \eta_n}\left(\frac{t}{\sigma\sqrt{n}}\right) = \left[\varphi_{\eta_1}\left(\frac{t}{\sigma\sqrt{n}}\right)\right]^n.$$

Using expansion (38.4), we obtain for a fixed t the relation

$$\varphi_{\zeta_n}(t) = \left[1 - \frac{1}{2}\sigma^2 \frac{t^2}{\sigma^2 n} + o\left(\frac{1}{n}\right)\right]^n = \left[1 - \frac{1}{2} \cdot \frac{t^2}{n} + o\left(\frac{1}{n}\right)\right]^n.$$

Instead of proving (38.2) it suffices to show that

(38.5) $$\lim_{n \to \infty} \log \varphi_{\zeta_n}(t) = -t^2/2.$$

From analysis we know the expansion

$$\log(1 + u) = u + o(u).$$

Taking, instead of u, the expression $-\frac{1}{2} \cdot \frac{t^2}{n} + o\left(\frac{1}{n}\right)$, we can write

$$\log \varphi_{\zeta_n}(t) = n \log\left(1 - \frac{1}{2} \cdot \frac{t^2}{n} + o\left(\frac{1}{n}\right)\right)$$

$$= n\left\{-\frac{1}{2} \cdot \frac{t^2}{n} + o\left(\frac{1}{n}\right)\right\} = -\frac{1}{2}t^2 + o(1).$$

This proves that for every fixed t we have (38.5), which completes the proof of Theorem 1.

Theorem 1 can also be formulated in a different way, in a manner analogous to that used in Theorem 2 of § 22. Namely we have

COROLLARY 1. *Under the assumptions of Theorem 1 the probability*

(38.6) $\quad P_n(a, b) = P(\{e: nm + a\sigma\sqrt{n} < \xi_1 + \ldots + \xi_n < nm + b\sigma\sqrt{n}\})$

that the sum $\xi_1 + \ldots + \xi_n$ will satisfy the inequalities indicated in the above formula tends, as $n \to \infty$, to the limit

$$\int_a^b \frac{1}{\sqrt{2\pi}} e^{-t^2/2} dt,$$

or

(38.7) $\qquad\qquad \lim_{n \to \infty} P_n(a, b) = \int_a^b \frac{1}{\sqrt{2\pi}} e^{-t^2/2} dt.$

The convergence is uniform with respect to a and b.

This corollary follows from relation (38.1), from the equivalence of the inequalities

$$nm + a\sigma\sqrt{n} < \xi_1 + \ldots + \xi_n < nm + b\sigma\sqrt{n} \quad \text{and} \quad a < \zeta_n < b$$

and from the fact that if the distribution functions $F_1(x), F_2(x), \ldots$ converge fundamentally to a continuous distribution function $F(x)$, the convergence is necessarily uniform.

We see here a precise formulation of the fact that sums or averages have distribution close to normal: the sum $\xi_1 + \ldots + \xi_n$ has expectation nm and variance $\sigma\sqrt{n}$, while relation (38.7) shows that this distribution is close to the normal distribution with the same parameters. Sometimes we formulate it as follows: if random variables $\xi_1, \xi_2, \ldots$ satisfy the assumptions of Theorem 1, then the sums $\xi_1 + \ldots + \xi_n$ are *asymptotically normal* $N(nm, \sigma\sqrt{n})$.

The Lindeberg–Lévy theorem formulated above is crucial for using the normal distribution in statistics. Indeed, this theorem asserts that when we perform independent repetitions of the same experiment, then the sample average will have a practically normal distribution. There is also another way of interpreting this type of limit theorems: it is claimed that the result of a random experiment is influenced by a large number of independent factors, and these influences are additive. Thus, one may expect that in many applications the characteristics under investigation will have normal distributions. Below we shall prove another theorem of this type, which will justify further the above interpretation: sums of independent terms will prove to be asymptotically normal even if the terms have different distributions. We point out, however, that in spite of the agreement between the limit theorems here discussed and the frequency of occurrence of the normal distribution in practical situations, the above interpretation has some weak points: it requires, namely, that the effects of different random factors be additive.

In some biological applications another mechanism of the phenomenon is assumed: it is postulated that the outcome depends not only on the magnitude of the random stimulus factor, but also on the magnitude of the organ on which it acts.

When the result of a stimulus is proportional to the magnitude of the object on which it acts, we observe as a limit distribution the so-called log-normal distribution, which occurs in biological problems concerning the magnitudes of certain organs in individuals of a given species. We say that random variable ξ has a *log-normal distribution* if there exists a constant a such that $\xi - a$ is positive with probability one and $\log(\xi - a)$ has a normal distribution.

We now prove

THEOREM 2 (Lapunov). *Consider a sequence* $\xi_1, \xi_2, \ldots, \xi_k, \ldots$ *of independent random variables with expectations* m_k, *variances* $\sigma_k^2 \neq 0$ *and finite absolute third central moments* $b_k^3 = E|\xi_k - m_k|^3$. *Further, let*

(38.5) $$\sigma^{(n)} = \sqrt{\sigma_1^2 + \ldots + \sigma_n^2}, \quad b^{(n)} = \sqrt[3]{b_1^3 + \ldots + b_n^3}.$$

If we have the relation

(38.6) $$\lim_{n \to \infty} \frac{b^{(n)}}{\sigma^{(n)}} = 0,$$

then the random variables

(38.7) $$\zeta_n = \frac{\xi_1 + \ldots + \xi_n - m_1 - \ldots - m_n}{\sigma^{(n)}}$$

have distributions weakly convergent to the normal distribution with the distribution function

$$\frac{1}{\sqrt{2\pi}} \int_{-\infty}^{x} e^{-t^2/2} dt.$$

PROOF. Denote by $\varphi_k(t)$ the characteristic function of the random variable $\xi_k - m_k$. Then the characteristic function of the random variable ζ_n will be

(38.8) $$\varphi_{\zeta_n}(t) = \prod_{k=1}^{n} \varphi_k\left(\frac{t}{\sigma^{(n)}}\right),$$

which follows from Theorems 3 of § 36 and 1 of § 37. To prove Theorem 2 it suffices to show that for every t we have

(38.9) $$\lim_{n\to\infty} \varphi_{\zeta_n}(t) = e^{-t^2/2}.$$

Using the expansion

$$e^{iz} = 1 + iz + \frac{(iz)^2}{2!} + \vartheta \frac{z^3}{3!}$$

valid for real z, where ϑ is a complex number smaller than unity in absolute value, we can write

$$\varphi_k(t) = E e^{-it(\xi_k - m_k)} = 1 - \tfrac{1}{2}\sigma_k^2 t^2 + \tfrac{1}{6}\vartheta_k b_k^3 t^3$$

where ϑ_k is a complex number such that $|\vartheta_k| \leqslant 1$.

Next, we have

$$\log \varphi_k\left(\frac{t}{\sigma^{(n)}}\right) = \log\left(1 - \frac{1}{2} \cdot \frac{\sigma_k^2}{(\sigma^{(n)})^2} t^2 + \frac{\vartheta_k}{6} \cdot \frac{b_k^3}{(\sigma^{(n)})^3} t^3\right) = \log(1 + z_{kn})$$

where

$$z_{kn} = -\frac{1}{2} \cdot \frac{\sigma_k^2}{(\sigma^{(n)})^2} t^2 + \frac{\vartheta_k}{6} \cdot \frac{b_k^3}{(\sigma^{(n)})^3} t^3.$$

In view of (38.6) we have for sufficiently large n

$$\frac{b_k}{\sigma^{(n)}} \leqslant \frac{b^{(n)}}{\sigma^{(n)}} < 1.$$

From the inequality given further on in the problem, we always have $\sigma_k \leqslant b_k$, whence we can write

$$z_{kn} = \vartheta'_{kn} \frac{b_k^2 t^2}{2(\sigma^{(n)})^2} + \vartheta_k \frac{b_k^3}{6(\sigma^{(n)})^3} t^3 = \vartheta''_{kn} \frac{b_k^2}{(\sigma^{(n)})^2}\left(\frac{t^2}{2} + \frac{|t|^3}{6}\right).$$

From (38.6) it follows that as $n \to \infty$ all z_{kn} tend to zero uniformly in k. For sufficiently large n we shall thus have $|z_{kn}| < 1/2$ for all k ($k = 1, 2, \ldots, n$). But for $|z| < 1/2$ we have

$$\log(1+z) = \frac{z}{1} - \frac{z^2}{2}\left(1 - \frac{2}{3}z + \frac{2}{4}z^2 - \ldots\right)$$

$$= z + \frac{1}{2}\theta z^2\left(1 + \frac{1}{2} + \frac{1}{2^2} + \ldots\right) = z + \theta z^2,$$

VII. CENTRAL LIMIT THEOREMS

where $|\theta| \leqslant 1$, hence

$$\log \varphi_k\left(\frac{t}{\sigma^{(n)}}\right) = -\frac{\sigma_k^2}{(\sigma^{(n)})^2} \cdot \frac{t^2}{2} + \frac{\vartheta_k}{6} \cdot \frac{b_k^3}{(\sigma^{(n)})^3} t^3 + \theta \vartheta_{kn}''^2 \frac{b_k^4}{(\sigma^{(n)})^4} \left(\frac{t^2}{2} + \frac{|t|^3}{6}\right)^2$$

$$= -\frac{\sigma_k^2}{(\sigma^{(n)})^2} \cdot \frac{t^2}{2} + \vartheta_{kn}''' \frac{b_k^3}{(\sigma^{(n)})^3} \left(\frac{|t|^3}{6} + \left(\frac{t^2}{2} + \frac{|t|^3}{6}\right)^2\right).$$

Adding with respect to k for $k = 1, 2, \ldots, n$ we obtain, in view of formula (38.8),

$$\log \varphi_{\zeta_n}(t) = -\frac{t^2}{2} + \vartheta_n' \frac{(b^{(n)})^3}{(\sigma^{(n)})^3} \left(\frac{|t|^3}{6} + \left(\frac{t^2}{2} + \frac{|t|^3}{6}\right)^2\right) \quad \text{where} \quad |\vartheta_n'| \leqslant 1.$$

We see that if (38.6) holds, we have for every t

$$\lim_{n \to \infty} \log \varphi_{\zeta_n}(t) = -\frac{t^2}{2},$$

which is equivalent to (38.9) and completes the proof of the theorem of Lapunov.

The most general result concerning the convergence of distributions of sums of independent random variables with finite variances is given by the following theorem:

THEOREM 3 (Lindeberg–Feller). *Let $\xi_1, \xi_2, \ldots, \xi_k, \ldots$ be independent random variables with probability distribution functions $F_k(x)$.*

The probability distributions of the normed sums

$$\zeta_n = \frac{\sum_{k=1}^{n} (\xi_k - E\xi_k)}{B_n},$$

where $B_n^2 = \sum_{k=1}^{n} D^2 \xi_k$ converge to the normal distribution with the distribution function

$$\Phi(x) = \int_{-\infty}^{x} \frac{1}{\sqrt{2\pi}} e^{-t^2/2} dt$$

and the terms are asymptotically negligible if and only if the Lindeberg condition holds, i.e., if for every $\varepsilon > 0$

$$\lim_{n \to \infty} \frac{1}{B_n^2} \sum_{k=1}^{n} \int_{|x| > \varepsilon B_n} x^2 \, dF_k(x + E\xi_k) = 0$$

(the terms $\xi_{nk} = (\xi_k - E\xi_k)/B_n$ are called *asymptotically negligible* if for every $\varepsilon > 0$

$$\lim_{n \to \infty} \sup_{1 \leqslant k \leqslant n} P\{|\xi_{nk}| > \varepsilon\} = 0).$$

We shall now show that the distributions of sums of independent random variables need not converge to the normal distribution; thus, to ensure this convergence one has to impose some conditions on the distributions of the terms. Indeed, let $\xi_1, \xi_2, \ldots$ be independent random variables having the Cauchy distribution with the probability density given by (37.14):

$$f(x) = \frac{1}{\pi} \cdot \frac{1}{1+x^2}, \quad -\infty < x < \infty.$$

Then, each of these variables has the characteristic function

$$\varphi(t) = e^{-|t|},$$

and the sum $\xi_1 + \ldots + \xi_n$ has the characteristic function

$$\varphi_{\zeta_n}(t) = e^{-n|t|}.$$

In view of Theorem 1 of § 37, the average $\frac{1}{n}(\xi_1 + \ldots + \xi_n) = \frac{1}{n}\zeta_n$ has the characteristic function

$$\varphi_{\zeta_n/n}(t) = \varphi_{\zeta_n}\left(\frac{t}{n}\right) = e^{-n|t|/n} = e^{-|t|},$$

and has, therefore, the same Cauchy distribution as each of the terms. Thus, the averages do not converge to the normal distribution.

Problems

1. Prove that if $\xi_1, \xi_2, \ldots$ are independent random variables and have the same distribution with a finite third moment then condition (38.6) is satisfied.

   (Theorem 1, however, does not follow from Lapunov's theorem, since the existence of third moments is not assumed in it.)

2. Prove that if the random variables $\xi_1, \xi_2, \ldots$ are independent and their distributions are given by

$$P\{\xi_k = 1\} = p_k, \quad P\{\xi_k = 0\} = q_k = 1 - p_k, \quad k = 1, 2, \ldots$$

then the random variables

$$\sum_{k=1}^{n} (\xi_k - p_k)/B_n,$$

where $B_n^2 = \sum_{k=1}^{n} p_k q_k$ have distributions convergent to the normal distribution

$$\frac{1}{\sqrt{2\pi}} \int_{-\infty}^{x} e^{-t^2/2} dt$$

if the series
$$\sum_{k=1}^{\infty} p_k q_k$$
diverges.

(Hint: Prove that the conditions of Lapunov's theorem are satisfied.)

3. Prove that the absolute moments of orders $v-1$, v and $v+1$ of the random variable ξ (if they exist) satisfy the relation
$$\beta_v^2 \leqslant \beta_{v-1}\beta_{v+1}.$$
Deduce that the absolute moments satisfy the inequalities
$$\beta_v^{1/v} \leqslant \beta_{v+1}^{1/(v+1)}, \quad v = 1, 2, \ldots$$
(Hint: Consider the expectation
$$E(u|\xi|^{(v+1)/2} + v|\xi|^{(v+1)/2})^2$$
as a function of u and v, and write the conditions for the non-negativeness of this function for all u and v.)

§ 39. DISTRIBUTIONS CONNECTED WITH THE NORMAL DISTRIBUTION WHICH OCCURS IN STATISTICS

We shall now discuss a few distributions used in mathematical statistics and connected with the normal distribution.

Chi-square distribution

DEFINITION 1. By the *chi-square distribution* with n degrees of freedom we mean the probability distribution of a sum $\xi_1^2 + \ldots + \xi_n^2$ where $\xi_1, \ldots, \xi_n$ are independent random variables with distribution $N(0, 1)$, i.e. with a normal distribution with expectation 0 and variance 1. The probability distribution function of this distribution will be denoted by $K_n(x)$, and the density by $k_n(x)$.

We shall now show that chi-square distributions are special cases of gamma distributions.

THEOREM 1. *The following relation holds*:

(39.1) $$k_n(x) = \gamma(x; \tfrac{1}{2}, \tfrac{1}{2}n) = \frac{1}{2^{n/2}\Gamma(n/2)} x^{n/2-1} e^{-x/2},$$

where $\gamma(x; a, p)$ is the density of the gamma distribution with parameters a and p given by formula (37.8).

We shall use the following fact as a lemma:

LEMMA 1. *If a random variable ξ has a probability distribution with density $f(x)$, then the random variable ξ^2 has a probability distribution with the density*

(39.2) $$f_{\xi^2}(x) = \begin{cases} 0 & \text{for } x < 0, \\ \dfrac{1}{2\sqrt{x}}(f(\sqrt{x})+f(-\sqrt{x})) & \text{for } x > 0. \end{cases}$$

Indeed, for $x < 0$ relation (39.2) is obvious. For $x > 0$ we have

$$P\{\xi^2 < x\} = P\{-\sqrt{x} < \xi < \sqrt{x}\} = \int_{-\sqrt{x}}^{\sqrt{x}} f_\xi(u)\,du;$$

differentiating with respect to x we obtain (39.2).

PROOF OF THEOREM 1. Applying (39.2) we see that if ξ has the normal distribution $N(0, 1)$, then its square, ξ^2, has a distribution with the density (for $x > 0$):

$$k_1(x) = \frac{1}{2\sqrt{x}} \cdot \frac{1}{\sqrt{2\pi}} (e^{-x/2}+e^{-x/2}) = \frac{1}{\sqrt{2\pi}} x^{-1/2} e^{-x/2}.$$

We have

$$\Gamma(\tfrac{1}{2}) = \int_0^\infty x^{-1/2} e^{-x}\,dx.$$

By substituting $x = t^2/2$ we get

$$\Gamma(\tfrac{1}{2}) = \sqrt{2} \int_0^\infty e^{-t^2/2}\,dt,$$

and finally, in view of (22.12), we have

$$\Gamma(\tfrac{1}{2}) = \sqrt{\pi}.$$

Comparing this with formula (37.8), we see that

$$k_1(x) = \gamma(x; \tfrac{1}{2}, \tfrac{1}{2}).$$

In view of the additivity of the gamma distribution (Corollary 4 of § 37) we obtain relation (39.1), which completes the proof.

Student's t distribution

DEFINITION 2. By *Student's t distribution* with n degrees of freedom we mean the probability distribution of the ratio

$$t_n = \frac{\xi}{\sqrt{\dfrac{1}{n}\chi_n^2}},$$

where ξ and χ_n^2 are independent random variables, ξ has the normal distribution $N(0, 1)$ and χ_n^2 has a chi-square distribution with n degrees of freedom. The probability distribution function of Student's t distribution will be denoted by $S_n(x)$, and its density by $s_n(x)$.

VII. CENTRAL LIMIT THEOREMS

THEOREM 2. *We have the following relation*

(39.3) $$s_n(x) = \frac{1}{\sqrt{n\pi}} \cdot \frac{\Gamma((n+1)/2)}{\Gamma(n/2)} \left(1 + \frac{x^2}{n}\right)^{-(n+1)/2}.$$

Before we prove Theorem 2, note the following two relations:

LEMMA 2. *If a random variable ξ has the probability distribution function $F_\xi(x)$ and the probability density $f_\xi(x)$, then the random variable $a\xi+b$, where $a \neq 0$, has the probability distribution function*

(39.4) $$F_{a\xi+b}(x) = \begin{cases} F_\xi\left(\dfrac{x-b}{a}\right) & \text{for } a > 0, \\ 1 - F_\xi\left(\dfrac{x-b}{a}\right) & \text{for } a < 0, \end{cases}$$

and the probability density

(39.5) $$f_{a\xi+b}(x) = \frac{1}{|a|} f_\xi\left(\frac{x-b}{a}\right).$$

Indeed, we have

$$F_{a\xi+b}(x) = P\{a\xi+b < x\}$$

$$= \begin{cases} P\left\{\xi < \dfrac{x-b}{a}\right\} = F_\xi\left(\dfrac{x-b}{a}\right) & \text{for } a > 0, \\ P\left\{\xi > \dfrac{x-b}{a}\right\} = 1 - F_\xi\left(\dfrac{x-b}{a}\right) & \text{for } a < 0 \end{cases}$$

which proves (39.4). Differentiating with respect to x, we find

$$f_{a\xi+b}(x) = \begin{cases} \dfrac{1}{a} f_\xi\left(\dfrac{x-b}{a}\right) & \text{for } a > 0, \\ -\dfrac{1}{a} f_\xi\left(\dfrac{x-b}{a}\right) & \text{for } a < 0, \end{cases}$$

which is equivalent to (39.5).

LEMMA 3. *If ξ is a non-negative random variable with the probability density $f_\xi(x)$, then $\sqrt{\xi}$ has the probability density*

(39.6) $$f_{\sqrt{\xi}}(x) = \begin{cases} 2x f_\xi(x^2) & \text{for } x > 0, \\ 0 & \text{for } x < 0. \end{cases}$$

In fact, for $x > 0$ we have
$$P\{\sqrt{\xi} < x\} = P\{\xi < x^2\} = \int_0^{x^2} f_\xi(u)\,du,$$
and (39.6) follows by differentiation.

PROOF OF THEOREM 2. In view of (39.1) and Lemma 2, the random variable $\eta^2 = \frac{1}{n}\chi_n^2$ has the probability density (for $x > 0$):
$$f_{\eta^2}(x) = \frac{n^{n/2}}{2^{n/2}\Gamma(n/2)} x^{n/2-1} e^{-nx/2};$$
applying Lemma 3 we see that the random variable $\eta = \sqrt{\frac{1}{n}\chi_n^2}$ has the probability density (for $x > 0$):
$$f_\eta(x) = \frac{n^{n/2}}{2^{(n-1)/2}\Gamma(n/2)} x^{n-1} e^{-nx^2/2}.$$

In view of the independence of ξ and η and the fact that the random variable η assumes positive values with probability one, we obtain
$$S_n(x) = P\left\{\frac{\xi}{\eta} < x\right\} = P\{\xi < x\eta\} = \iint_G f_\xi(u) f_\eta(v)\,du\,dv,$$
where the domain G is determined by the inequalities $0 < v < \infty$, $-\infty < u < xv$. Substituting $v = v'$, $u = u'v'$, or $v' = v$, $u' = u/v$, we transform G into G', defined by the inequalities $0 < v' < \infty$, $-\infty < u' < x$. We have
$$\frac{\partial(u,v)}{\partial(u',v')} = \begin{vmatrix} v' & u' \\ 0 & 1 \end{vmatrix} = v'.$$
Thus, in new variables we can write
$$S_n(x) = \iint_{G'} v' f_\xi(u', v') f_\eta(v')\,du'\,dv'.$$
Omitting the symbol $'$ and substituting the appropriate expressions in the above formula, we obtain
$$S_n(x) = \int_{-\infty}^{x} \int_0^{\infty} c_n v^n \exp\{-\tfrac{1}{2}v^2(u^2+n)\}\,dv\,du, \quad c_n = \frac{n^{n/2}}{\sqrt{\pi}\,2^{(n-1)/2}\Gamma(n/2)}.$$
By differentiating with respect to x we get
$$s_n(x) = c_n \int_0^{\infty} v^n \exp\{-\tfrac{1}{2}v^2(x^2+n)\}\,dv.$$

VII. CENTRAL LIMIT THEOREMS

Substituting $\frac{1}{2}v^2(x^2+n) = r$, or

$$v = \frac{2^{1/2}}{(x^2+n)^{1/2}} r^{1/2}, \quad dv = \frac{1}{2^{1/2}(x^2+n)^{1/2} r^{1/2}} dr,$$

we obtain

$$s_n(x) = c_n \frac{2^{(n-1)/2}}{(x^2+n)^{(n+1)/2}} \int_0^\infty r^{(n-1)/2} e^{-r} dr$$

$$= c_n 2^{(n-1)/2} \Gamma\left(\frac{n+1}{2}\right) \frac{1}{(n+x^2)^{(n+1)/2}}$$

$$= \frac{n^{(n+1)/2} \Gamma((n+1)/2) 2^{(n-1)/2}}{\sqrt{n\pi} 2^{(n-1)/2} \Gamma(n/2) n^{(n+1)/2}} \cdot \frac{1}{(1+x^2/n)^{(n+1)/2}}$$

$$= \frac{\Gamma((n+1)/2)}{\sqrt{n\pi} \Gamma(n/2)} \cdot \frac{1}{(1+x^2/n)^{(n+1)/2}},$$

which was to be proved.

We see that Student's distribution is symmetric with respect to the origin; it has a finite expectation for $n > 1$, a finite variance for $n > 2$, and in general, one can prove that if t_n is a random variable having Student's distribution with n degrees of freedom, then

(39.7) $$\mu_2(t_n) = \frac{n}{n-2} \quad \text{for} \quad n > 2,$$

and, more generally,

(39.8) $$\mu_{2\nu}(t_n) = \frac{1 \cdot 3 \cdot \ldots \cdot (2\nu-1) n^\nu}{(n-2)(n-4)\ldots(n-2\nu)} \quad \text{for} \quad 2\nu < n.$$

Fisher's F distribution

DEFINITION 3. By *Fisher's F distribution*, or shortly the *Fisher's distribution*, with (m, n) degrees of freedom (m for the numerator and n for the denominator) we mean the probability distribution of the ratio

$$\varkappa = \frac{\xi}{\eta},$$

where ξ and η are independent random variables, each having a chi-square distribution, ξ with m degrees of freedom, and η with n degrees of freedom. We shall denote the distribution function of this random variable by $F_{mn}(x)$ and its density by $f_{mn}(x)$.

THEOREM 3. *We have*

(39.9) $$f_{mn}(x) = \begin{cases} 0 & \text{for } x < 0, \\ \dfrac{\Gamma((m+n)/2)}{\Gamma(m/2)\Gamma(n/2)} \cdot \dfrac{x^{m/2-1}}{(x+1)^{(m+n)/2}} & \text{for } x > 0. \end{cases}$$

PROOF. According to Definition 3 we may write for $x > 0$

$$F_{mn}(x) = P\{\varkappa < x\} = P\{\xi < x\eta\} = \iint_G k_m(u) k_n(v) \, du \, dv,$$

where G is the domain determined by the inequalities $0 < v < \infty$, $0 < u < xv$. Introducing new variables $v = v'$, $u = u'v'$, we transform the domain G into the domain G' determined by the inequalities $0 < v' < \infty$, $0 < u' < x$. Since

$$\frac{\partial(u, v)}{\partial(u', v')} = \begin{vmatrix} v' & u' \\ 0 & 1 \end{vmatrix} = v',$$

we may write, omitting the symbol $'$:

$$F_{mn}(x) = \iint_{G'} v k_m(uv) k_n(v) \, du \, dv = \int_0^x \int_0^\infty v k_m(uv) k_n(v) \, dv \, du.$$

By differentiating we obtain

$$f_{mn}(x) = \int_0^\infty v k_m(xv) k_n(v) \, dv.$$

Substituting for k_m and k_n the expressions given in (39.1), we find

$$f_{mn}(x) = \int_0^\infty \frac{1}{2^{(m+n)/2} \Gamma(m/2) \Gamma(n/2)} \, v(xv)^{m/2-1} e^{-xv/2} v^{n/2-1} e^{-v/2} \, dv$$

$$= \frac{x^{m/2-1}}{2^{(m+n)/2} \Gamma(m/2) \Gamma(n/2)} \int_0^\infty v^{(m+n)/2-1} e^{-(x+1)v/2} \, dv.$$

Substituting $\dfrac{(x+1)}{2} v = u$, we get $v = \dfrac{2u}{x+1}$, $dv = \dfrac{2du}{x+1}$ and

$$\int_0^\infty v^{(m+n)/2-1} e^{-(x+1)v/2} \, dv$$

$$= \frac{2^{(m+n)/2}}{(x+1)^{(m+n)/2}} \int_0^\infty u^{(m+n)/2-1} e^{-u} \, du = \frac{2^{(m+n)/2} \Gamma((m+n)/2)}{(x+1)^{(m+n)/2}}.$$

Finally,
$$f_{mn}(x) = \frac{x^{m/2-1}}{2^{(m+n)/2}\Gamma(m/2)\Gamma(n/2)} \cdot \frac{2^{(m+n)/2}\Gamma((m+n)/2)}{(x+1)^{(m+n)/2}}$$
which is equivalent to formula (39.9).

Instead of the Fisher distribution discussed above, we usually deal with some distributions related to it, namely the distributions called Snedecor's T distribution and Fisher's z distribution. T and z are traditional symbols for random variables with Snedecor or Fisher distribution, which we shall presently define.

Snedecor's T distribution

DEFINITION 4. By the *Snedecor distribution* with the degrees of freedom (m, n) we mean the probability distribution of the ratio

$$T = \frac{\frac{1}{m}\xi}{\frac{1}{n}\eta}$$

where ξ and η are independent random variables, ξ has a chi-square distribution with m degrees of freedom, and η has a chi-square distribution with n degrees of freedom. This distribution is also called *Snedecor's T distribution*.

There is obviously a linear relation between the random variables T and $\varkappa$ appearing in Definition 3, namely $T = \frac{n}{m}\varkappa$. Thus, using formula (39.9) and Lemma 2, we obtain

COROLLARY 1. *If T is a random variable with the Snedecor distribution with (m, n) degrees of freedom, then its probability density is*

(39.10) $\quad f_T(x) = \begin{cases} 0 & \text{for } x < 0, \\ \dfrac{\Gamma((m+n)/2)}{\Gamma(m/2)\Gamma(n/2)} \left(\dfrac{n}{m}\right)^{n/2} \dfrac{x^{m/2-1}}{\left(x+\dfrac{n}{m}\right)^{(m+n)/2}} & \text{for } x > 0. \end{cases}$

Fisher's z distribution

There exists also tables of the distribution called Fisher's z distribution, defined as follows:

DEFINITION 5. By *Fisher's z distribution* with (m, n) degrees of freedom we mean the probability distribution of the random variable

$$z = \tfrac{1}{2}\log T,$$

where T is a random variable having the Snedecor distribution with (m, n) degrees of freedom.

The distributions of Fisher's z and Snedecor's T are related as follows:

$$P\{z < x\} = P\{\tfrac{1}{2}\log T < x\} = P\{\log T < 2x\} = P\{T < e^{2x}\}.$$

Thus, by differentiation we find the expression for the probability density of z:

COROLLARY 2. *If the random variable z has Fisher's z distribution with (m, n) degrees of freedom, then*

$$(39.11) \quad f_z(x) = 2e^{2x} f_T(e^{2x}) = \frac{\Gamma((m+n)/2)}{\Gamma(m/2)\Gamma(n/2)} \left(\frac{n}{m}\right)^{n/2} 2e^{mx} \left(e^{2x} + \frac{n}{m}\right)^{-(m+n)/2}.$$

We have already found the probability density of the chi-square distribution and of Student's distribution. Now we shall show that when the number of degrees of freedom increases to infinity, these two distributions converge to the normal distribution. To formulate this convergence it will be convenient to use the following definition:

DEFINITION 6. Suppose we are given random variables $\xi_1, \xi_2, \ldots$ If the probability distributions of the random variables

$$\frac{\xi_1 - m_1}{\sigma_1}, \frac{\xi_2 - m_2}{\sigma_2}, \ldots$$

converge to the normal distribution

$$\Phi(x) = \int_{-\infty}^{x} \frac{1}{\sqrt{2\pi}} e^{-t^2/2} dt,$$

then we say that the random variables $\xi_1, \xi_2, \ldots$ are *asymptotically normal* $N(m_i, \sigma_i)$.

When the random variables $\xi_1, \xi_2, \ldots$ have finite variances, then m_i is usually equal to the expectation of ξ_i, and σ_i^2 to its variance; this, however, is not necessary.

THEOREM 4. *Let $t_1, t_2, \ldots$ be random variables such that t_n has Student's distribution with n degrees of freedom. Then the random variables $t_1, t_2, \ldots$ are asymptotically normal $N(0, 1)$.*

In other words, when the number of degrees of freedom increases, the Student distributions converge to the normal distribution with expectation 0 and variance 1.

VII. CENTRAL LIMIT THEOREMS

PROOF. The probability distribution of the random variable t_n has density given by (39.3). We can write it in the form

$$s_n(x) = \frac{\Gamma((n+1)/2)}{\sqrt{n/2}\,\Gamma(n/2)} \cdot \frac{1}{\sqrt{2\pi}} \left(1 + \frac{x^2}{n}\right)^{-(n+1)/2}.$$

We shall prove first that for every x we have

(39.12) $$\lim_{n \to \infty} s_n(x) = \frac{1}{\sqrt{2\pi}} e^{-x^2/2}.$$

We know from analysis that

$$\lim_{p \to \infty} \frac{\Gamma(p+h)}{p^h \Gamma(p)} = 1.$$

It follows that we must have

$$\lim_{n \to \infty} \frac{\Gamma((n+1)/2)}{\sqrt{n/2}\,\Gamma(n/2)} = 1.$$

To prove relation (39.12) it suffices to show that

(39.13) $$\lim_{n \to \infty} \left(1 + \frac{x^2}{n}\right)^{-(n+1)/2} = e^{-x^2/2}.$$

Passing to logarithms and applying the expansion

$$\log(1+u) = u + o(u),$$

we obtain

$$-\frac{n+1}{2} \log\left(1 + \frac{x^2}{n}\right) = -\frac{n+1}{2}\left(\frac{x^2}{n} + o\left(\frac{1}{n}\right)\right) = -\frac{x^2}{2}\left(\frac{n+1}{n} + o(1)\right).$$

It follows that

$$\lim_{n \to \infty} \left(-\frac{n+1}{2} \log\left(1 + \frac{x^2}{n}\right)\right) = -\frac{x^2}{2},$$

hence we obtain (39.13), which completes the proof of (39.12).

Let r denote the largest integer not exceeding $(n+1)/2$. Then $r \geq n/2$ and for all $n \geq 1$ and real x we get

$$\left(1 + \frac{x^2}{n}\right)^{(n+1)/2} \geq \left(1 + \frac{x^2}{n}\right)^r \geq 1 + r\frac{x^2}{n} \geq 1 + \frac{x^2}{2}.$$

Thus, all functions $s_n(x)$ can be majorized by the integrable function of the form $A(1+x^2/2)^{-1}$. From relation (39.12) it follows therefore that for every x

$$\lim_{n \to \infty} \int_{-\infty}^{x} s_n(u)\,du = \int_{-\infty}^{x} \frac{1}{\sqrt{2\pi}} e^{-u^2/2}\,du,$$

or for every x

$$\lim_{n \to \infty} S_n(x) = \Phi(x),$$

which completes the proof of Theorem 4.

THEOREM 5. *Let $\chi_1^2, \chi_2^2, \ldots$ be random variables such that χ_n^2 has a chi-square distribution with n degrees of freedom. Then these random variables are asymptotically normal $N(n, \sqrt{2n})$.*

PROOF. By definition, the random variable χ_n^2 has the same distribution as the sum of the squares $\xi_1^2 + \ldots + \xi_n^2$ of the independent random variables $\xi_1, \ldots, \xi_n$, each having a normal distribution with expectation 0 and variance 1. The random variable ξ_1^2 has a gamma distribution with the density $\gamma(x; \frac{1}{2}, \frac{1}{2})$, hence it has a characteristic function $\varphi(t) = (1-2it)^{-1/2}$. Differentiating, we obtain

$$\varphi'(t) = i(1-2it)^{-3/2},$$

$$\varphi''(t) = 3i^2(1-2it)^{-5/2},$$

hence, in view of Theorem 3 of § 36, we have $E\xi_1^2 = 1$, $E(\xi_1^2)^2 = 3$, which implies that $D^2\xi_1^2 = 2$. By Theorem 1 of § 38 the sums $\xi_1^2 + \ldots + \xi_n^2$, and hence the variables χ_n^2 are asymptotically normal $N(n, \sqrt{2n})$, which completes the proof of Theorem 5.

THEOREM 6. *Suppose that $\chi_1^2, \chi_2^2, \ldots$ are random variables such that χ_n^2 has a chi-square distribution with n degrees of freedom. Then the random variables $\sqrt{2\chi_1^2}, \sqrt{2\chi_2^2}, \ldots$ are asymptotically normal $N(\sqrt{2n}, 1)$.*

PROOF. Consider the inequality

$$\sqrt{2\chi_n^2} < \sqrt{2n} + x.$$

Squaring both sides and ordering the expression obtained, we may write

$$2\chi_n^2 < 2n + 2x\sqrt{2n} + x^2, \quad \chi_n^2 < n + x\sqrt{2n} + x^2/2,$$

$$\frac{\chi_n^2 - n}{\sqrt{2n}} < x + \frac{x^2}{2\sqrt{2n}}.$$

By Theorem 5 we have

$$\lim_{n \to \infty} P\left\{\frac{\chi_n^2 - n}{\sqrt{2n}} < x\right\} = \frac{1}{\sqrt{2\pi}} \int_{-\infty}^{x} e^{-t^2/2} dt.$$

Since the limit distribution is continuous, and $x^2/\sqrt{2n}$ tends to zero as $n \to \infty$ we have

$$\lim_{n\to\infty} P\left\{\frac{\chi_n^2-n}{\sqrt{2n}} < x + \frac{x^2}{2\sqrt{2n}}\right\} = \frac{1}{\sqrt{2\pi}} \int_{-\infty}^{x} e^{-t^2/2}dt.$$

This proves Theorem 6.

Fisher proved that the convergence to the normal distribution is even better if $2n$ is replaced by $2n-1$. In applications, for $n > 30$, the random variable $\sqrt{2\chi_n^2}-\sqrt{2n-1}$ may be treated as a random variable with the normal distribution $N(0, 1)$.

Problems

1. Prove that a random variable ξ with a Poisson distribution with parameter λ is for $\lambda \to \infty$ asymptotically normal $N(\lambda, \sqrt{\lambda})$.

2. In analysis, one considers the beta function, defined for $a > 0$ and $b > 0$ by the equation

$$B(a, b) = \int_0^1 x^{a-1}(1-x)^{b-1}dx.$$

The probability distribution with the density

$$\beta(x; a, b) = \begin{cases} 0 & \text{for } x < 0, \\ \frac{1}{B(a, b)} x^{a-1}(1-x)^{b-1} & \text{for } 0 < x < 1, \\ 0 & \text{for } 1 < x \end{cases}$$

is called the *beta distribution*.

Prove that if $\xi_1, \ldots, \xi_{m+n}$ are independent random variables with the normal distribution $N(0, 1)$, then the random variable

$$\frac{\xi_1^2 + \ldots + \xi_m^2}{\xi_1^2 + \ldots + \xi_{m+n}^2}$$

has a beta distribution with parameters $a = m/2$, $b = n/2$.

3. Using the relation

$$B(a, b) = \frac{\Gamma(a)\Gamma(b)}{\Gamma(a+b)}$$

show that the expectation of the beta distribution $\beta(x; a, b)$ equals $a/(a+b)$ and the variance equals $ab/[(a+b)^2(a+b+1)]$.

4. Prove that random variables $\xi_1, \xi_2, \ldots$ such that ξ_n has a beta distribution with the density $\beta(x; na, nb)$ are asymptotically normal $N(a/(a+b), \sqrt{ab/[(a+b)^2(a+b+1)]})$.

§ 40. COVARIANCE MATRIX OF A PROBABILITY DISTRIBUTION IN A MULTIDIMENSIONAL EUCLIDEAN SPACE

Before we can continue our considerations concerning tests used in classical mathematical statistics based on the assumption of the normality of distribution in a population, we have to make a few digressions. One of them concerns the role of the matrix of second central moments for the characterization of multidimensional distributions; this will be discussed in the present section. In the next section we shall discuss the characteristic functions of multidimensional distributions.

The Čebyšev inequality, proved in § 29, showed the relation between the variance, i.e., the central second moment, and the probability that the random variable will assume a value differing from its expectation by more than a given number $\varepsilon > 0$. From the definition of variance given in § 27 we see that if the variance is equal to zero, then the random variable assumes with probability one a value equal to its expectation; in this case the whole probability distribution is concentrated at one point.

Let us consider now a system of random variables $\xi_1, \ldots, \xi_n$ or, which is equivalent, a multidimensional random variable, or *random vector*, defined by the equality $\xi = (\xi_1, \ldots, \xi_n)$. In the case of a multidimensional random vector, to the concept of expectation corresponds the concept of vector whose components are equal to the expectations of the components of our random vector, in accordance with the formula

(40.1) $$E\xi = (E\xi_1, \ldots, E\xi_n)$$

where

$$E\xi_i = \int \xi_i(e)\,dP(e) = \int \ldots \int x_i\,dF(x_1, \ldots, x_n).$$

The first of these integrals defines the expectation of the component ξ_i of the random vector as an integral with respect to the probability measure on the space of elementary events e; the second integral equals the integral with respect to the n-dimensional distribution function:

$$F(x_1, \ldots, x_n) \stackrel{df}{=} P\{\xi_1 < x_1, \ldots, \xi_n < x_n\}.$$

What is the analogue of variance here? It is the matrix of second order central moments, which can conveniently be denoted by

(40.2) $$\Lambda = [\lambda_{ij}] = \begin{bmatrix} \lambda_{11} & \ldots & \lambda_{1n} \\ \ldots & \ldots & \ldots \\ \lambda_{n1} & \ldots & \lambda_{nn} \end{bmatrix}.$$

Here

(40.3) $$\lambda_{ij} \stackrel{df}{=} E(\xi_i - E\xi_i)(\xi_j - E\xi_j) = \mathrm{Cov}(\xi_i, \xi_j),$$

equals the expectation of the product of the deviations of the random variables ξ_i and ξ_j from their expectations. We can see that λ_{ii} is simply the variance of the random variable ξ_i, i.e., $\lambda_{ii} = \mathrm{Cov}(\xi_i, \xi_i) = D^2\xi_i$.

VII. CENTRAL LIMIT THEOREMS

For $i \neq j$, the coefficient λ_{ij} is called the *covariance* of the random variables ξ_i and ξ_j. For simplicity of formulations, it will be convenient to treat the variance $D^2\xi_i$ of the random variable ξ_i as the covariance of two identical random variables ξ_i and ξ_i, as suggested formally by (40.3). Thus, the matrix Λ will be called the *covariance matrix*. Our first aim will be to indicate the relation between the properties of the covariance matrix and the properties of the joint probability distribution of those random variables. Also, we shall investigate the different properties of covariance matrices, aiming to discover those which, as will be seen, characterize such matrices (see Theorem 2 of § 42).

LEMMA 1. *The covariance matrix is symmetric.*

In fact, by definition (40.3) we always have $\lambda_{ij} = \lambda_{ji}$.

Suppose we are given a multidimensional random variable $\boldsymbol{\xi} = (\xi_1, \ldots, \xi_n)$ with the expectation

$$E\boldsymbol{\xi} = (E\xi_1, \ldots, E\xi_n) = (m_1, \ldots, m_n) = \boldsymbol{m}$$

and a covariance matrix given by (40.2). What is the expectation and variance of the random variable $\tau_1 = t_1\xi_1 + \ldots + t_n\xi_n$? It can easily be seen that

$$E\tau_1 = t_1 E\xi_1 + \ldots + t_n E\xi_n = t_1 m_1 + \ldots + t_n m_n,$$

and next,

(40.4) $\quad D^2\tau_1 = E(\tau_1 - E\tau_1)^2$
$$= E\big(t_1\xi_1 + \ldots + t_n\xi_n - (t_1 m_1 + \ldots + t_n m_n)\big)^2$$
$$= E\big(t_1(\xi_1 - m_1) + \ldots + t_n(\xi_n - m_n)\big)^2$$
$$= \sum_{i=1}^{n} \sum_{j=1}^{n} t_i t_j E(\xi_i - m_i)(\xi_j - m_j) = \sum_{i=1}^{n} \sum_{j=1}^{n} t_i t_j \lambda_{ij}.$$

The relation which we have just obtained shows two things. First, with the covariance matrix Λ there is related in a natural way a quadratic form of n real variables $t_1, \ldots, t_n$, given by the expression appearing after the last equality sign in (40.4); this quadratic form expresses the variance of the linear combination $t_1\xi_1 + \ldots + t_n\xi_n$ of the random variables $\xi_1, \ldots, \xi_n$. Secondly, as this quadratic form expresses the variance of a certain random variable, it must be non-negative for all the possible values $t_1, \ldots, t_n$. A quadratic form with this property is called *positive semidefinite*. If the quadratic form assumes a positive value unless all numbers $t_1, \ldots, t_n$ are zero, it is called *positive definite*. Thus, we have

LEMMA 2. *If $\Lambda = [\lambda_{ij}]$, where $i, j = 1, 2, \ldots, n$, is the covariance matrix of the random variables $\xi_1, \ldots, \xi_n$, then the quadratic form*

(40.5) $$\sum_{i=1}^{n}\sum_{j=1}^{n} \lambda_{ij} t_i t_j$$

of the variables $t_1, \ldots, t_n$ is positive semidefinite; the same is true for all forms obtained from (40.5) by substituting 0 for some of the variables $t_1, \ldots, t_n$ (indeed, this means that we consider a linear combination of some of the random variables $\xi_1, \ldots, \xi_n$ only).

From the second part of Lemma 2 it follows in particular that every quadratic form of two variables u and v given by the formula

(40.6) $$\lambda_{ii} u^2 + 2\lambda_{ij} uv + \lambda_{jj} v^2$$

($i, j = 1, 2, \ldots, n$, $i \neq j$, we write $2\lambda_{ij}$ instead of $\lambda_{ij} + \lambda_{ji}$ in view of symmetry) must be positive semidefinite. It follows that we must have $\lambda_{ii} \geq 0$ and $\lambda_{jj} \geq 0$. This condition is automatically satisfied if Λ is a covariance matrix, since then λ_{ii} is the variance of the random variable ξ_i. It follows also that if $\lambda_{ii} = \lambda_{jj} = 0$, then $\lambda_{ij} = 0$.

Consider now the case where one of the numbers $\lambda_{ii}, \lambda_{jj}$ is positive. Without loss of generality we may assume that $\lambda_{ii} > 0$. Then the transformation

$$\lambda_{ii} u^2 + 2\lambda_{ij} uv + \lambda_{jj} v^2$$
$$= \lambda_{ii}\left\{u^2 + 2\frac{\lambda_{ij}}{\lambda_{ii}} uv + \frac{\lambda_{jj}}{\lambda_{ii}} v^2\right\} = \lambda_{ii}\left\{\left(u + \frac{\lambda_{ij}}{\lambda_{ii}} v\right)^2 + \frac{\lambda_{ii}\lambda_{jj} - \lambda_{ij}^2}{\lambda_{ii}^2} v^2\right\}$$

shows that the form (40.6) is positive definite or semidefinite depending on whether the discriminant $W = \lambda_{ii}\lambda_{jj} - \lambda_{ij}^2$ is positive or equal to zero.

Thus, we have the following lemma:

LEMMA 3. *If $\Lambda = [\lambda_{ij}]$ is a covariance matrix, then for arbitrary distinct i, j*

(40.7) $$W = \lambda_{ii}\lambda_{jj} - \lambda_{ij}^2 \geq 0$$

i.e.,

(40.8) $$\lambda_{ij}^2 \leq \lambda_{ii}\lambda_{jj}.$$

The last inequality is known as the *Schwarz inequality* for covariances (see problem 3 of §38). It shows that covariances cannot exceed certain limits; in particular, a random variable with variance zero has covariance zero with any other random variable.

The quadratic form (40.5) determines, as we have seen, the variance of the linear combination $\tau_1 = t_1\xi_1 + \ldots + t_n\xi_n$. If this variance is equal to zero then the random variable τ_1 has a distribution concentrated at one point $E\tau = t_1 m_1 + \ldots + t_n m_n$. If, in addition, not all numbers $t_1, \ldots, t_n$ vanish, it follows

VII. CENTRAL LIMIT THEOREMS

that the random variables $\xi_1, \ldots, \xi_n$ satisfy, with probability one, the linear relation

$$t_1 \xi_1 + \ldots + t_n \xi_n = t_1 m_1 + \ldots + t_n m_n$$

or

(40.9) $$t_1(\xi_1 - m_1) + \ldots + t_n(\xi_n - m_n) = 0.$$

Geometrically, this means that with probability one the value of the multidimensional random variable $\xi = (\xi_1, \ldots, \xi_n)$ lies on the hyperplane

$$t_1(x_1 - m_1) + \ldots + t_n(x_n - m_n) = 0,$$

and $t = (t_1, \ldots, t_n)$ is a vector perpendicular to this hyperplane.

Suppose now that relation (40.9) is satisfied with probability one if $t_1, \ldots, t_n$ do not all vanish. Then we can multiply it by $\xi_1 - m_1$ and take the expectation. It follows that we must have $\lambda_{11} t_1 + \ldots + \lambda_{1n} t_n = 0$. In a similar way, multiplying relation (40.9) by $\xi_2 - m_2, \ldots, \xi_n - m_n$ and taking the expectations, we can see that the following system of n linear equations with the unknowns $t_1, \ldots, t_n$ and the coefficients from the covariance matrix

(40.10)
$$\lambda_{11} t_1 + \ldots + \lambda_{1n} t_n = 0,$$
$$\ldots\ldots\ldots\ldots\ldots\ldots\ldots\ldots$$
$$\lambda_{n1} t_1 + \ldots + \lambda_{nn} t_n = 0$$

must be satisfied.

We see that if relation (40.9) is satisfied with probability one, the system of equations (40.10) is also satisfied. Conversely, if system (40.10) is satisfied, then multiplying the first equation by t_1, the second by $t_2, \ldots$, and the last by t_n and adding the left-hand sides, we obtain

$$\sum_{i=1}^{n} \sum_{j=1}^{n} \lambda_{ij} t_i t_j = 0.$$

Thus, the value of the quadratic form expressing the variance of the linear combination $t_1(\xi_1 - m_1) + \ldots + t_n(\xi_n - m_n)$ is zero. This proves that relation (40.9) must hold. Thus, we have shown that relation (40.9) holds with probability one if and only if system (40.10) is satisfied. From algebra we know that if this system has a non-vanishing solution, then its determinant is zero. Moreover, system (40.10) has r linearly independent solutions if and only if its matrix is of the rank $n-r$. In this case, the random variables $\xi_1, \ldots, \xi_n$ satisfy with probability one r independent linear equations, i.e., the whole distribution of the multidimensional random vector $\xi = (\xi_1, \ldots, \xi_n)$ is concentrated on an $(n-r)$-dimensional subspace of an n dimensional Euclidean space. In this case we say that this distribution is $(n-r)$-dimensional. Replacing $n-r$ by r we may formulate the following

THEOREM 1 (Frisch). *The joint probability distribution of random variables $\xi_1, \ldots, \xi_n$ is r-dimensional if and only if the covariance matrix $\Lambda = [\lambda_{ij}]$ is of the rank r.*

This is an analogue of the theorem stated at the beginning of this section, which asserts that a random variable with variance zero is concentrated at one point with probability one. In the particular case where the variances λ_{ii} of the random variables $\xi_1, \ldots, \xi_n$ are all zero, the joint probability distribution is concentrated at one point $(E\xi_1, \ldots, E\xi_n)$. According to formula (40.8), the covariance matrix consists of zeros only, and is of the rank 0, which is consistent with Frisch's theorem.

Suppose now that we are given a system of random variables $\xi = (\xi_1, \ldots, \xi_n)$ with the covariance matrix $\Lambda = [\lambda_{ij}]$. Let us consider new random variables defined by the equations

(40.11)
$$\eta_1 = c_{11}\xi_1 + \ldots + c_{1n}\xi_n,$$
$$\ldots\ldots\ldots\ldots\ldots\ldots\ldots\ldots\ldots$$
$$\eta_n = c_{n1}\xi_1 + \ldots + c_{nn}\xi_n.$$

Let us find the covariance matrix of the random variables $\eta_1, \ldots, \eta_n$. We have

(40.12) $\varkappa_{ij} = \text{Cov}(\eta_i, \eta_j) = E(\eta_i - E\eta_i)(\eta_j - E\eta_j)$
$$= E(c_{i1}\xi_1 + \ldots + c_{in}\xi_n - (c_{i1}E\xi_1 + \ldots + c_{in}E\xi_n)) \times$$
$$\times (c_{j1}\xi_1 + \ldots + c_{jn}\xi_n - (c_{j1}E\xi_1 + \ldots + c_{jn}E\xi_n))$$
$$= E(c_{i1}(\xi_1 - E\xi_1) + \ldots + c_{in}(\xi_n - E\xi_n)) \times$$
$$\times (c_{j1}(\xi_1 - E\xi_1) + \ldots + c_{jn}(\xi_n - E\xi_n))$$
$$= Ec_{i1}(\xi_1 - E\xi_1)(c_{j1}(\xi_1 - E\xi_1) + \ldots + c_{jn}(\xi_n - E\xi_n)) + \ldots$$
$$+ Ec_{in}(\xi_n - E\xi_n)(c_{j1}(\xi_1 - E\xi_1) + \ldots + c_{jn}(\xi_n - E\xi_n))$$
$$= \sum_{s=1}^{n} c_{is} \sum_{k=1}^{n} \lambda_{sk} c_{jk}.$$

Consider now the matrix $C = [c_{ij}]$ of coefficients in equations (40.11) and the matrix $K = [\varkappa_{ij}]$ of covariances of the new random variables $\eta_1, \ldots, \eta_n$. The calculation performed shows that the matrix K is obtained from the matrices C and Λ by multiplication. Indeed, if we denote by $C' = [c'_{ij}]$ the matrix C transposed, i.e. the matrix defined as $c'_{ij} = c_{ji}$, the last sum in (40.12) may be written as follows:

$$\sum_{k=1}^{n} \lambda_{sk} c_{jk} = \sum_{k=1}^{n} \lambda_{sk} c'_{kj}.$$

VII. CENTRAL LIMIT THEOREMS

On the right-hand side we recognize the element appearing in the sth row and jth column of the product AC' of the matrices A and C'. Denoting this element by u_{sj}, we can write for the extreme members of (40.12):

$$\varkappa_{ij} = \sum_{i=1}^{n} c_{is} u_{sj},$$

which shows that $\varkappa_{ij}$ is the element appearing in the ith row and jth column of the matrix $C(AC')$. Thus, we have proved the following important:

THEOREM 2. *If $\xi_1, \ldots, \xi_n$ are random variables with covariance matrix $A = [\lambda_{ij}]$, the random variables $\eta_1, \ldots, \eta_n$ are defined by equations* (40.11) *with the matrix of coefficients $C = [c_{ij}]$, and C' denotes the transpose of the matrix C, then the covariance matrix $K = [\varkappa_{ij}]$ of the random variables $\eta_1, \ldots, \eta_n$ is given by*
(40.13) $$K = CAC'.$$

REMARK 1. Formula (40.13) remains valid also if the number of the variables η_i is different from the number of the variables ξ_j; then the matrices C and C' are rectangular.

REMARK 2. If the matrix C of the coefficients c_{ij} in equations (40.11) is not singular, i.e. if the determinant $|C|$ of this matrix is different from zero, then—as is known from algebra—multiplication by C does not change the rank of the matrix. From this remark, and from Theorem 1 it follows that the dimension of the joint probability distribution of the random vector $\eta_1, \ldots, \eta_n$ is the same as the dimension of the joint probability distribution of the random vector $\xi_1, \ldots, \xi_n$. In other words, the transformation (40.11) with the non-singular matrix C does not change the dimension of the distribution.

In § 28 we proved the additivity of the variance for independent random variables. Below we show that a similar property holds for the addition of the covariances of independent random vectors. Indeed, suppose that $\boldsymbol{\xi} = (\xi_1, \xi_2)$ and $\boldsymbol{\eta} = (\eta_1, \eta_2)$ are two independent random vectors, and let $\boldsymbol{\zeta} = (\zeta_1, \zeta_2)$ be their sum, i.e. let $\boldsymbol{\zeta} = (\xi_1 + \eta_1, \xi_2 + \eta_2)$. We then have

(40.14) $$\mathrm{Cov}(\zeta_1, \zeta_2) = \mathrm{Cov}(\xi_1, \xi_2) + \mathrm{Cov}(\eta_1, \eta_2).$$

Indeed, we have the following relations

$$\begin{aligned}\mathrm{Cov}(\zeta_1, \zeta_2) &= E(\zeta_1 - E\zeta_1)(\zeta_2 - E\zeta_2) \\ &= E\big(\xi_1 + \eta_1 - E(\xi_1 + \eta_1)\big)\big(\xi_2 + \eta_2 - E(\xi_2 + \eta_2)\big) \\ &= E\big((\xi_1 - E\xi_1) + (\eta_1 - E\eta_1)\big)\big((\xi_2 - E\xi_2) + (\eta_2 - E\eta_2)\big) \\ &= E(\xi_1 - E\xi_1)(\xi_2 - E\xi_2) + E(\xi_1 - E\xi_1)(\eta_2 - E\eta_2) \\ &\quad + E(\eta_1 - E\eta_1)(\xi_2 - E\xi_2) + E(\eta_1 - E\eta_1)(\eta_2 - E\eta_2).\end{aligned}$$

The random variables ξ_1 and η_2 are independent by assumption; thus, the second term in the last sum vanishes. Similarly, in view of the independence of η_1 and ξ_2 the third term vanishes. The first term equals $\mathrm{Cov}(\xi_1, \xi_2)$ and the fourth equals $\mathrm{Cov}(\eta_1, \eta_2)$, which proves relation (40.14).

The result which we have just obtained may be formulated as follows:

THEOREM 3. *If $\xi = (\xi_1, \ldots, \xi_n)$ is a random vector with a covariance matrix $\Lambda = [\lambda_{ij}]$, $\eta = (\eta_1, \ldots, \eta_n)$ is a random vector with a covariance matrix $K = [\varkappa_{ij}]$, vectors ξ and η are stochastically independent, and $\zeta = (\zeta_1, \ldots, \zeta_n)$ is their sum, then the covariance matrix $U = [u_{ij}]$ of the random vector ζ equals to the sum of the covariance matrices of vectors ξ and η:*

(40.15) $$U = \Lambda + K$$

(by the sum of the matrices Λ and K we mean the matrix U whose elements are equal to the sums of the corresponding elements of the matrices Λ and K, i.e. $u_{ij} = \lambda_{ij} + \varkappa_{ij}$).

This theorem shows that there is a complete analogy between the variance and the covariance matrix. Let us also note without proof the following relation of this series:

THEOREM 4. *If the random vector $\xi = (\xi_1, \ldots, \xi_n)$ has covariance matrix $\Lambda = [\lambda_{ij}]$, then the vector $a\xi = (a\xi_1, \ldots, a\xi_n)$ has a covariance matrix $a^2 \Lambda = [a^2 \lambda_{ij}]$.*

Problems

The problems presented below contain complementary material concerning linear regression (called also *regression of the second kind*) and linear correlation.

1. a) Prove that if ξ_1 and ξ_2 are random variables with expectations $m_1 = E\xi_1$ and $m_2 = E\xi_2$ and a covariance matrix

$$\Lambda = \begin{bmatrix} \lambda_{11} & \lambda_{12} \\ \lambda_{21} & \lambda_{22} \end{bmatrix},$$

then the expectation $E[(\xi_1 - m_1) + \alpha(\xi_2 - m_2)]^2$ attains its minimum equal to

$$\frac{|\Lambda|}{\lambda_{22}} = \frac{\lambda_{11}\lambda_{22} - \lambda_{12}^2}{\lambda_{22}}$$

for $\alpha = -\lambda_{12}/\lambda_{22}$.

The line with the equation

$$x_1 - m_1 = \frac{\lambda_{12}}{\lambda_{22}}(x_2 - m_2)$$

is called the *regression* of the random variable ξ_1 on the random variable ξ_2. The number

$$\varrho = \frac{\lambda_{12}}{\sqrt{\lambda_{11}\lambda_{22}}}$$

VII. CENTRAL LIMIT THEOREMS

is called the *correlation coefficient* between ξ_1 and ξ_2. The number

$$\beta_{1;2} = \frac{\lambda_{12}}{\lambda_{22}}$$

is called the *regression coefficient* of ξ_1 on ξ_2, and the number

$$\sigma_{1;2}^2 = \frac{|\Lambda|}{\lambda_{22}}$$

is called the *residual variance*.

b) Prove that the residual variance $\sigma_{1;2}^2$ can be expressed by the correlation coefficient ϱ and the variance λ_{11} of ξ_1 as follows:

$$\sigma_{1;2}^2 = \lambda_{11}(1-\varrho^2).$$

c) Prove that the random variables $\xi_2 - m_2$ and $(\xi_1 - m_1) - \lambda_{12}(\xi_2 - m_2)/\lambda_{22}$ are uncorrelated (the remainder of $\xi_1 - m_1$ after subtracting from it the linear function of $\xi_2 - m_2$ which gives its best approximation is uncorrelated with $\xi_2 - m_2$).

d) Prove that the expectation $E(\xi_1 + \alpha \xi_2 + \beta)^2$ attains its minimum for

$$\alpha = -\frac{\lambda_{12}}{\lambda_{22}}, \quad \beta = \frac{\lambda_{12} m_2 - \lambda_{22} m_1}{\lambda_{22}},$$

and verify that the equation

$$x_1 = \frac{\lambda_{12}}{\lambda_{22}} x_2 - \frac{\lambda_{12} m_2 - \lambda_{22} m_1}{\lambda_{22}}$$

coincides with the regression line of ξ_1 on ξ_2.

2. a) Prove that if $\xi_0, \xi_1, \ldots, \xi_n$ are random variables with expectations $m_0, m_1, \ldots, m_n$ and a covariance matrix $\Lambda = [\lambda_{ij}]$, $i, j = 0, 1, \ldots, n$, then the system of coefficients for which the expectation

$$E[(\xi_0 - m_0)^2 + \alpha_1(\xi_1 - m_1)^2 + \ldots + \alpha_n(\xi_n - m_n)^2]$$

attains its minimum equal to $|\Lambda|/|\Lambda_{11}|$ is given by the formula

(40.16) $$\alpha_i = \frac{\Lambda_{i0}}{\Lambda_{00}}, \quad i = 1, 2, \ldots, n;$$

here Λ_{ij} denotes the algebraic complement of the element λ_{ij} of the matrix Λ, and $|\Lambda|$ is the determinant of the matrix Λ.

The hyperplane with the equation

$$x_0 - m_0 = -\alpha_1(x_1 - m_1) - \ldots - \alpha_n(x_n - m_n)$$

with $\alpha_1, \ldots, \alpha_n$ given by (40.16) is called the *regression hyperplane* of ξ_0 on $\xi_1, \xi_2, \ldots, \xi_n$. The coefficients $\beta_{0;1,2,\ldots,n} = -\Lambda_{i0}/\Lambda_{00}$ are called the *regression coefficients* of ξ_0 on $\xi_1, \ldots, \xi_n$. The number

$$r_{0;1,2,\ldots,n} = \sqrt{1 - \frac{|\Lambda|}{\lambda_{00}\Lambda_{00}}}$$

is called the *multiple correlation coefficient* between ξ_0 and the random variables $\xi_1, \ldots, \xi_n$. The number

$$\sigma_{0;1,2,\ldots,n}^2 = \frac{|\Lambda|}{\Lambda_{00}}$$

is called the *residual variance*.

b) Prove that the expectation
$$E(\xi_0 + \alpha_1 \xi_1 + \ldots + \alpha_n \xi_n + \beta)^2$$
attains its minimum for
$$\alpha_i = \frac{\Lambda_{i0}}{\Lambda_{00}}, \quad i = 1, 2, \ldots, n,$$
$$\beta = -\frac{\Lambda_{00} m_0 + \Lambda_{10} m_1 + \ldots + \Lambda_{n0} m_n}{\Lambda_{00}},$$
and verify that the equation
$$x_0 = -\alpha_1 x_1 - \ldots - \alpha_n x_n - \beta,$$
where $\alpha_1, \ldots, \alpha_n$ and β are defined above, coincides with the equation of the regression hyperplane of ξ_0 on the system $\xi_1, \ldots, \xi_n$.

c) Prove that for an arbitrary $k = 1, 2, \ldots, n$ the random variables
$$\xi_0 - m_0 + \frac{\Lambda_{10}}{\Lambda_{00}}(\xi_1 - m_1) + \ldots + \frac{\Lambda_{n0}}{\Lambda_{00}}(\xi_n - m_n) \quad \text{and} \quad \xi_k - m_k$$
are uncorrelated (the remainder of $\xi_0 - m_0$ after subtracting from it that linear combination of random variables $\xi_1 - m_1, \ldots, \xi_n - m_n$ which gives best approximation is uncorrelated with these random variables).

H i n t. The expectation of the above random variables is proportional to the sum of the products of the elements of the kth row of the covariance matrix Λ by the algebraic complements of the elements of the first row of the matrix Λ.

3. Let ξ_1, ξ_2 be a pair of random variables with expectations m_1 and m_2 and a non-degenerate covariance matrix
$$\Lambda = \begin{bmatrix} \lambda_{11} & \lambda_{12} \\ \lambda_{21} & \lambda_{22} \end{bmatrix}.$$

Find the coefficients α_1 and α_2 for which
(40.17) $$\alpha_1^2 + \alpha_2^2 = 1$$
and the expectation
(40.18) $$E(\alpha_1(\xi_1 - m_1) + \alpha_2(\xi_2 - m_2))^2 = \lambda_{11}\alpha_1^2 + 2\lambda_{12}\alpha_1\alpha_2 + \lambda_{22}\alpha_2^2$$
attains its extremum.

a) Applying the method of Lagrange multipliers to find the extremum of functions of several variables under side conditions we find that in order that α_1 and α_2 satisfying (40.17), give the extremum of (40.18) it is necessary that (40.17) hold, and that the partial derivatives with respect to α_1, α_2 and the multiplier λ of the function
$$\lambda_{11}\alpha_1^2 + 2\lambda_{12}\alpha_1\alpha_2 + \lambda_{22}\alpha_2^2 - \lambda(\alpha_1^2 + \alpha_2^2 - 1)$$
vanish. By considering partial derivatives with respect to α_1 and α_2, we obtain the system of homogeneous equations
(40.19)
$$(\lambda_{11} - \lambda)\alpha_1 + \lambda_{12}\alpha_2 = 0,$$
$$\lambda_{21}\alpha_1 + (\lambda_{22} - \lambda)\lambda_2 = 0.$$
This system can have a non-zero solution α_1, α_2 only if its determinant vanishes. Thus, λ must satisfy the secular equation
$$\begin{vmatrix} \lambda_{11} - \lambda & \lambda_{12} \\ \lambda_{21} & \lambda_{22} - \lambda \end{vmatrix} = 0.$$

VII. CENTRAL LIMIT THEOREMS

b) If λ is a root of the secular equation, then α_1 and α_2 satisfy the proportion

$$\alpha_1 : \alpha_2 = (\lambda_{22} - \lambda) : \lambda_{12}.$$

c) If λ' and λ'' are distinct roots of the secular equation, then the vectors $(\lambda_{22} - \lambda', \lambda_{12})$ and $(\lambda_{22} - \lambda'', \lambda_{12})$, whence also the vectors (α'_1, α'_2) and (α''_1, α''_2), equal to the solutions of the homogeneous equation (40.19) corresponding to these roots, are perpendicular.

This can easily be checked: we have

$$\lambda' + \lambda'' = \lambda_{11} + \lambda_{22}, \quad \lambda'\lambda'' = \lambda_{11}\lambda_{22} - \lambda_{12}^2,$$

hence

$$(\lambda_{22} - \lambda')(\lambda_{22} - \lambda'') + \lambda_{12}^2 = \lambda_{22}^2 + \lambda_{12}^2 - \lambda_{22}(\lambda' + \lambda'') + \lambda'\lambda''$$
$$= \lambda_{22}^2 + \lambda_{12}^2 - \lambda_{22}(\lambda_{11} + \lambda_{22}) + \lambda_{11}\lambda_{22} - \lambda_{12}^2 = 0.$$

d) If λ is a root of the secular equation and α_1 and α_2 are solutions of system (40.19) corresponding to these roots, satisfying (40.17), then the expectation of (40.18) equals λ.

Indeed, we have

$$\lambda_{11}\alpha_1^2 + 2\lambda_{12}\alpha_1\alpha_2 + \lambda_{22}\alpha_2^2$$
$$= \alpha_1(\lambda_{11}\alpha_1 + \lambda_{12}\alpha_2) + \alpha_2(\lambda_{12}\alpha_1 + \lambda_{22}\alpha_2)$$
$$= \alpha_1[(\lambda_{11} - \lambda)\alpha_1 + \lambda_{12}\alpha_2] + \alpha_2[\lambda_{12}\alpha_1 + (\lambda_{22} - \lambda)\alpha_2] + \lambda(\alpha_1^2 + \alpha_2^2) = \lambda.$$

e) If λ' and λ'' are distinct roots of the secular equation and (α'_1, α'_2) and (α''_1, α''_2) are two solutions of the homogeneous system (40.19) corresponding to these roots, then the random variables $\alpha'_1(\xi_1 - m_1) + \alpha'_2(\xi_2 - m_2)$ and $\alpha''_1(\xi_1 - m_1) + \alpha''_2(\xi_2 - m_2)$ are uncorrelated.

Indeed, we have

$$E[\alpha'_1(\xi_1 - m_1) + \alpha'_2(\xi_2 - m_2)][\alpha''_1(\xi_1 - m_1) + \alpha''_2(\xi_2 - m_2)]$$
$$= \alpha'_1[\lambda_{11}\alpha''_1 + \lambda_{12}\alpha''_2] + \alpha'_2[\lambda_{12}\alpha''_1 + \lambda_{22}\alpha''_2]$$
$$= \alpha'_1\lambda''\alpha''_1 + \alpha'_2\lambda''\alpha''_2 = \lambda''(\alpha'_1\alpha''_1 + \alpha'_2\alpha''_2) = 0.$$

In passing to the third equality from the end we used the fact that equations (40.19) are satisfied, and in passing to the last equality we used the fact that, in view of c), the vectors (α'_1, α'_2) and (α''_1, α''_2) are orthogonal.

The orthogonal lines

$$\alpha'_1(x_1 - m_1) + \alpha'_2(x_2 - m_2) = 0,$$
$$\alpha''_1(x_1 - m_1) + \alpha''_2(x_2 - m_2) = 0$$

are called the *correlation axes*.

We saw in e) that the distances of the random point (ξ_1, ξ_2) from the correlation axis, given by

$$\alpha'_1(\xi_1 - m_1) + \alpha'_2(\xi_2 - m_2) \quad \text{and} \quad \alpha''_1(\xi_1 - m_1) + \alpha''_2(\xi_2 - m_2)$$

are uncorrelated random variables with variances λ' and λ''. Thus, the transformation of the plane given by the equations

$$y_1 = \alpha'_1(x_1 - m_1) + \alpha'_2(x_2 - m_2),$$
$$y_2 = \alpha''_1(x_1 - m_1) + \alpha''_2(x_2 - m_2)$$

is called the *transformation to uncorrelated coordinates* (or *variables*).

The ellipse with the equation

$$\frac{y_1^2}{\lambda'} + \frac{y_2^2}{\lambda''} = 1$$

which has its centre at the point (m_1, m_2) and orthogonal axes with half lengths $\sqrt{\lambda'}$ and $\sqrt{\lambda''}$ parallel to the vectors (α_1', α_2') and (α_1'', α_2''), is called the *correlation ellipse*.

4. What we said in problem 3 about the axes and ellipse of correlation for two random variables extends to arbitrary systems of random variables.

Let $\xi_1, \ldots, \xi_n$ be a system of random variables with expectations $m_1, \ldots, m_n$ and a non-singular covariance matrix

$$\varLambda = [\lambda_{ij}], \quad i, j = 1, \ldots, n.$$

Then the values of the coefficients satisfying the side condition

$$\alpha_1^2 + \ldots + \alpha_n^2 = 1,$$

for which the expectation

$$E(\alpha_1(\xi_1 - m_1) + \ldots + \alpha_n(\xi_n - m_n))^2$$

can have an extremum, is determined for λ satisfying the secular equation

$$\begin{vmatrix} \lambda_{11} - \lambda & \lambda_{12} & \ldots & \lambda_{1n} \\ \lambda_{21} & \lambda_{22} - \lambda & \ldots & \lambda_{2n} \\ \vdots & & & \\ \lambda_{n1} & \lambda_{n2} & \ldots & \lambda_{nn} - \lambda \end{vmatrix} = 0$$

from the system of homogeneous linear equations

$$(\lambda_{11} - \lambda)\alpha_1 + \lambda_{12}\alpha_2 + \ldots + \lambda_{1n}\alpha_n = 0,$$
$$\lambda_{21}\alpha_1 + (\lambda_{22} - \lambda)\alpha_2 + \ldots + \lambda_{2n}\alpha_n = 0,$$
$$\ldots$$
$$\lambda_{n1}\alpha_1 + \lambda_{n2}\alpha_2 + \ldots + (\lambda_{nn} - \lambda)\alpha_n = 0.$$

If $\lambda^{(1)}, \lambda^{(2)}, \ldots, \lambda^{(n)}$ are distinct roots of the secular equation, and the vectors $(\alpha_1^{(1)}, \ldots, \alpha_n^{(1)})$, $\ldots, (\alpha_1^{(n)}, \ldots, \alpha_n^{(n)})$ are solutions of the homogeneous equations satisfying side condition and corresponding to these roots, then the vectors in question are orthogonal, and the random variables $\alpha_1^{(1)}(\xi_1 - m_1) + \ldots + \alpha_n^{(1)}(\xi_n - m_n), \ldots, \alpha_1^{(n)}(\xi_1 - m_1) + \ldots + \alpha_n^{(n)}(\xi_n - m_n)$ are uncorrelated and have variances equal to $\lambda^{(1)}, \ldots, \lambda^{(n)}$.

The hyperplanes

$$\alpha_1^{(1)}(x_1 - m_1) + \ldots + \alpha_n^{(1)}(x_n - m_n) = 0,$$
$$\ldots$$
$$\alpha_1^{(n)}(x_1 - m_1) + \ldots + \alpha_n^{(n)}(x_n - m_n) = 0$$

are called *correlation hyperplanes*, and the transformation

$$y_1 = \alpha_1^{(1)}(x_1 - m_1) + \ldots + \alpha_n^{(1)}(x_n - m_n),$$
$$\ldots$$
$$v = \alpha_1^{(n)}(x_1 - m_1) + \ldots + \alpha_n^{(n)}(x_n - m_n)$$

is called the *transformation to orthogonal (uncorrelated) coordinates*. The hyperellipsoid in the space of points $(x_1, \ldots, x_n)$ defined by the equation

$$\frac{y_1^2}{\lambda^{(1)}} + \ldots + \frac{y_n^2}{\lambda^{(n)}} = 1$$

which has the centre at the point $(m_1, \ldots, m_n)$ and orthogonal axes parallel to the vectors $(\alpha_1^{(j)}, \ldots, \alpha_n^{(j)})$ with half lengths $\sqrt{\lambda^{(j)}}$, $j = 1, \ldots, n$, is called the *correlation hyperellipsoid*.

5. Let random variables ξ, η, ζ have zero expectations and the covariance matrix

$$\varLambda = \begin{bmatrix} 4 & 1 & -1 \\ 1 & 4 & 2 \\ -1 & 2 & 4 \end{bmatrix}.$$

a) Find the roots of the secular equation and the correlation planes.

A n s w e r: $\lambda' = 3 - \sqrt{3}$, $\lambda'' = 3 + \sqrt{3}$, $\lambda''' = 6$, $(3+\sqrt{3})x+(-3-2\sqrt{3})y+(3+2\sqrt{3})z = 0$, $(3-\sqrt{3})x+(-3+2\sqrt{3})y+(3-2\sqrt{3})z = 0$, $y+z = 0$.

b) Find the regression lines of each of these random variables on the remaining two and compute the residual variances.

A n s w e r: regression lines: $x = \frac{1}{2}y - \frac{1}{2}z$, $y = \frac{2}{5}x + \frac{3}{5}z$, $z = -\frac{2}{5}x + \frac{3}{5}y$, residual variances: 3, 12/5, 12/5.

§41. CHARACTERISTIC FUNCTIONS OF MULTIDIMENSIONAL RANDOM VARIABLES

In § 36 we discussed the characteristic functions of one dimensional random variables, and in subsequent paragraphs we showed their use in studying limit distributions and in studying some important properties of probability distributions. In the sequel we shall have to deal with the multidimensional version of the normal distribution, and the multidimensional version of the central limit theorem. Here again the tool of characteristic functions is very convenient. Thus, in this section we shall put together the most important information concerning the characteristic function of multidimensional probability distributions. We shall omit most of the proofs, restricting ourselves to definitions and theorems.

DEFINITION 1. By the *characteristic function* of an n-dimensional random vector $\boldsymbol{\xi} = (\xi_1, \ldots, \xi_n)$ we mean the function of n real variables (i.e., the function of an n-dimensional vector $\boldsymbol{t} = (t_1, \ldots, t_n)$) defined by the following equation

(41.1) $\qquad \varphi_{\boldsymbol{\xi}}(\boldsymbol{t}) = \varphi_{(\xi_1, \ldots, \xi_n)}(t_1, \ldots, t_n) = E\exp\bigl(i(t_1\xi_1 + \ldots + t_n\xi_n)\bigr).$

If we agree to denote by $\boldsymbol{t}\boldsymbol{\xi}$ the scalar product $t_1\xi_1 + \ldots + t_n\xi_n$ of the vectors $\boldsymbol{t} = (t_1, \ldots, t_n)$ and $\boldsymbol{\xi} = (\xi_1, \ldots, \xi_n)$ we may write (41.1) in a form graphically identical with the formula used in Definition 1 of § 36: $\varphi_{\boldsymbol{\xi}}(\boldsymbol{t}) = Ee^{i\boldsymbol{t}\boldsymbol{\xi}}$.

Here are some of the properties of characteristic functions:

THEOREM 1. *If $\varphi(t_1, ..., t_n)$ is the characteristic function of a random vector* $\xi = (\xi_1, ..., \xi_n)$, *then*

(a) $\varphi(0, ..., 0) = 1$;

(b) $|\varphi(t_1, ..., t_n)| \leq 1$ *for any* $t_1, ..., t_n$;

(c) $\varphi(t_1, ..., t_n) = \overline{\varphi(-t_1, ..., -t_n)}$, *where the bar denotes the complex conjugate;*

(d) $\varphi(t_1, ..., t_n)$ *is uniformly continuous in the whole n-dimensional space of the variables* $t_1, ..., t_n$.

If we have the characteristic function $\varphi(t_1, ..., t_n)$ of the random vector $\xi = (\xi_1, ..., \xi_n)$, we can easily find the characteristic function of any k-dimensional vector $(k \leq n)$ resulting from the vector ξ by omitting from it $n-k$ of its components; in order to obtain such a characteristic function it suffices to substitute value 0 for all the variables t_s which correspond to the coordinates not included in the k-dimensional vector under consideration. Thus, $\varphi(t_1, 0, ..., 0)$ is the characteristic function of the first coordinate.

Now we give a theorem which relates the independence of the components of vector $\xi = (\xi_1, ..., \xi_n)$ and the form of its characteristic function.

THEOREM 2. *The components of a random vector* $\xi = (\xi_1, ..., \xi_n)$ *are independent random variables if and only if its characteristic function equals the product of the characteristic functions of the components, i.e. if*

(41.2) $$\varphi_{(\xi_1, ..., \xi_n)}(t_1, ..., t_n) = \varphi_{\xi_1}(t_1) ... \varphi_{\xi_n}(t_n).$$

As in the case of one-dimensional random variables, the moments of a random vector can be expressed in terms of the derivatives of the characteristic function:

THEOREM 3. *If $\varphi(t_1, ..., t_n)$ is the characteristic function of a random vector* $\xi = (\xi_1, ..., \xi_n)$ *and there exists a finite moment*

$$\mu_{k_1, ..., k_n} = E\xi_1^{k_1} ... \xi_n^{k_n},$$

then there exists a continuous partial derivative of the order $k_1 + ... + k_n$ of the function $\varphi(t_1, ..., t_n)$ and

(41.3) $$\mu_{k_1, ..., k_n} = (-i)^{k_1 + ... + k_n} \left[\frac{\partial^{k_1 + ... + k_n} \varphi(t_1, ..., t_n)}{\partial t_1^{k_1} ... \partial t_n^{k_n}} \right]_{(t_1, ..., t_n) = (0, ..., 0)}.$$

Formula (36.5) extends to the case of the addition of independent random vectors:

VII. CENTRAL LIMIT THEOREMS

THEOREM 4. *If $\xi = (\xi_1, \ldots, \xi_n)$ is a random vector with a characteristic function $\varphi_\xi(t_1, \ldots, t_n)$, and $\eta = (\eta_1, \ldots, \eta_n)$ is a random vector with a characteristic function $\varphi_\eta(t_1, \ldots, t_n)$, and if the vectors ξ and η are independent, then the characteristic function of their sum $\zeta = \xi + \eta$ is*

(41.4) $\qquad \varphi_\zeta(t_1, \ldots, t_n) = \varphi_\xi(t_1, \ldots, t_n) \varphi_\eta(t_1, \ldots, t_n).$

Indeed, we have the following equalities:

$$\varphi_\zeta(t_1, \ldots, t_n) = E\exp[i(t_1\zeta_1 + \ldots + t_n\zeta_n)]$$
$$= E\exp\{i[t_1(\xi_1+\eta_1) + \ldots + t_n(\xi_n+\eta_n)]\}$$
$$= E\exp[i(t_1\xi_1 + \ldots + t_n\xi_n)]\exp[i(t_1\eta_1 + \ldots + t_n\eta_n)].$$

In view of the independence of the vectors ξ and η the random variables $\exp[i(t_1\xi_1 + \ldots + t_n\xi_n)]$ and $\exp[i(t_1\eta_1 + \ldots + t_n\eta_n)]$ appearing under the last expectation sign are independent, hence the expectation of their product equals to the product of their expectations. This proves Theorem 4.

The usefulness of characteristic functions in studying probability distributions lies mainly in the fact that there exists a one-to-one correspondence between characteristic functions and probability distributions, and that there exists a relation between convergence of probability distributions and convergence of characteristic functions. The theorems which assert these properties for one-dimensional distributions remain true in the multidimensional case almost without change of formulation:

THEOREM 5. *The probability distribution function $F(x_1, \ldots, x_n)$ of an n-dimensional distribution is determined uniquely by its characteristic function.*

THEOREM 6. *If the n-dimensional distribution functions $F_1(x_1, \ldots, x_n)$, $F_2(x_1, \ldots, x_n), \ldots$ converge to the distribution function $F(x_1, \ldots, x_n)$ at every point of continuity of the latter then their characteristic functions $\varphi_1(t_1, \ldots, t_n)$, $\varphi_2(t_1, \ldots, t_n), \ldots$ converge at every point $(t_1, \ldots, t_n)$ to the characteristic function of the distribution $F(x_1, \ldots, x_n)$. If the characteristic functions $\varphi_1(t_1, \ldots, t_n)$, $\varphi_2(t_1, \ldots, t_n), \ldots$ of certain n-dimensional distributions converge at every point $(t_1, \ldots, t_n)$ to a certain continuous function $\varphi(t_1, \ldots, t_n)$, then this function is the characteristic function of a certain probability distribution, and the corresponding distribution function is, at every point of continuity, the limit of the distribution functions corresponding to the characteristic functions $\varphi_1, \varphi_2, \ldots$*

Finally, we give a few easily verifiable formulas, which will be used in the sequel.

THEOREM 7. *If $\varphi(t_1, \ldots, t_n)$ is the characteristic function of a random vector $\xi = (\xi_1, \ldots, \xi_n)$, then*

(a) $\exp[i(b_1 t_1 + \ldots + b_n t_n)]\varphi(a_1 t_1, \ldots, a_n t_n)$ is the characteristic function of a random vector $(a_1 \xi_1 + b_1, \ldots, a_n \xi_n + b_n)$;

(b) $\varphi(t, t, \ldots, t)$ is the characteristic function of the sum $\xi_1 + \ldots + \xi_n$, and, more generally,

(c) $\varphi(t_1 t, t_2 t, \ldots, t_n t)$ is the characteristic function of the linear combination $t_1 \xi_1 + \ldots + t_n \xi_n$.

§ 42. MULTIDIMENSIONAL NORMAL DISTRIBUTION

We have already discussed the one-dimensional normal distribution; it is the distribution with the density

$$f(x; m, \sigma) = \frac{1}{\sigma \sqrt{2\pi}} e^{-(x-m)^2/2\sigma^2},$$

and the characteristic function

$$\varphi(t; m, \sigma) = e^{-imt - \sigma^2 t^2/2}.$$

The parameters m and $\sigma > 0$ represent the expectation and the standard deviation of this distribution.

In the sequel we shall discuss the multidimensional analogue of this distribution. We shall see that, as regard the limit distributions of sums of independent random vectors, it plays the same role as the normal distribution for sums of independent random variables. We shall also use the properties of the multidimensional normal distribution in discussing tests and distributions important in statistics.

Suppose that $\xi_1, \ldots, \xi_n$ are independent random variables, each with a normal distribution with expectation m_i and standard deviation $\sigma_i > 0$. Then the probability density of the random vector $\xi = (\xi_1, \ldots, \xi_n)$ is

$$(42.1) \quad f(x_1, \ldots, x_n) = \frac{1}{(\sqrt{2\pi})^n \sigma_1 \ldots \sigma_n} \exp\left[-\frac{1}{2} \sum_{i=1}^n \left(\frac{x_i - m_i}{\sigma_i}\right)^2\right],$$

and, according to Theorem 2 of § 41, the characteristic function of this distribution has the form

$$(42.2) \quad \varphi(t_1, \ldots, t_n) = \exp\left[i(m_1 t_1 + \ldots + m_n t_n) - \frac{1}{2}(\sigma_1^2 t_1^2 + \ldots + \sigma_n^2 t_n^2)\right].$$

Let us look more closely at the exponent of the right-hand side of formula (42.1). We recognize in it the quadratic form of the variables $y_i = x_i - m_i$ (multiplied by $-1/2$) with coefficients given by the diagonal matrix

$$M = \begin{bmatrix} \frac{1}{\sigma_1^2} & 0 & \cdots & 0 \\ 0 & \frac{1}{\sigma_2^2} & \cdots & 0 \\ \cdots & \cdots & \cdots & \cdots \\ 0 & 0 & \cdots & \frac{1}{\sigma_n^2} \end{bmatrix}.$$

Let us compare it with the exponent of the right-hand side of formula (42.2). We see in it the term $i(m_1 t_1 + \ldots + t_n m_n)$ depending on the expectations of the components of the random vector and the quadratic form

(42.3) $$\sigma_1^2 t_1^2 + \ldots + \sigma_n^2 t_n^2$$

(multiplied by $-1/2$) of the real variables $t_1, \ldots, t_n$ with the following diagonal matrix of the coefficients Λ:

(42.4) $$\Lambda = \begin{bmatrix} \sigma_1^2 & 0 & \cdots & 0 \\ 0 & \sigma_2^2 & \cdots & 0 \\ \cdots & \cdots & \cdots & \cdots \\ 0 & 0 & \cdots & \sigma_n^2 \end{bmatrix}.$$

Clearly, the quadratic form (42.3) is positive definite. In view of the independence of the components $\xi_1, \ldots, \xi_n$, we recognize in the matrix Λ the covariance matrix of the random vector ξ. Next, we see that the matrix M is the inverse of the matrix Λ, in the sense of matrix multiplication. Indeed, we have $M\Lambda = I$, where I is the unit matrix:

$$I = \begin{bmatrix} 1 & 0 & \cdots & 0 \\ 0 & 1 & \cdots & 0 \\ \cdots & \cdots & \cdots & \cdots \\ 0 & 0 & \cdots & 1 \end{bmatrix}.$$

Finally, in the product of standard deviations $\sigma_1, \ldots, \sigma_n$ appearing in the denominator on the right-hand side of formula (42.1) we recognize the square root of the determinant of the matrix Λ:

$$\sigma_1 \ldots \sigma_n = \sqrt{|\Lambda|}.$$

The question arises whether the relations which we have observed can be generalized to a larger class of distributions, not restricted to distributions of random vectors with independent components having normal one-dimensional distributions. The following theorem gives a positive answer to this question:

THEOREM 1. *Let*

$$(42.5) \quad \sum_{i=1}^{n}\sum_{j=1}^{n} a_{ij} y_i y_j$$

be a positive definite quadratic form of the variables $y_1, \ldots, y_n$ *with a symmetric matrix* $A = [a_{ij}]$ *of the coefficients* a_{ij}. *Then for an arbitrary vector* $\mathbf{m} = (m_1, \ldots, m_n)$

(a) *the function*

$$(42.6) \quad f(x_1, \ldots, x_n) = \frac{\sqrt{|A|}}{(\sqrt{2\pi})^n} \exp\left\{-\frac{1}{2}\sum_{i=1}^{n}\sum_{j=1}^{n} a_{ij}(x_i - m_i)(x_j - m_j)\right\}$$

is an n-dimensional probability density; *if, moreover,* $\Lambda = [\lambda_{ij}]$ *is the matrix inverse to* A, *then*

(b) *the function*

$$(42.7) \quad \varphi(t_1, \ldots, t_n) = \exp\left\{i(m_1 t_1 + \ldots + m_n t_n) - \frac{1}{2}\sum_{i=1}^{n}\sum_{j=1}^{n} \lambda_{ij} t_i t_j\right\}$$

is the characteristic function of an n-dimensional normal distribution with density (42.6), *and*

(c) *the matrix* Λ *is the covariance matrix of this distribution, and the vector* $\mathbf{m} = (m_1, \ldots, m_n)$ *is the vector of the expectations of the components.*

PROOF. To prove point (a) it suffices to show that

$$I = \int \ldots \int f(x_1, \ldots, x_n) dx_1 \ldots dx_n = 1.$$

By substituting $y_i = x_i - m_i$, or $x_i = y_i + m_i$, whose Jacobian equals one, we see that

$$(42.8) \quad I = \frac{\sqrt{|A|}}{\sqrt{(2\pi)^n}} \int \ldots \int \exp\left\{-\frac{1}{2}\sum_{i=1}^{n}\sum_{j=1}^{n} a_{ij} y_i y_j\right\} dy_1 \ldots dy_n = \frac{\sqrt{|A|}}{\sqrt{(2\pi)^n}} J$$

We have denoted the last integral by J. Now we compute its value. Let us introduce new variables by orthogonal transformation:

$$(42.9) \quad \begin{aligned} u_1 &= c_{11} y_1 + \ldots + c_{1n} y_n, \\ & \ldots\ldots\ldots\ldots\ldots\ldots\ldots\ldots \\ u_n &= c_{n1} y_1 + \ldots + c_{nn} y_n \end{aligned}$$

(the transformation is orthogonal if for every i we have $c_{i1}^2 + \ldots + c_{in}^2 = 1$ and for any distinct i, j we have $c_{i1} c_{j1} + \ldots + c_{in} c_{jn} = 0$). Then the inverse transformation has the transpose of $C = [c_{ij}]$ as its matrix; thus we have

$$(42.10) \quad \begin{aligned} y_1 &= c'_{11} u_1 + \ldots + c'_{1n} u_n, \\ & \ldots\ldots\ldots\ldots\ldots\ldots\ldots\ldots \\ y_n &= c'_{n1} u_1 + \ldots + c'_{nn} u_n, \end{aligned}$$

where the matrix $C' = [c'_{ij}]$ is defined as $c'_{ij} = c_{ji}$. The calculations essentially identical with those which we presented in the proof of Theorem 2 of § 40 show that the quadratic form in the exponent of the integrand in the integral J will assume the form

(42.11)
$$\sum_{i=1}^{n}\sum_{j=1}^{n}\varkappa_{ij}u_{i}u_{j}.$$

Thus, (42.11) is a quadratic form of the variables $u_1, \ldots, u_n$ with the matrix of coefficients $K = [\varkappa_{ij}]$ equal to

(42.12) $$K = CAC'.$$

Indeed, we have

$$\sum_{i=1}^{n}\sum_{j=1}^{n}a_{ij}y_{i}y_{j} = \sum_{i=1}^{n}\sum_{j=1}^{n}a_{ij}\left(\sum_{s=1}^{n}c'_{is}u_{s}\right)\left(\sum_{k=1}^{n}c'_{jk}u_{k}\right),$$

which, after changing the order of summation and using the equality $c'_{is} = c_{si}$, will give the expression

$$\sum_{i=1}^{n}\sum_{j=1}^{n}c_{si}a_{ij}c'_{jk}$$

as the coefficient at the product $u_s u_k$. This proves the equality $K = CAC'$.

Next, we know from algebra that the transformation (42.9) can always be chosen in such a way that the matrix K is diagonal; this means reducing the quadratic form (42.5) to a sum of squares. In view of the positive definiteness of form (42.5), the elements on the main diagonal of the matrix K will be positive, and will be equal to the roots of the secular equation

(42.13)
$$\begin{vmatrix} a_{11}-\lambda & a_{12} & \cdots & a_{1n} \\ a_{21} & a_{22}-\lambda & \cdots & a_{2n} \\ \cdots & \cdots & \cdots & \cdots \\ a_{n1} & a_{n2} & \cdots & a_{nn}-\lambda \end{vmatrix} = 0.$$

Denote $\varkappa_{ii} = 1/\sigma_i^2$. We then have

(42.14)
$$K = \begin{bmatrix} \frac{1}{\sigma_1^2} & 0 & \cdots & 0 \\ 0 & \frac{1}{\sigma_2^2} & \cdots & 0 \\ \cdots & \cdots & \cdots & \cdots \\ 0 & 0 & \cdots & \frac{1}{\sigma_n^2} \end{bmatrix}.$$

The determinant of a matrix of orthogonal transformation is always equal to $+1$ or -1, and we can choose our transformation so that its value will be $+1$. Then we shall be able to write

(42.15)
$$J = \int \cdots \int \exp\left\{-\frac{1}{2}\sum_{i=1}^{n}\sum_{j=1}^{n} a_{ij} y_i y_j\right\} dy_1 \ldots dy_n$$

$$= \int \cdots \int \exp\left\{-\frac{1}{2}\sum_{i=1}^{n} \frac{u_i^2}{\sigma_i^2}\right\} du_1 \ldots du_n$$

$$= \int \exp\left\{-\frac{1}{2}\cdot\frac{u_1^2}{\sigma_1^2}\right\} du_1 \ldots \int \exp\left\{-\frac{1}{2}\cdot\frac{u_n^2}{\sigma_n^2}\right\} du_n.$$

Thus, the integral J is represented as a product of integrals of one variable. The latter can be calculated according to the formula

$$\int_{-\infty}^{+\infty} \exp\left[-\frac{t_1^2}{2\sigma_1^2}\right] dt_1 = \sqrt{2\pi}\,\sigma_1$$

(since, as we know, $\dfrac{1}{\sigma_1\sqrt{2\pi}} \exp\left[-\dfrac{t_1^2}{2\sigma_1^2}\right]$ is a probability density). Thus, we have

(42.16)
$$J = (\sqrt{2\pi})^n \sigma_1 \ldots \sigma_n.$$

Let us find the relation between $\sigma_1, \ldots, \sigma_n$ and $|A|$. In view of equality (42.12) we have $|K| = |C| \cdot |A| \cdot |C'|$. Since the matrix C has been chosen in such a way that its determinant is equal to 1, we have $|K| = |A|$. In view of (42.15) we obtain therefore $\sqrt{|A|} = 1/(\sigma_1 \ldots \sigma_n)$. Combining this with relations (42.16) and (42.8), we see that $I = 1$, which completes the proof of assertion (a).

We shall now prove the remaining assertions of Theorem 1.

Let $\xi = (\xi_1, \ldots, \xi_n)$ denote a random vector with probability density (42.6), and let $\eta = (\eta_1, \ldots, \eta_n)$ be related with ξ by formula $\eta = \xi - m$, i.e., $(\eta_1, \ldots, \eta_n) = (\xi_1 - m_1, \ldots, \xi_n - m_n)$. Then the probability density of η will be given by the formula

$$f_\eta(y_1, \ldots, y_n) = \frac{\sqrt{|A|}}{(\sqrt{2\pi})^n} \exp\left\{-\frac{1}{2}\sum_{i=1}^{n}\sum_{j=1}^{n} a_{ij} y_i y_j\right\},$$

which explains the substitution preceding formula (42.8). If we now regard formula (42.8) as a transformation of random variables with the matrix C used in the preceding calculations, we shall obtain instead of (42.9) the formula

VII. CENTRAL LIMIT THEOREMS

(42.17)
$$\zeta_1 = c_{11}\eta_1 + \ldots + c_{1n}\eta_n,$$
$$\ldots\ldots\ldots\ldots\ldots\ldots\ldots\ldots\ldots$$
$$\zeta_n = c_{n1}\eta_1 + \ldots + c_{nn}\eta_n.$$

Instead of (42.10) we obtain the inverse transformation

(42.18)
$$\eta_1 = c'_{11}\zeta_1 + \ldots + c'_{n1}\zeta_n,$$
$$\ldots\ldots\ldots\ldots\ldots\ldots\ldots\ldots\ldots$$
$$\eta_n = c'_{n1}\zeta_1 + \ldots + c'_{nn}\zeta_n.$$

From the form of the integrand in the n-dimensional integral in (42.15) (equal, up to a constant to the probability density of the vector ζ, in view of (42.8) we see that the components of the vector ζ are stochastically independent, and the component ζ_i has a normal distribution with expectation m_i and variance σ_i^2. Thus, the matrix

(42.19)
$$S = \begin{bmatrix} \sigma_1^2 & 0 & \ldots & 0 \\ 0 & \sigma_2^2 & \ldots & 0 \\ \ldots\ldots\ldots\ldots\ldots\ldots \\ 0 & 0 & \ldots & \sigma_n^2 \end{bmatrix}$$

is the covariance matrix of the random variables $\zeta_1, \ldots, \zeta_n$. In view of (42.18) and Theorem 2 of § 40, the covariance matrix Λ of the random variables $\eta_1, \ldots, \eta_n$ can be expressed by S and C according to the formula

(42.20) $\qquad\qquad\qquad \Lambda = C'SC.$

Combining (42.14) and (42.19) we see that the matrix S is the inverse of the matrix K, whence

(42.21) $\qquad\qquad\qquad SK = I.$

Next, multiplying (42.12) by C' on the left and by C on the right, using the associative law for matrix multiplication and the fact that for orthogonal matrices we have $CC' = C'C = I$, we obtain the following expression for the matrix A by the matrices K and C:

(42.22) $\qquad\qquad\qquad A = C'KC.$

From formulas (42.20), (42.21) and (42.22) it follows that the covariance matrix of the random variables $\eta_1, \ldots, \eta_n$ is the inverse of the matrix A. Indeed, we have

$$\Lambda \cdot A = C'SC \cdot C'KC = C'S(C \cdot C')KC = C'(SK)C = C'C = I.$$

This proves the first part of assertion (c). The second part follows from the remark that, in view of relations (42.18), the random variables $\eta_1, \ldots, \eta_n$ have expectations zero, and the vector ξ is translated by m with respect to η.

It remains to prove assertion (b).

In view of the relation $\xi = \eta + m$ it suffices to verify that the function
$$\exp\left\{-\frac{1}{2}\sum_{i=1}^{n}\sum_{j=1}^{n}\lambda_{ij}t_it_j\right\},$$
where $\Lambda = [\lambda_{ij}]$ is the inverse of the matrix A, is the characteristic function of the random vector η. We must therefore show that

(42.23) $\quad \int \ldots \int \exp\{i(t_1y_1 + \ldots + t_ny_n)\}\dfrac{\sqrt{|A|}}{(\sqrt{2\pi})^n} \times$

$$\times \exp\left\{-\frac{1}{2}\sum_{i=1}^{n}\sum_{j=1}^{n}a_{ij}y_iy_j\right\}dy_1\ldots dy_n$$

$$= \exp\left\{-\frac{1}{2}\sum_{i=1}^{n}\sum_{j=1}^{n}\lambda_{ij}t_it_j\right\}.$$

We shall use transformation (42.9) again. The integral appearing on the left-hand side of equality (42.23) can be written as follows:

$$B = \int \ldots \int \exp[i(v_1u_1 + \ldots + v_nu_n)]\prod_{i=1}^{n}\frac{1}{\sigma_i\sqrt{2\pi}}\exp\left\{-\frac{1}{2}\left(\frac{u_i}{\sigma_i}\right)^2\right\}du_1\ldots du_n$$

$$= \int_{-\infty}^{+\infty}\exp(iv_1u_1)\frac{1}{\sigma_1\sqrt{2\pi}}\exp\left\{-\frac{1}{2}\left(\frac{u_1}{\sigma_1}\right)^2\right\}du_1 \times$$

$$\times \ldots \times \int_{-\infty}^{+\infty}\exp(iv_nu_n)\frac{1}{\sigma_n\sqrt{2\pi}}\exp\left\{-\frac{1}{2}\left(\frac{u_n}{\sigma_n}\right)^2\right\}du_n,$$

where
$$v_1 = c_{11}t_1 + \ldots + c_{1n}t_n,$$
$$\ldots\ldots\ldots\ldots\ldots\ldots\ldots\ldots\ldots\ldots$$
$$v_n = c_{n1}t_1 + \ldots + c_{nn}t_n.$$

But the integrals of one variable appearing on the right-hand side are simply the values of the characteristic functions of the random variables ζ_i at the points $v_1, \ldots, v_n$. From Theorem 4 of § 37 and from the fact that ζ_i has a normal distribution with expectation 0 and variance σ_i^2 it follows that the characteristic function of ζ_i equals $\exp[-\frac{1}{2}\sigma_i^2 t_i^2]$. Finally,

$$B = \exp\left[-\frac{1}{2}\sum_{i=1}^{n}\sigma_i^2 v_i^2\right].$$

VII. CENTRAL LIMIT THEOREMS

It remains to show that

$$\sum_{i=1}^{n} \sigma_i^2 v_i^2 = \sum_{i=1}^{n} \sum_{j=1}^{n} \lambda_{ij} t_i t_j.$$

We have

$$v_i = \sum_{s=1}^{n} c_{is} t_s,$$

whence

$$v_i^2 = \sum_{s=1}^{n} \sum_{k=1}^{n} c_{is} t_s c_{ik} t_k,$$

and next

$$\sum_{i=1}^{n} \sigma_i^2 v_i^2 = \sum_{i=1}^{n} \sum_{s=1}^{n} \sum_{k=1}^{n} \sigma_i^2 c_{is} c_{ik} t_s t_k = \sum_{s=1}^{n} \sum_{k=1}^{n} t_s t_k \sum_{i=1}^{n} c'_{si} \sigma_i^2 c_{ik}.$$

In the last expression we recognize the quadratic form of the variables $t_1, \ldots, t_n$, whose coefficients are elements of the product $C'SC$, equal to the matrix Λ according to formula (42.20).

Thus, the proof of Theorem 1 is complete.

Theorem 1 shows that to every positive definite quadratic form of n variables there corresponds a probability density of an n-dimensional probability distribution (the n-dimensional probability distribution in the sense of Frisch's theorem). This justifies the following definition:

DEFINITION 1. *The probability distribution in an n-dimensional Euclidean space corresponding to the probability density (42.6) with the positive definite quadratic form (42.5) will be called a non-degenerate n-dimensional normal distribution.*

When we were discussing normal distributions on the real line, it proved convenient to treat degenerate distributions which have the whole probability concentrated at one point as special cases of normal distributions.

Indeed, such distributions may appear as limits of non-degenerate normal distributions and we may say therefore that the limit of normal distributions is a normal distribution, though possibly degenerate. The same applies to an even greater degree for normal distributions in n-dimensional Euclidean spaces. Suppose that $\xi_1, \ldots, \xi_n$ are independent random variables with normal distributions with expectation 0 and variances $\sigma_i > 0$. Then the joint probability distribution of the random variables $\xi_1, \ldots, \xi_n$, i.e. of the random vector $\xi = (\xi_1, \ldots, \xi_n)$, will have the characteristic function

$$\exp\left\{-\frac{1}{2}\sum_{i=1}^{n}\sigma_i^2 t_i^2\right\}.$$

Now, if $\sigma_n \to 0$, then the characteristic functions will converge to the continuous function

$$\exp\left\{-\frac{1}{2}\sum_{i=1}^{n-1}\sigma_i^2 t_i^2\right\}.$$

In view of Theorem 6 of § 41, this limit function will be the characteristic function of a certain probability distribution which we also want to include in the class of normal distributions. This distribution, however, will be degenerate in the sense that the whole probability will be concentrated on the hyperplane $x_n = 0$. Thus, this distribution will be $(n-1)$-dimensional in an n-dimensional Euclidean space, and will have no density in this space.

What we said above may be generalized in the form of a theorem, which—in the domain of non-degenerate distributions—is the converse of Theorem 1.

THEOREM 2. *If*

(42.23) $$\sum_{i=1}^{n}\sum_{j=1}^{n}\lambda_{ij}t_i t_j$$

is a positive semidefinite quadratic form with a symmetric matrix $\Lambda = [\lambda_{ij}]$ *then for every vector* $\boldsymbol{m} = (m_1, \ldots, m_n)$ *the function*

(42.24) $$\varphi(t_1, \ldots, t_n) = \exp\left\{i(m_1 t_1 + \ldots + m_n t_n) - \frac{1}{2}\sum_{i=1}^{n}\sum_{j=1}^{n}\lambda_{ij}t_i t_j\right\}$$

is the characteristic function of a certain probability distribution in an n-dimensional Euclidean space; here $m_1, \ldots, m_n$ *are the expectations of the components, and* Λ *is the covariance matrix. If, in addition, the matrix* Λ *is non-singular, and* $A = [a_{ij}]$ *is the matrix inverse to* Λ, *then the probability distribution with characteristic function* (42.24) *has the probability density*

(42.25) $$f(x_1, \ldots, x_n) = \frac{\sqrt{|A|}}{(\sqrt{2\pi})^n}\exp\left\{-\frac{1}{2}\sum_{i=1}^{n}\sum_{j=1}^{n}a_{ij}(x_i - m_i)(x_j - m_j)\right\}.$$

The proof of this theorem is virtually contained in the proof of Theorem 1. First of all, we have to show that the function

(42.26) $$\exp\left\{-\frac{1}{2}\sum_{i=1}^{n}\sum_{j=1}^{n}\lambda_{ij}t_i t_j\right\}$$

VII. CENTRAL LIMIT THEOREMS

is the characteristic function of a distribution with the covariance matrix Λ and zero expectations for the components; according to Theorem 7(a) of § 41, the distribution with the characteristic function (42.24) is obtained through a translation by the vector $\boldsymbol{m} = (m_1, \ldots, m_n)$. Suppose that the matrix Λ is of the rank $n' \leqslant n$. By the well-known theorem from algebra, we can find an orthogonal matrix $C = [c_{ij}]$ with determinant 1 such that the transformation

(42.27)
$$u_1 = c_{11} t_1 + \ldots + c_{1n} t_n,$$
$$\ldots\ldots\ldots\ldots\ldots\ldots\ldots\ldots\ldots$$
$$u_n = c_{n1} t_1 + \ldots + c_{nn} t_n$$

transforms the quadratic form (42.23) into the sum of squares

(42.28)
$$\sum_{i=1}^{n} \sigma_i^2 u_i^2$$

with the diagonal matrix of coefficients $S = [\sigma_{ij}]$, where $\sigma_{ii} = \sigma_i^2$, $\sigma_{ij} = 0$ for $i \neq j$, and only n' among the numbers $\sigma_1^2, \ldots, \sigma_n^2$ will be different from zero; we may assume that the first n' terms will be different from zero, i.e. $\sigma_i^2 > 0$ for $i = 1, 2, \ldots, n'$.

Then function (42.26) will be transformed into

$$\exp\left\{-\frac{1}{2} \sum_{i=1}^{n} \sigma_i^2 u_i^2\right\}$$

and we recognize the characteristic function of the random vector $\zeta = (\zeta_1, \ldots, \zeta_n)$, with normal components, independent, with expectations 0 and variances σ_i^2. The matrix S, equal to the covariance matrix of the random variables $\zeta_1, \ldots, \zeta_n$, is related to the matrices Λ and C by the relation

$$S = C\Lambda C',$$

which implies that

(42.29)
$$\Lambda = C'SC.$$

Using Theorem 2 of § 40, we see now that Λ is the covariance matrix of the random variables $\eta_1, \ldots, \eta_n$ defined by the equalities

$$\eta_1 = c'_{11} \zeta_1 + \ldots + c'_{1n} \zeta_n,$$
$$\ldots\ldots\ldots\ldots\ldots\ldots\ldots\ldots\ldots$$
$$\eta_n = c'_{n1} \zeta_1 + \ldots + c'_{nn} \zeta_n.$$

It follows in particular that the random variables η_i have expectations equal to zero. We shall now show that function (42.26) is the characteristic function of the random vector $\eta = (\eta_1, \ldots, \eta_n)$. Indeed, we have

$$\varphi_\eta(t_1, \ldots, t_n) = E\exp[i(t_1\eta_1 + \ldots + t_n\eta_n)]$$

$$= E\exp\left\{i\sum_{j=1}^n t_j \sum_{s=1}^n c'_{js}\zeta_s\right\} = E\exp\left\{i\sum_{s=1}^n \zeta_s \sum_{j=1}^n c_{sj}t_j\right\}.$$

In view of the independence of the random variables ζ_i and relation (42.27) we obtain

$$\varphi_\eta(t_1, \ldots, t_n) = \exp\left\{-\frac{1}{2}\sum_{j=1}^n \sigma_j^2 u_j^2\right\}.$$

Returning to the variables $t_1, \ldots, t_n$ by another application of (42.27), we find

$$\sum_{j=1}^n \sigma_j^2 u_j^2 = \sum_{j=1}^n \sigma_j^2 \left(\sum_{k=1}^n c_{jk}t_k\right)^2$$

$$= \sum_{j=1}^n \sigma_j^2 \sum_{k=1}^n \sum_{s=1}^n c_{jk}t_k c_{js}t_s = \sum_{k=1}^n \sum_{s=1}^n t_k t_s \sum_{j=1}^n c'_{kj}\sigma_j^2 c_{js},$$

and we recognize the quadratic form of the variables $t_1, \ldots, t_n$ with the matrix of coefficients equal to $C'SC$. Using equality (42.29) we finally get

$$\varphi_\eta(t_1, \ldots, t_n) = \exp\left\{-\frac{1}{2}\sum_{i=1}^n \sum_{j=1}^n \lambda_{ij} t_i t_j\right\},$$

which completes the proof of the main assertion of Theorem 2.

The remaining part of the assertion of Theorem 2 follows directly from Theorem 1 and Lemma 1 given below.

LEMMA 1. *If*

(42.30) $$\sum_{i=1}^n \sum_{j=1}^n \lambda_{ij} t_i t_j$$

is a positive definite quadratic form with a symmetric matrix of coefficients $\Lambda = [\lambda_{ij}]$, then there exists a matrix $A = [a_{ij}]$ inverse to matrix Λ, and the quadratic form

(42.31) $$\sum_{i=1}^n \sum_{j=1}^n a_{ij} t_i t_j$$

is positive definite.

PROOF. We can find an orthogonal matrix $C = [c_{ij}]$ with a determinant equal to 1 such that after introducing new variables $u_1, \ldots, u_n$ according to formulas

(42.27), form (42.30) will become equal to

$$\sum_{i=1}^{n}\sum_{j=1}^{n}\sigma_{ij}u_iu_j,$$

with a diagonal matrix $S = [\sigma_{ij}]$ such that $\sigma_{ii} = \sigma_i^2 > 0$ for $i = 1, 2, \ldots, n$ and $\sigma_{ij} = 0$ for $i \neq j$. As already computed in the proof of Theorem 2, we shall have

$$S = C\Lambda C',$$

whence

$$\Lambda = C'SC.$$

Clearly, the quadratic form

$$\frac{u_1^2}{\sigma_1^2} + \ldots + \frac{u_n^2}{\sigma_n^2}$$

is positive definite, similarly to the form

$$\sigma_1^2 u_1^2 + \ldots + \sigma_n^2 u_n^2.$$

The second of these forms has the matrix S, and the first—a matrix inverse to S, which we will denote by S^{-1}. If in the first of these forms we return from the variables $u_1, \ldots, u_n$ to the variables $t_1, \ldots, t_n$ according to (42.27), we obtain

$$\frac{u_1^2}{\sigma_1^2} + \ldots + \frac{u_n^2}{\sigma_n^2} = \sum_{j=1}^{n}\frac{u_j^2}{\sigma_j^2} = \sum_{k=1}^{n}\sum_{s=1}^{n}t_k t_s \sum_{j=1}^{n}c'_{kj}\frac{1}{\sigma_i^2}c_{js}.$$

On the right-hand side we recognize the quadratic form of the coefficients $t_1, \ldots, t_n$ with the matrix $C'S^{-1}C$. This equality, the positive definiteness of the quadratic form appearing on the left-hand side, and fact that transformation (42.27) assigns to any system $t_1, \ldots, t_n$ non-vanishing identically a system $u_1, \ldots, u_n$, non-vanishing identically, shows that the quadratic form on the right-hand side of the equality in question is positive definite. Note that the matrix $C'S^{-1}C$ is the required matrix inverse to Λ. Indeed,

$$\Lambda(C'S^{-1}C) = (C'SC)(C'S^{-1}C) = C'S(CC')S^{-1}C$$
$$= C'(SS^{-1})C = C'C = I,$$

which completes the proof of Lemma 1.

In connection with Theorem 2 we introduce the following definition.

DEFINITION 2. The probability distribution in an n-dimensional Euclidean space with the characteristic function of form (42.24) described in Theorem 2 will be called a *normal distribution*. If the rank of the matrix Λ is smaller than n, we say that this distribution is *degenerate*.

It should be remarked that Theorem 2 is, in a sense, converse to Lemma 2 of § 40. There we showed that a quadratic form whose coefficients are taken from the covariance matrix of a certain system of random variables is positive semidefinite. In Theorem 2 we showed that the converse is also true: for every positive semidefinite quadratic form with a symmetric matrix of coefficients one can find a system of random variables for which that matrix will be their covariance matrix. These two facts characterize covariance matrices, and we state this in the form of a corollary:

COROLLARY 1. *A symmetric square matrix* $\Lambda = [\lambda_{ij}]$, $i, j = 1, 2, ..., n$, *is a covariance matrix of a certain system of random variables if and only if the quadratic form*

$$\sum_{i=1}^{n} \sum_{j=1}^{n} \lambda_{ij} t_i t_j$$

is positive semidefinite.

Let us also mention the following property of the multidimensional normal distribution, implicit in the proof of Theorem 2:

COROLLARY 2. *Every k-dimensional ($k \leqslant n$) normal distribution in an n-dimensional Euclidean space can be treated as a joint distribution of k linear combinations of n independent random variables with non-degenerate normal distributions.*

(This is the distribution of random variables $\eta_1, ..., \eta_n$ which we expressed as linear combinations of independent random variables $\zeta_1, ..., \zeta_n$.)

We now present some more properties of multidimensional normal distributions.

THEOREM 3. *If $\xi = (\xi_1, ..., \xi_n)$ is a random vector with a joint normal distribution, and the vector $\eta = (\eta_1, ..., \eta_m)$ is defined by the linear equations*

(42.32)
$$\eta_1 = c_{11}\xi_1 + ... + c_{1n}\xi_n,$$
$$\dots\dots\dots\dots\dots\dots\dots\dots\dots$$
$$\eta_m = c_{m1}\xi_1 + ... + c_{mn}\xi_n,$$

then the vector η also has a normal distribution.

PROOF. We shall show that the characteristic function of the joint distribution of the random variables $\eta_1, ..., \eta_m$ has the form required in the definition of the normal distribution. Indeed, we may write

$$\varphi_\eta(t_1, \ldots, t_m) = E\exp\{i(t_1\eta_1 + \ldots + t_m\eta_m)\}$$

$$= E\exp\left\{i\sum_{j=1}^{n} t_j \sum_{k=1}^{n} c_{jk}\xi_k\right\} = E\exp\left\{i\sum_{k=1}^{n} \xi_k \sum_{j=1}^{n} c_{jk}t_j\right\}$$

$$= \varphi_\xi(u_1, \ldots, u_n) = \exp\left\{i(m_1 u_1 + \ldots + m_n u_n) - \frac{1}{2}\sum_{k=1}^{n}\sum_{s=1}^{n}\lambda_{ks} u_k u_s\right\},$$

where $u_1, \ldots, u_n$ are given by

$$u_1 = c_{11}t_1 + \ldots + c_{m1}t_m,$$
$$\ldots\ldots\ldots\ldots\ldots\ldots\ldots\ldots\ldots\ldots$$
$$u_n = c_{1n}t_1 + \ldots + c_{mn}t_m,$$

$m_1, \ldots, m_n$ are the expectations of the random variables $\xi_1, \ldots, \xi_n$, and Λ is their covariance matrix. It remains to return to variables $t_1, \ldots, t_m$ in the last expression and verify that we obtain for $\varphi_\eta(t_1, \ldots, t_m)$ the function of the required form.

§ 43. MULTIDIMENSIONAL FORM OF THE CENTRAL LIMIT THEOREM OF PROBABILITY THEORY

Let $\xi^{(1)} = (\xi_1^{(1)}, \xi_2^{(1)})$, $\xi^{(2)} = (\xi_1^{(2)}, \xi_2^{(2)})$, ... be a sequence of independent vectors having the same probability distribution with expectation $m = (0, 0)$ and covariance matrix

$$\Lambda = \begin{bmatrix} \lambda_{11} & \lambda_{12} \\ \lambda_{21} & \lambda_{22} \end{bmatrix}.$$

Denote by $\varphi(t_1, t_2)$ the common characteristic function. Then the sum

$$\zeta^{(n)} = \xi^{(1)} + \ldots + \xi^{(n)}$$

has the characteristic function $[\varphi(t_1, t_2)]^n$, expectation $(0, 0)$ and covariance matrix

$$n\Lambda = \begin{bmatrix} n\lambda_{11} & n\lambda_{12} \\ n\lambda_{21} & n\lambda_{22} \end{bmatrix}$$

(see Theorem 4 of § 41 and Theorem 3 of § 40). Similarly, we find that the normed sums

$$\eta^{(n)} = \frac{1}{\sqrt{n}}\zeta^{(n)} = \frac{1}{\sqrt{n}}(\xi^{(1)} + \ldots + \xi^{(n)})$$

have the characteristic function

(43.1) $$\varphi_n(t_1, t_2) = \left[\varphi\left(\frac{t_1}{\sqrt{n}}, \frac{t_2}{\sqrt{n}}\right)\right]^n,$$

expectation $(0, 0)$ and the same covariance matrix as $\xi^{(1)}$, i.e. equal to Λ.

Using Theorem 3 of § 41, we may write the equalities

$$E\xi_1^{(1)} = (-i)\left[\frac{\partial \varphi(t_1, t_2)}{\partial t_1}\right]_{t_1=t_2=0} = 0,$$

$$E\xi_2^{(1)} = (-i)\left[\frac{\partial \varphi(t_1, t_2)}{\partial t_2}\right]_{t_1=t_2=0} = 0,$$

$$E(\xi_1^{(1)})^2 = (-i)^2\left[\frac{\partial^2 \varphi(t_1, t_2)}{\partial t_1^2}\right]_{t_1=t_2=0} = \lambda_{11},$$

$$E(\xi_1^{(1)}\xi_2^{(1)}) = (-i)^2\left[\frac{\partial^2 \varphi(t_1, t_2)}{\partial t_1 \partial t_2}\right]_{t_1=t_2=0} = \lambda_{12} = \lambda_{21},$$

$$E(\xi_2^{(1)})^2 = (-i)^2\left[\frac{\partial^2 \varphi(t_1, t_2)}{\partial t_2^2}\right]_{t_1=t_2=0} = \lambda_{22}.$$

Expanding the function $\varphi(t_1, t_2)$ into a Taylor series in the neighbourhood of the point $(t_1, t_2) = (0, 0)$ and using terms up to the second order, we may write

(43.2) $$\varphi(t_1, t_2) = 1 - \tfrac{1}{2}(\lambda_{11}t_1^2 + 2\lambda_{12}t_1t_2 + \lambda_{22}t_2^2) + o(t_1^2 + t_2^2)$$

(we have used here the symmetry $\lambda_{12} = \lambda_{21}$). Let us fix the point (t_1, t_2) and ask about the value of the characteristic function of the vector $\eta^{(n)}$ at this point. We shall have

$$\varphi\left(\frac{t_1}{\sqrt{n}}, \frac{t_2}{\sqrt{n}}\right) = 1 - \frac{1}{2n}(\lambda_{11}t_1^2 + 2\lambda_{12}t_1t_2 + \lambda_{22}t_2^2) + o\left(\frac{1}{n}\right).$$

Using this fact, and denoting for convenience $\lambda_{11}t_1^2 + 2\lambda_{12}t_1t_2 + \lambda_{22}t_2^2$ by T, we get, in view of the expansion $\log(1+u) = u + o(u)$, valid in the neighbourhood of $u = 0$,

$$\log \varphi\left(\frac{t_1}{\sqrt{n}}, \frac{t_2}{\sqrt{n}}\right) = -\frac{1}{2n}T + o\left(\frac{1}{n}\right).$$

Therefore, using (43.1), we obtain

$$\log \varphi_n(t_1, t_2) = \log \varphi\left[\left(\frac{t_1}{\sqrt{n}}, \frac{t_2}{\sqrt{n}}\right)\right]^n$$

$$= n \log \varphi\left(\frac{t_1}{\sqrt{n}}, \frac{t_2}{\sqrt{n}}\right) = n\left(-\frac{1}{2n}T + o\left(\frac{1}{n}\right)\right) = -\frac{1}{2}T + o(1).$$

VII. CENTRAL LIMIT THEOREMS

This relation shows that at every point (t_1, t_2) we have the convergence

$$\lim_{n \to \infty} \log \varphi_n(t_1, t_2) = -\tfrac{1}{2} T,$$

which is equivalent to the convergence

$$\lim_{n \to \infty} \varphi_n(t_1, t_2) = \exp(-\tfrac{1}{2} T) = \exp\{-\tfrac{1}{2}(\lambda_{11} t_1^2 + 2\lambda_{12} t_1 t_2 + \lambda_{22} t_2^2)\}$$

at every point (t_1, t_2).

We have thus proved that the characteristic functions of normed averages η_n converge to a continuous function equal to the characteristic function of the bivariate normal distribution with expectation $(0, 0)$ and covariance matrix Λ.

Obviously, the calculations performed may easily be generalized to the case of distributions in Euclidean spaces of an arbitrary number of dimensions.

We have assumed for simplicity that the random vectors $\xi^{(i)}$ have expectation $(0, 0)$. In the general case we would have to consider vectors $\xi^{(i)} - E\xi^{(i)}$.

Thus, we have the following analogue of Theorem 1 of § 38 concerning the limit distribution of sums of independent terms having the same distribution and second moments:

THEOREM 1. *If random vectors* $\xi^{(1)} = (\xi_1^{(1)}, \ldots, \xi_k^{(1)})$, $\xi^{(2)} = (\xi_1^{(2)}, \ldots, \xi_k^{(2)}), \ldots$ *are stochastically independent and have identical probability distributions with the expectation* $\boldsymbol{m} = (m_1, \ldots, m_k)$ *and covariance matrix* $\Lambda = [\lambda_{ij}]$, *then the normed sums*

$$\boldsymbol{\eta}^{(n)} = \frac{1}{\sqrt{n}} \{(\xi^{(1)} - \boldsymbol{m}) + \ldots + (\xi^{(n)} - \boldsymbol{m})\}$$

have in the limit a normal distribution with the expectation $(0, 0, \ldots, 0)$ *and covariance matrix* Λ.

Sometimes it is more convenient to formulate the assertion of Theorem 1 as follows: *the sums* $\zeta^{(n)} = \xi^{(1)} + \ldots + \xi^{(n)}$ *are asymptotically normal with the expectation* $n\boldsymbol{m}$ *and covariance matrix* $n\Lambda$.

This theorem may be generalized in a similar way as Theorem 1 of § 38. The sums of random vectors prove to be asymptotically normal under very weak conditions concerning their distributions. Thus, the role of normal distributions in the case of multidimensional distributions is completely analogous to the role of one-dimensional normal distribution discussed in § 38.

§ 44. LIMIT DISTRIBUTIONS OF SAMPLE STATISTICS

In the chapters concerning the laws of large numbers, and in particular in § 33 of Chapter VI, we showed (using statistical terminology) that samples may be treated as miniatures of the populations from which they were taken, in the sense that as the sample size increases, all the important parameters of empirical distributions converge to the corresponding parameters of the distribution of the population. This assertion, however, has certain disadvantages: it does not tell us how to assert the accuracy of the approximation. In the present paragraph we return to this subject and we shall show that on the basis of limit theorems we can estimate how exact is the miniature of the population presented by the sample; in other words, we shall learn how to judge the accuracy with which the sample informs us about the parameters of the distribution of the population. What we shall discuss here will be essentially a strenghtening of the laws of large numbers. They asserted, in general, that for every fixed $\varepsilon > 0$, the sample parameter differed from the corresponding population parameter by more than ε with probability tending to zero as the sample size increased. The theorems which we shall now formulate give the relation between this probability and the sample size, hence they allow us (approximately, to be sure) to evaluate the numerical value of these probabilities.

We shall try to present the theorems in an order stressing their parallelism with those of § 33.

THEOREM 1. *If* $\xi_1 = \xi_1(e), \xi_2 = \xi_2(e), \ldots$ *is a sequence of independent random variables with identical distributions and if there exists a finite moment of order* 2ν, *i.e.* $\alpha_{2\nu} = E\xi_1^{2\nu} < \infty$, *then the sample moments of order* ν, *defined by the equality*

$$a_\nu^{(n)} = \frac{1}{n}\left(\xi_1^\nu(e) + \ldots + \xi_n^\nu(e)\right)$$

are asymptotically normal $N\left(\alpha_\nu, \frac{1}{\sqrt{n}}\sqrt{\alpha_{2\nu} - \alpha_\nu^2}\right)$, *where* $\alpha_{2\nu}$ *and* α_ν *are moments of the random variable* ξ_1.

PROOF. This theorem is a direct consequence of Theorem 1 of § 38; it suffices to apply the latter to the νth powers $\xi_1^\nu, \xi_2^\nu, \ldots$ of the random variables $\xi_1, \xi_2, \ldots$

In other words, this theorem asserts that the probability of $a_\nu^{(n)}$ differing from α_ν by at least $\frac{\varepsilon}{\sqrt{n}}\sqrt{\alpha_{2\nu} - \alpha_\nu^2}$ tends to a limit equal to the probability that a normal random variable with expectation 0 and variance 1 will exceed ε in absolute value, i.e.

$$\lim_{n\to\infty} P\left(\left\{e: |a_\nu^{(n)} - \alpha_\nu| > \frac{\varepsilon}{\sqrt{n}}\sqrt{\alpha_{2\nu} - \alpha_\nu^2}\right\}\right) = 2\int_\varepsilon^\infty \frac{1}{\sqrt{2\pi}} e^{-t^2/2} dt.$$

If we replace here the unknown number $\sqrt{\alpha_{2\nu} - \alpha_\nu^2}$ by the number $s = \sqrt{a_{2\nu}^{(n)} - (a_\nu^{(n)})^2}$ computed from the sample, we will be able to approximate the probabilities of

VII. CENTRAL LIMIT THEOREMS

the deviations of the sample moments from the population moments. For instance, if it should turn out after taking the sample of 50 elements that $s = 4.18$, and we wanted the difference $|a_v^{(n)} - \alpha_v|$ to exceed 0.5 with a probability at most 0.05, we would find in the tables of normal distribution that $\varepsilon = 1.96$, and we would solve with respect to n the inequality

$$\frac{\varepsilon s}{\sqrt{n}} \leqslant 0.5 \quad \text{or} \quad n \geqslant \frac{\varepsilon^2 s^2}{0.5^2} = 268.4.$$

Hence the indication that, with the above requirements, we should increase the sample to almost 300 elements.

THEOREM 2. *Let $\xi_1 = \xi_1(e), \xi_2 = \xi_2(e), \ldots$ be independent random variables with identical distributions. Suppose that there exists a finite moment of order $2v$, i.e. $\alpha_{2v} = E\xi_1^{2v} < \infty$, and let $f(x_1, \ldots, x_m)$ be a continuous function defined in the Euclidean space of m dimensions, with continuous partial derivatives of the first and second order at the point $x_1 = \alpha_{i_1}, \ldots, x_m = \alpha_{i_m}$, where α_{i_k} is the moment of order i_k of the random variables in question (i.e. $\alpha_{i_k} = E\xi_1^{i_k}$) where $i_k \leqslant v$ for $k = 1, 2, \ldots, m$. Then the random variables*

$$\zeta_n(e) = f(a_{i_1}^{(n)}(e), \ldots, a_{i_m}^{(n)}(e)),$$

where $a_{i_k}^{(n)} = \frac{1}{n}(\xi_1^{i_k}(e) + \ldots + \xi_n^{i_k}(e))$, are asymptotically normal $N(m, \sigma_n)$, where

$$m = f(\alpha_{i_1}, \ldots, \alpha_{i_m}),$$

$$\sigma_n^2 = \frac{1}{n} \sum_{j=1}^{m} \sum_{k=1}^{m} c_j c_k (\alpha_{i_j + i_k} - \alpha_{i_j} \alpha_{i_k}),$$

$$c_j = \left[\frac{\partial}{\partial x_j} f(x_1, \ldots, x_m)\right]_{(x_1, \ldots, x_m) = (\alpha_{i_1}, \ldots, \alpha_{i_m})}.$$

Outline of the proof. We have to prove that for every x the probability of the inequality $\frac{\zeta_n - m}{\sigma_n} < x$ tends to $\Phi(x)$, where

$$\Phi(x) = \frac{1}{\sqrt{2\pi}} \int_{-\infty}^{x} e^{-t^2/2} dt.$$

Note first the that m-dimensional random variables

$$(\eta_1^{(j)}, \ldots, \eta_m^{(j)}) = (\xi_j^{i_1}, \ldots, \xi_j^{i_m})$$

have the vector of expectations equal to $(\alpha_{i_1}, \ldots, \alpha_{i_m})$, and the covariance matrix

$$\Lambda = [\lambda_{jk}] = \alpha_{i_j + i_k} - \alpha_{i_j} \alpha_{i_k}.$$

Thus, in view of Theorem 1 of § 43, the system of sample moments $(a_{i_1}^{(n)}, \ldots, a_{i_m}^{(n)})$ is an asymptotically normal random variable with the expectation $(\alpha_{i_1}, \ldots, \alpha_{i_m})$ and covariance matrix $\dfrac{1}{n} \Lambda$.

Next, applying the Čebyšev inequality

$$P(\{e: (\xi - E\xi)^2 \geq \varepsilon^2\}) \leq \frac{D^2 \xi}{\varepsilon^2}$$

to successive sample moments, we see that for every j separately

$$P(\{e: |a_{i_j}^{(n)} - \alpha_{i_j}| \geq \varepsilon\}) \leq \frac{D^2 a_{i_j}^{(n)}}{\varepsilon^2} = \frac{D^2 \xi_i^{i_j}}{n\varepsilon^2},$$

hence the vector $(a_{i_1}^{(n)}, \ldots, a_{i_m}^{(n)})$ will assume values from outside the domain

$$A = \{(x_1, \ldots, x_m): |x_j - \alpha_{i_j}| < \varepsilon \text{ for } j = 1, \ldots, m\}$$

with a probability not exceeding $c/\varepsilon^2 n$, where $c = \sum_{j=1}^{m} D^2 \xi_i^{i_j}$. Substituting $\varepsilon = n^{-3/8}$. we obtain a sequence of increasingly smaller domains A_n with the property that the probability of the vector $(a_{i_1}^{(n)}, \ldots, a_{i_m}^{(n)})$ assuming any value from outside A_n tends to zero.

Let us now consider the expansion of the function $f(x_1, \ldots, x_m)$ in the neighbourhood of the point $(\alpha_{i_1}, \ldots, \alpha_{i_m})$ and, more precisely, in the domain A_n. We have

(44.1) $\quad f(x_1, \ldots, x_m) = f(\alpha_{i_1}, \ldots, \alpha_{i_m}) + c_1(x_1 - \alpha_{i_1}) + \ldots + c_m(x_m - \alpha_{i_m}) + R,$

where, according to the Taylor formula, $c_1, \ldots, c_m$ denote the first partial derivatives of $f(x_1, \ldots, x_m)$ with respect to the individual variables, while

$$R = \sum_{j=1}^{m} \sum_{k=1}^{m} H_{jk}(x_j - \alpha_{i_j})(x_k - \alpha_{i_k}),$$

where H_{jk} are mixed derivatives of the second order computed at some intermediate point between $(\alpha_{i_1}, \ldots, \alpha_{i_m})$ and $(x_1, \ldots, x_m)$. Denote by M the number which bounds from above the absolute values of the derivatives H_{jk}. We can write

$$|R| \leq \sum_{j=1}^{m} \sum_{k=1}^{m} |H_{jk}(x_j - \alpha_{i_j})(x_k - \alpha_{i_k})| \leq \sum_{j=1}^{m} \sum_{k=1}^{m} M n^{-3/4} = m^2 M n^{-3/4}.$$

Let us now write expansion (44.1) in the form

$$\sqrt{n}\left(f(x_1, \ldots, x_m) - f(\alpha_{i_1}, \ldots, \alpha_{i_m})\right) = \sqrt{n}\left(c_1(x_1 - \alpha_{i_1}) + \ldots + c_m(x_m - \alpha_{i_m})\right) + \sqrt{n}\,R,$$

and substitute the sample moments $a_{i_1}^{(n)}, \ldots, a_{i_m}^{(n)}$ for the variables $x_1, \ldots, x_m$.

We have shown that with a probability arbitrarily close to one the random variable $\sqrt{n}(\zeta_n - m)$ differs from the linear combination

$$\sqrt{n}\left(c_1(a_{i_1}^{(n)} - \alpha_{i_1}) + \ldots + c_m(a_{i_m}^{(n)} - \alpha_{i_m})\right)$$

of the random variables $\sqrt{n}(a_{i_1}^{(n)} - \alpha_{i_1}), \ldots, \sqrt{n}(a_{i_m}^{(n)} - \alpha_{i_m})$ by a random variable $\sqrt{n}\,R$ tending to zero in probability. It follows that $\sqrt{n}(\zeta_n - m)$ has the same limit distribution as the above linear combination. Since sample moments have an asymptotically normal joint distribution, the linear combination in question has also a limiting normal distribution, which completes the proof of Theorem 2.

Theorem 2 asserts that central moments and their continuously differentiable functions are asymptotically normally distributed.

We shall now present the theorem concerning the limit distribution of a sample quantile.

THEOREM 3. *Let $\xi_1 = \xi_1(e), \xi_2 = \xi_2(e), \ldots$ be a sequence of independent identically distributed random variables, and let $F(x)$ be their common probability distribution function. Next, let z_p, $0 < p < 1$, be the quantile of order p of the distribution $F(x)$, and let $\zeta_p^{(n)}$ be the largest quantile of order p in a sample of n elements (i.e. $\zeta_p^{(n)} = \zeta_p^{(n)}(e)$ is the largest number z satisfying the inequality*

$$F_n(z) \leqslant p \leqslant F_n(z+0)$$

where

$$F_n(x) = \frac{1}{n}\,\mathrm{Card}\{i\colon \xi_i < x, i = 1, 2, \ldots, n\}$$

is the empirical distribution function).

If $F(x)$ is differentiable in the neighbourhood of the point $x = z_p$, and has a continuous derivative $f(x) = F'(x)$ such that $f(z_p) > 0$, then the sample quantiles of order p are asymptotically normal with the expectation z and standard deviation

$$\frac{1}{f(z_p)}\sqrt{\frac{p(1-p)}{n}}.$$

PROOF. For fixed n and e, let $x_k^{(n)}$ denote the kth smallest of the numbers $\xi_1(e), \xi_2(e), \ldots, \xi_n(e)$. We know from § 33, that for the largest sample quantile $\zeta_p^{(n)}$ of order p we have

$$\zeta_p^{(n)} = x_{[np]+1}^{(n)}.$$

Thus, our problem is to find the limit distribution for the statistic $x_k^{(n)}$, where $k = [np]+1$.

Let us first find the probability distribution function of the statistic $x_k^{(n)}$. We have $x_k^{(n)} < x$ if and only if at least k of the random variables $\xi_1(e), \ldots, \xi_n(e)$ assume values smaller than x. Hence

$$P(\{e: x_k^{(n)} < x\}) = \sum_{i=k}^{n} \binom{n}{i} F^i(x)(1-F(x))^{n-i}.$$

By assumption, the distribution function $F(x)$ has a continuous derivative $f(x)$ in the neighbourhood of the point $x = z_p$. Thus, by differentiating we find that the statistic $x_k^{(n)}$ has, in the neighbourhood of the point $x = z_p$, a continuous probability density $f_k^{(n)}(x)$ given by the formula

$$f_k^{(n)}(x) = \frac{d}{dx}\left\{\binom{n}{k}F^k(x)(1-F(x))^{n-k} + \binom{n}{k+1}F^{k+1}(x)(1-F(x))^{n-k-1}\right.$$

$$\left. + \ldots + \binom{n}{n}F^n(x)\right\}$$

$$= f(x)\left\{\binom{n}{k}kF^{k-1}(1-F)^{n-k}\right.$$

$$- \binom{n}{k}(n-k)F^k(1-F)^{n-k-1} + \binom{n}{k+1}(k+1)F^k(1-F)^{n-k-1}$$

$$\left. - \binom{n}{k+1}(n-k-1)F^{k+1}(1-F)^{n-k-2} + \ldots + \binom{n}{n}nF^{n-1}\right\}.$$

Since

$$\binom{n}{k}(n-k) = \frac{n!}{(n-k)!k!}(n-k) = \frac{n!}{(n-k-1)!k!}$$

$$= \frac{n!}{(n-k-1)!(k+1)!}(k+1) = \binom{n}{k+1}(k+1),$$

most terms in the above formula cancel each other out, and we finally obtain the following expression for the density $f_k^{(n)}(x)$ of the statistic $x_k^{(n)}$ in the neighbourhood of the point $x = z_p$:

(44.2) $$f_k^{(n)}(x) = \binom{n}{k}kf(x)F^{k-1}(x)(1-F(x))^{n-k}.$$

We want to prove that the statistic $x_k^{(n)}$ with $k = [np]+1$ is asymptotically normal with the expectation z_p and standard deviation $\frac{1}{f(z_p)}\sqrt{\frac{p(1-p)}{n}}$. It suffices to show therefore that the statistic

$$\eta_n = \frac{x_k^{(n)} - z_p}{\frac{1}{f(z_p)}\sqrt{\frac{p(1-p)}{n}}}$$

has a distribution asymptotically normal with expectation 0 and variance 1.

VII. CENTRAL LIMIT THEOREMS

Note that if the random variable ξ has density $h(x)$ in the neighbourhood $z-\varepsilon < x < z+\varepsilon$ of the point z, then the random variable $A\xi+B$ where $A > 0$ has density $\frac{1}{A} h\left(\frac{x-B}{A}\right)$ in the neighbourhood $B+Az-A\varepsilon < x < B+Az+A\varepsilon$ of the point $x = B+Az$. Applying this to the statistic $x_k^{(n)}$ for

$$A = \frac{f(z_p)}{\sqrt{\frac{p(1-p)}{n}}}, \quad B = -\frac{z_p f(z_p)}{\sqrt{\frac{p(1-p)}{n}}}$$

we obtain the following result: if $x_k^{(n)}$ has density $f_k^{(n)}(x)$ in the neighbourhood $z_p - \varepsilon < x < z_p + \varepsilon$ of the point z_p, then the statistic η_n has the probability density

$$g_n(x) = \frac{\sqrt{p(1-p)}}{\sqrt{n} f(z_p)} f_k^{(n)}\left(z_p + x \frac{\sqrt{p(1-p)}}{f(z_p)\sqrt{n}}\right)$$

in the neighbourhood

(44.3) $$-\frac{f(z_p)\sqrt{n}}{\sqrt{p(1-p)}} < x < \frac{f(z_p)\sqrt{n}}{\sqrt{p(1-p)}}$$

of the point $x = 0$.

We shall prove that for every x

(44.4) $$\lim_{n \to \infty} g_n(x) = \frac{1}{\sqrt{2\pi}} e^{-x^2/2}.$$

Let us fix x. In view of inequality (44.3) the value of x will lie in the interval in which $g_n(x)$ is defined starting from some n. Using (44.2), we may write

$$g_n(x) = \frac{\sqrt{p(1-p)}}{\sqrt{n} f(z_p)} f(y) \binom{n}{k} k F^{k-1}(y)(1-F(y))^{n-k}$$

$$= \frac{f(y)}{f(z_p)} \binom{n}{k} (1-p)^{n-k} \sqrt{np(1-p)} \frac{k}{np} \left(\frac{F(y)}{p}\right)^{k-1} \left(\frac{1-F(y)}{1-p}\right)^{n-k},$$

where we put for brevity

$$y = z_p + x \frac{\sqrt{p(1-p)}}{f(z_p)\sqrt{n}}.$$

The density $f(x)$ is continuous by assumption, whence

$$\lim_{n \to \infty} \frac{f(y)}{f(z_p)} = 1.$$

Next, in view of Theorem 1 of § 22 and the fact that for $k = [np]+1$ we have
$$\lim_{n\to\infty} \frac{k-np}{\sqrt{np(1-p)}} = \lim_{n\to\infty} \frac{[np]+1-np}{\sqrt{np(1-p)}} = 0,$$
we obtain the relation
$$\lim_{n\to\infty} \binom{n}{k} p^k (1-p)^{n-k} \sqrt{np(1-p)} = \frac{1}{\sqrt{2\pi}}.$$
Since $k = [np]+1$, we may also write
$$\lim_{n\to\infty} \frac{k}{np} = 1.$$

To complete the proof of our theorem it remains to show that

(44.5) $$\lim_{n\to\infty} \left(\frac{F(y)}{p}\right)^{k-1} \left(\frac{1-F(y)}{1-p}\right)^{n-k} = e^{-x^2/2}.$$

Applying the Lagrange theorem on the mean value to $F(y)$, and using the continuity of the density $f(x)$, we may write
$$F(y) = F\left(z_p + \frac{x}{\sqrt{n}} \cdot \frac{\sqrt{p(1-p)}}{f(z_p)}\right) = p + a \frac{x}{\sqrt{n}},$$
where

(44.6) $$\lim_{n\to\infty} a = \sqrt{p(1-p)}.$$

Thus
$$\left(\frac{F(y)}{p}\right)^{k-1} \left(\frac{1-F(y)}{1-p}\right)^{n-k} = \left(\frac{p+a\frac{x}{\sqrt{n}}}{p}\right)^{k-1} \left(\frac{q-a\frac{x}{\sqrt{n}}}{q}\right)^{n-k},$$
where $q = 1-p$. Now, instead of proving (44.5), we show that

(44.7) $$\lim_{n\to\infty} \log \left(\frac{F(y)}{p}\right)^{k-1} \left(\frac{1-F(y)}{1-p}\right)^{n-k} = -\frac{1}{2} x^2.$$

We have the following expansion:
$$\log(1+u) = u - \frac{u^2}{2} + o(u^2).$$

Applying this expansion to our case, we obtain
$$\log \left(\frac{F(y)}{p}\right)^{k-1} \left(\frac{1-F(y)}{1-p}\right)^{n-k}$$
$$= (k-1)\log\left(1 + \frac{a}{p} \cdot \frac{x}{\sqrt{n}}\right) + (n-k)\log\left(1 - \frac{a}{q} \cdot \frac{x}{\sqrt{n}}\right)$$

$$= (k-1)\frac{a}{p}\cdot\frac{x}{\sqrt{n}} - (n-k)\frac{a}{q}\cdot\frac{x}{\sqrt{n}} - \frac{1}{2}(k-1)\frac{a^2}{p^2}\cdot\frac{x^2}{n}$$
$$-\frac{1}{2}(n-k)\frac{a^2}{q^2}\cdot\frac{x^2}{n} + o(1)$$
$$= a\frac{x}{\sqrt{n}}\left\{\frac{k-1}{p} - \frac{n-k}{q}\right\} - \frac{a^2}{2}\cdot\frac{x^2}{n}\left\{\frac{k-1}{p^2} + \frac{n-k}{q^2}\right\} + o(1).$$

From the equality
$$\lim_{n\to\infty}\frac{1}{\sqrt{n}}\left\{\frac{k-1}{p} - \frac{n-k}{q}\right\} = \lim_{n\to\infty}\frac{1}{\sqrt{n}}\cdot\frac{(k-1)q-(n-k)p}{pq}$$
$$= \lim_{n\to\infty}\frac{1}{\sqrt{n}}\cdot\frac{[np]q\{-n-[np]-1\}p}{pq}$$
$$= \lim_{n\to\infty}\frac{1}{\sqrt{n}}\cdot\frac{[np]-np+p}{pq} = 0,$$

and
$$\lim_{n\to\infty}\frac{1}{n}\left\{\frac{k-1}{p^2} + \frac{n-k}{q^2}\right\} = \lim_{n\to\infty}\frac{1}{n}\cdot\frac{(k-1)q^2+(n-k)p^2}{p^2q^2}$$
$$= \lim_{n\to\infty}\frac{1}{n}\cdot\frac{k(q^2-p^2)+np^2-q^2}{p^2q^2}$$
$$= \lim_{n\to\infty}\frac{1}{n}\cdot\frac{np(q-p)+np^2}{p^2q^2} = \frac{p(q-p)+p^2}{p^2q^2} = \frac{1}{pq},$$

by use of formula (44.6) relation (44.7) follows, which completes the proof of Theorem 3.

Finally, we shall present without proof the famous theorem of Kolmogorov (1933), supplementing Glivenko's theorem of § 33 by giving the limit distribution.

THEOREM 4 (Kolmogorov). *Let $\xi_1 = \xi_1(e), \xi_2 = \xi_2(e), \ldots$ be a sequence of independent identically distributed random variables with a common continuous distribution function $F(x)$. Let $F_n(x)$ be the empirical distribution function, i.e. let*

$$F_n(x) = F_n(x; e) = \frac{1}{n}\mathrm{Card}\{i\colon \xi_i(e) < x,\ i = 1, 2, \ldots, n\}$$

(*as before, Card A denotes the number of elements of the set A*). *Finally, let*

$$\delta_n = \sup_{-\infty < x < \infty} |F_n(x) - F(x)|.$$

Then for every x we have the convergence

$$\lim_{n\to\infty} P(\{e\colon \sqrt{n}\delta_n < x\}) = K(x)$$

where $K(x)$ is the probability distribution function defined by the formula

$$K(x) = \begin{cases} 0 & \text{for } x \leqslant 0, \\ \sum_{k=-\infty}^{+\infty} (-1)^k e^{-2k^2 x^2} & \text{for } x > 0. \end{cases}$$

Problems

1. Prove that if the distribution in the population has a continuous density $f(x)$, and for $0 < p_1 < p_2 < 1$ we have $f(z') > 0$ and $f(z'') > 0$, where $z' = z_{p_1}$ and $z'' = z_{p_2}$ are quantiles of orders p_1 and p_2, then the pair $(\zeta_{p_1}^{(n)}, \zeta_{p_2}^{(n)})$ of sample quantiles has an asymptotically normal bivariate distribution with expectations z' and z'' and the covariance matrix

$$\begin{bmatrix} \dfrac{p_1(1-p_1)}{nf^2(z')} & \dfrac{p_1(1-p_2)}{nf(z')f(z'')} \\ \dfrac{p_1(1-p_2)}{nf(z')f(z'')} & \dfrac{p_2(1-p_2)}{nf^2(z'')} \end{bmatrix}.$$

2. *Angular transformation.* The fraction m/n of the number of successes in n independent trials with the probability of success equal to p has an asymptotically normal distribution with the expectation p and variance $\dfrac{1}{n}p(1-p)$. Find the function $g(u)$ such that the variance of the asymptotic distribution of the random variable $g(m/n)$ is independent of p.

Answer: $g(u) = \arcsin\sqrt{u}$.

CHAPTER VIII

Statistical Inference

§ 45. EXAMPLES OF TESTS RELATED TO THE NORMAL DISTRIBUTION

In the laws of large numbers and limit theorems of probability theory which we have discussed so far we assumed a fixed (though perhaps unknown) probabilistic description of the scheme of a random experiment in the form of a probability distribution, and we assumed the independence of successive repetitions of the experiment. Under these assumptions we explained why and in what sense the sample may be regarded as a miniature of the population; more precisely, we explained in what sense sample characteristics such as mean, variance, quantiles, empirical distribution function, etc., converge to the corresponding characteristics of the distribution of the population. These considerations show, therefore, how exactly we estimate the characteristics of the population on the basis of the sample. However, we did not make use of the fact that one usually has some prior information about the population. For instance, we may want to estimate the expectation of the population on the basis of a sample, and we know that the distribution of the population is, say, normal, or Poisson, or—more generally—belongs to a certain family of distributions. It is one of the features distinguishing problems of mathematical statistics from problems of probability theory that in the former we stress the use of information not contained in the sample when estimating the parameters of the population on the basis of the sample. The mathematical structure of problems of mathematical statistics, which we shall presently discuss, is, generally speaking, the following: we know that the distribution of a population belongs to a certain family of distributions, and we want to decide, on the basis of a sample, which distribution of that family is the actual distribution of the population in question.

Thus, the main feature distinguishing the problems of mathematical statistics lies in fact that they are concerned with the best possible identification of the distribution of the population, and using the prior knowledge of what class of distributions comprises the distribution in question.

According to the desired degree of identification of the distribution, we distinguish two large classes of problems. The requirements are strictest in *estimation problems*. They are concerned with evaluating on the basis of a sample some parameters of a distribution which usually determine this distribution uniquely. At the other extreme we have problems of *verification*, or *hypothesis testing*, where the class of distributions occurring in a given our problems is partitioned into two subclasses, and we want to decide on the basis of a sample which of these subclasses contains the distribution of the population in question.

We shall discuss first the statistical tests, starting from some examples connected with the normal distribution.

Below we present a description of empirical situations leading to the verification of a hypothesis concerning the expectation of the normal distribution.

EXAMPLE 1. *One-sided test for the hypothesis of zero expectation in the normal distribution with a known variance.* Suppose that an experimental station has succeeded in growing a new variety of tomatoes, apparently yielding higher crops. How can we verify this fact? We know that the crop depends on a number of factors, many of them beyond control, such is for instance the local soil conditions. Thus, experiments are often conducted with the following way: we grow tomato plants in pairs, one member of each pair being of an old variety and the other of a new variety of tomatoes; we observe the differences in crops between the members of every pair. As a result we obtain a system of pairs of numbers, on the basis of which we are to decide whether the new variety is better than the old as regards the magnitude of crops. The trouble in answering this question arises from the fact that plants of different varieties may occasionally give crops within the same interval, and we cannot exclude the possibility that it is only by accident that the new variety has given a slightly better average crops than the old. What difference in the average crops can be regarded as accidental and how great must it be to constitute a reliable indication that the new variety is better?

Suppose that $x_1, ..., x_n$ are the observed differences of crops in pairs of neighbouring plants, and suppose that we know that these differences have normal distribution with a known standard deviation σ. Then our reasoning goes as follows: if there were no difference between the tomato varieties in question, the observed differences would have a normal distribution with the expectation 0. The average

$$\bar{x} = \frac{1}{n}(x_1 + ... + x_n)$$

can thus be treated as the observed average of n independent observations of random variables with a normal distribution with expectation 0 and standard deviation σ, i.e., as an observation of a random variable with a normal distribu-

VIII. STATISTICAL INFERENCE

tion with expectation 0 and standard deviation $\sigma/\sqrt{n}$. We think that the greater is the value of $\bar{x}$, the more strongly our observation indicates that the new variety is better. We should determine the limit starting from which observation $\bar{x}$ can be regarded as inconsistent with the assumption that there is no difference between the two varieties. As this limit we take a number λ such that the probability, under the null hypothesis of there being no difference between the two varieties, that the average $\bar{x}$ exceeds λ is small (say equals 0.05). In tables of the normal distribution we find that the function

$$\Phi(x) = \frac{1}{\sqrt{2\pi}} \int_{-\infty}^{x} e^{-t^2/2} dt$$

assumes value 0.95 for $x = 1.645$, whence as λ we may take the number $1.645\sigma/\sqrt{n}$. Thus, we have found a method of deciding, on the basis of the sample, whether the new variety is better or not:

If $\bar{x} \leqslant \lambda$ we decide that there is no difference between the two varieties;

If $\bar{x} > \lambda$ we decide that the new variety is better than the old one.

Let us look at our construction formally. We have assumed that the points $x_1, \ldots, x_n$ of the n-dimensional Euclidean space represent the results of our random experiment. They constitute elementary events, or—as we usually say in statistics—sample points. Next, we have assumed that our random experiment may be described by a probability distribution $P(A)$ defined on the space X of all sample points, this distribution depending on one parameter m, equal to the expected difference of crops in a pair of plants of the new and the old varieties. More precisely, we have assumed that we deal with a probability distribution with the density

$$f(x_1, \ldots, x_n; m) = \frac{1}{(\sigma\sqrt{2\pi})^n} \exp\left[-\frac{(x_1-m)^2}{2\sigma^2}\right] \ldots \exp\left[-\frac{(x_n-m)^2}{2\sigma^2}\right],$$

where m is a non-negative number, and σ is constant and known. Each value of m specifies the probability distribution in the space X, hence it constitutes a certain *statistical hypothesis*. We call such a hypothesis *simple*, since it specifies exactly the probability distribution, and *parametric*, since the specification of the probability distribution in space X has the form of a specification of a numerical value of a certain parameter. As admissible hypotheses we take here those probability distribution which correspond to the values of m from the interval $0 \leqslant m < \infty$, or, which amounts to the same, the value of the parameter m from this interval. We single out the *null hypothesis* $m = 0$; it is distinguished by the interpretation which we assign to it: there is no increment of crops. The hypotheses corresponding to the positive values of the parameter m are called *alternatives* of the null hypothesis; these hypotheses are *simple*. The supposition $m > 0$ constitutes a *composite* hypothesis, since it consists of a set of points in the space of parameters: $K = \{m: 0 < m < \infty\}$. We do not admit negative values for the parameter m, as assuming that, since we have tried to obtain a better variety, the new variety is no worse than the old one. Thus, we test the hypothesis $m = 0$ against the alternative $m > 0$. The test is given in the form of a function defined on the space of all sample points and assuming two values: "accept the null hypothesis" and "reject the null hypothesis". The set of points $(x_1, \ldots, x_n)$ connected with the second of these two decisions is called the *critical region*. In our case

$$W = \{(x_1, \ldots, x_n): \lambda < \bar{x}\}.$$

Naturally, the test is completely determined by the critical region. Thus, in the sequel, we shall speak of tests W, instead of tests with critical regions W. Since, in turn, the critical region W is determined by the condition $\lambda < \bar{x}$, we often say that this condition constitutes the test. Thus, the test which we have just defined is a one-sided test for the hypothesis of zero expectation in a normal distribution with a known variance.

The probability $\alpha = P\{W|m = 0\}$, or, as we say, the probability of rejecting the null hypothesis when it is true, is called the *significance level of test W*. For $n = 10$ and $\alpha = 0.05$ we would have to take for λ the value $1.645\sigma/\sqrt{10}$.

Below, we give other variants of tests connected with comparing the average crops of different varieties of tomatoes from Example 1.

EXAMPLE 2. *Two-sided test of the hypothesis concerning the expectation in a normal distribution with a known variance.* In Example 1 we described the situation in which the new variety was no worse than the old one. Imagine now another situation, where the conditions are as before: we observe differences in crops in the pairs of plants of the two varieties, but we have no fixed prior opinion as to which variety is better. We simply want to compare the crops of the two varieties.

Now the space of parameters changes: instead of the space $\{m: 0 \leq m < \infty\}$ we have $\{m: -\infty < m < \infty\}$. Positive values of m signify that the first variety is better than the second, while negative values of m signify the opposite. The null hypothesis is still $m = 0$, but the new alternative hypothesis is now represented by the set $\{m: m \neq 0\}$. In the present situation, the values of $\bar{x}$ differing only in the sign will constitute equally strong indications for or against the null hypothesis, since the set of alternative hypotheses is symmetric with respect to the point $m = 0$. Using the fact that under the null hypothesis the value $\bar{x}$

may be treated as the value assumed by a random variable with the normal distribution $N(0, \sigma/\sqrt{n})$, we build a test with a significance level α in a different way, namely by choosing λ so that for the critical set of the form
$$W = \{(x_1, ..., x_n): \lambda < |\bar{x}|\}$$
we have
$$P\{W|m = 0\} = \alpha.$$
We have thus obtained a two-sided test for the hypothesis of zero expectation in a normal distribution with a known variance. For $\alpha = 0.05$ we find in tables of the normal distribution
$$2(1-\Phi(x)) = 0.05$$
for $x = 1.9600$, whence $\lambda = 1.96\sigma/\sqrt{n}$.

EXAMPLE 3. *Testing the hypothesis of the equality of the means in normal populations with the same known variance.* Suppose now that we want to compare the average crops of two varieties of tomatoes, but the setting of the experiment is different: taking care that all side conditions affecting the crop are identical, we observe separately the crops $x_1, ..., x_n$ on n plants of the first variety, and the crops $y_1, ..., y_k$ on k plants of the second variety. Suppose, moreover, that these crops can be treated as values of independent random variables having a normal distribution with a known variance σ^2 and unknown expectations m_1 for the first variety and m_2 for the second variety. Then we have the space of sample points $(x_1, ..., x_n, y_1, ..., y_k)$ of an $(n+k)$-dimensional Euclidean space, and the probability distribution in this space, depending on the two parameters m_1 and m_2, has the density

$$f(x_1, ..., x_n, y_1, ..., y_k)$$
$$= \frac{1}{(\sigma\sqrt{2\pi})^{n+k}} \prod_{j=1}^{n} \exp\left\{-\frac{(x_j-m_1)^2}{2\sigma^2}\right\} \prod_{j=1}^{k} \exp\left\{-\frac{(y_j-m_2)^2}{2\sigma^2}\right\}.$$

Now the parameter space is the Euclidean plane
$$P = \{(m_1, m_2): -\infty < m_1 < \infty, -\infty < m_2 < \infty\}.$$
The null hypothesis is now composite: it consists of the set $H_0 = \{(m_1, m_2): m_1 = m_2\}$, the alternative hypothesis is also composite: it equals to the set $\{(m_1, m_2): m_1 \neq m_2\}$. The test for the null hypothesis at the significance level α may be constructed by using the fact that the difference of means $\bar{x}-\bar{y}$, where $\bar{x} = \frac{1}{n}(x_1 + ... + x_n)$, $\bar{y} = \frac{1}{k}(y_1 + ... + y_k)$ can be treated as a value of a random variable

with a normal distribution with the expectation m_1-m_2 and standard deviation

$$\sigma\sqrt{\frac{1}{n}+\frac{1}{k}}=\sigma\sqrt{\frac{n+k}{nk}}.$$

Under the null hypothesis we have $m_1-m_2=0$, and the distribution of the difference $\bar{x}-\bar{y}$ is independent of the particular point (m_1, m_2) of H_0; for this reason we may speak here of the test for hypothesis H_0 at the significance level α.

The critical region may be defined by choosing λ so that for the critical set of the form

$$W=\{(x_1,\ldots,x_n,y_1,\ldots,y_k)\colon \lambda<|\bar{x}-\bar{y}|\}$$

we have

$$P\{W|(m_1,m_2)\in H_0\}=\alpha.$$

For $\alpha=0.01$ we find from the tables of the normal distribution that $\lambda=2.56\sigma\sqrt{(n+k)/nk}$.

EXAMPLE 4. *Student's test; testing the hypothesis of zero expectation in a normal distribution with unknown variance.* Suppose now that we are in the same situation as in Example 1, but we do not know the value σ. In this case we must consider the sample space X of points $(x_1,\ldots,x_n)$ and a two-parameter family of distributions with the densities

$$f(x_1,\ldots,x_n)=\frac{1}{(\sqrt{2\pi})^n\sigma^n}\prod_{i=1}^{n}\exp\left\{-\frac{(x_i-m)^2}{2\sigma^2}\right\}.$$

Now the parameter space will have the form

$$P=\{(m,\sigma)\colon 0\leqslant m<\infty, 0<\sigma<\infty\},$$

and the null hypothesis will be composite:

$$H_0=\{(m,\sigma)\colon m=0, 0<\sigma<\infty\}.$$

The alternative hypothesis will be

$$\{(m,\sigma)\colon m>0, 0<\sigma<\infty\}.$$

How can we build a critical region so as to obtain a test for H_0 at the significance level α? The point is that the same value $\bar{x}$ will constitute an indication for or against the null hypothesis, but the strength of this indication will depend on σ. Thus, if we took a critical region such as in Example 1, the probability of rejecting the null hypothesis would depend on σ, and we could not speak of a test at the significance level α. We have to replace the average $\bar{x}$ by another function of the sample $(x_1,\ldots,x_n)$. It turns out that one can obtain a function of a sample, i.e., a *statistic*, whose distribution is independent of the particular point of H_0

VIII. STATISTICAL INFERENCE

in the parameter space. By the results of § 33 we know that the sample standard deviation

$$(45.1) \qquad s = \sqrt{\frac{1}{n}\{(x_1-\bar{x})^2 + \ldots + (x_n-\bar{x})^2\}}$$

can be treated as an approximation of the population standard deviation σ. We also know that the statistic $\bar{x}/(\sigma/\sqrt{n})$ would have, under the null hypothesis, a normal distribution with expectation 0 and variance 1, independently of the choice of the point of H_0. This suggests using in our case *Student's t statistic*, defined as

$$(45.2) \qquad t_{n-1} = \frac{\bar{x}}{s}\sqrt{n-1}.$$

We shall prove in Theorem 1 that the statistic t has Student's t distribution with $n-1$ degrees of freedom, independently of the choice of the point from H_0. Thus, we may choose the number λ so that for the critical region of the form

$$W = \left\{(x_1, \ldots, x_n) : \lambda < \frac{\bar{x}}{s}\sqrt{n-1}\right\}$$

we have

$$P\{W|(m, \sigma) \in H_0\} = \alpha.$$

Then W will be the required test at the significance level α.

For example, if $n = 10$ and $\alpha = 0.05$, we have to take for λ the value 1.833, i.e., by 0.188 more than the number 1.645 by which we had to multiply the standard deviation in Example 1.

We now present a theorem which is basic for the construction given in Example 4.

THEOREM 1. *If $\xi_1, \ldots, \xi_n$ are independent random variables having a normal distribution with the expectation m and standard deviation σ, then*

(a) *the sample mean*

$$\bar{x} = \frac{1}{n}(x_1 + \ldots + x_n)$$

and the sample standard deviation

$$s = \sqrt{\frac{1}{n}\{(x_1-\bar{x})^2 + \ldots + (x_n-\bar{x})^2\}}$$

are stochastically independent;

(b) *the statistic ns^2/σ^2 has a chi-square distribution with $n-1$ degrees of freedom;*

(c) *the statistic*

$$t = \frac{\bar{x}-m}{s}\sqrt{n-1}$$

has Student's t-distribution with $n-1$ *degrees of freedom.*

PROOF. Without loss of generality we may assume that $m = 0$. To prove assertion (a) it suffices to show (see Theorem 7 of § 28) that the random variables $\sqrt{n}\bar{x} = (\xi_1 + \ldots + \xi_n)/\sqrt{n}$ and $ns^2 = (\xi_1-\bar{x})^2 + \ldots + (\xi_n-\bar{x})^2$ are stochastically independent.

Note that we have the equality

$$ns^2 = (\xi_1-\bar{x})^2 + \ldots + (\xi_n-\bar{x})^2$$
$$= \xi_1^2 + \ldots + \xi_n^2 - n\bar{x}^2 = \xi_1^2 + \ldots + \xi_n^2 - (\sqrt{n}\bar{x})^2.$$

Next, the expression

$$\sqrt{n}\bar{x} = \frac{1}{\sqrt{n}}\xi_1 + \ldots + \frac{1}{\sqrt{n}}\xi_n = c_{11}\xi_1 + \ldots + c_{1n}\xi_n$$

may be treated as the first row of the orthogonal transformation

$$\eta_1 = c_{11}\xi_1 + \ldots + c_{1n}\xi_n,$$
$$\ldots\ldots\ldots\ldots\ldots\ldots\ldots\ldots\ldots\ldots$$
$$\eta_n = c_{n1}\xi_1 + \ldots + c_{nn}\xi_n,$$

where $\eta_1 = \sqrt{n}\bar{x}$. We use here the theorem, known from algebra, asserting that if we have an arbitrary number of rows satisfying the requirements for orthogonal matrices, we can always complete them to an orthogonal matrix by adding other rows. From Theorem 3 of § 42 we know that the random variables $\eta_1, \ldots, \eta_n$ as linear combinations of normally distributed random variables, have themselves normal distribution. Their expectations are zero, their variances $E\eta_j^2$ are equal to σ^2, and the covariances are $E\eta_i\eta_j = 0$. This shows that the random variables $\eta_1, \ldots, \eta_n$ are stochastically independent and have identical normal distributions. Next, in view of the orthogonality of the matrix $C = [c_{jk}]$ we have for every elementary event e:

$$\eta_1^2 + \ldots + \eta_n^2 = \sum_{j=1}^{n}\left(\sum_{k=1}^{n}c_{jk}\xi_k\right)^2 = \sum_{j=1}^{n}\sum_{k=1}^{n}\sum_{s=1}^{n}c_{jk}c_{js}\xi_k\xi_s$$
$$= \sum_{k=1}^{n}\sum_{s=1}^{n}\xi_k\xi_s\sum_{j=1}^{n}c_{jk}c_{js} = \xi_1^2 + \ldots + \xi_n^2.$$

VIII. STATISTICAL INFERENCE

It follows that in the whole space of elementary events we have

$$\sqrt{n}\bar{x} = \eta_1,$$

$$ns^2 = \xi_1^2 + \ldots + \xi_n^2 - (\sqrt{n}\bar{x})^2 = \eta_1^2 + \ldots + \eta_n^2 - \eta_1^2 = \eta_2^2 + \ldots + \eta_n^2.$$

In view of the independence of the random variables $\eta_1, \ldots, \eta_n$ the assertion (a) of Theorem 1 follows.

Further, we have found that ns^2 equals the sum of the squares of $n-1$ independent random variables with the same normal distribution $N(0, \sigma)$. We may therefore write

$$\frac{ns^2}{\sigma^2} = \left(\frac{\eta_2}{\sigma}\right)^2 + \ldots + \left(\frac{\eta_n}{\sigma}\right)^2,$$

and now the random variables $\eta_2/\sigma, \ldots, \eta_n/\sigma$ have variance 1. The assertion (b) of Theorem 1 follows.

Finally, the formula defining the statistic t, after taking into account the assumption $m = 0$ can be written as follows:

$$t = \frac{\bar{x}\sqrt{n-1}}{s} = \frac{\dfrac{\bar{x}}{\sigma/\sqrt{n}}}{\sqrt{\dfrac{1}{n-1} \cdot \dfrac{ns^2}{\sigma^2}}}.$$

In view of Theorem 1, assertion (a), the numerator and the denominator are stochastically independent, the numerator has a normal distribution $N(0, 1)$, and in view of (b), the denominator has a distribution such as $\sqrt{\chi^2/(n-1)}$, where χ^2 has a chi-square distribution with $n-1$ degrees of freedom. In view of Definition 2 of § 39, assertion (c) follows, which completes the proof of Theorem 1.

The assertion (a), expressing a peculiar property of the normal distribution is worthy of notice: the mean and the standard deviation, though computed from the same sample, are stochastically independent, as if they were computed from two independent samples.

EXAMPLE 5. *Student's test; testing the hypothesis of the equality of expectations in two normal populations with the same unknown standard deviation.* Consider now a situation differing from that described in Example 3 only in the assumption that we do not know the standard deviation σ, knowing only that this standard deviation is the same for both varieties of tomatoes. How can we decide which variety is better?

In the sample space consisting of points $(x_1, \ldots, x_n, y_1, \ldots, y_k)$ we now have a family of distributions depending on three numerical parameters m_1, m_2 and σ,

with a probability density given by
$$f(x_1, \ldots, x_n, y_1, \ldots, y_k)$$
$$= \frac{1}{(\sigma\sqrt{2\pi})^{n+k}} \prod_{j=1}^{n} \exp\left(-\frac{(x_j-m_1)^2}{2\sigma^2}\right) \prod_{j=1}^{k} \exp\left(-\frac{(y_j-m_2)^2}{2\sigma^2}\right).$$

The parameter space has the form
$$P = \{(m_1, m_2, \sigma): -\infty < m_1 < \infty, -\infty < m_2 < \infty, 0 < \sigma < \infty\}.$$
The role of the composite null hypothesis is played by the set
$$H_0 = \{(m_1, m_2, \sigma): m_1 = m_2, \sigma > 0\}$$
while the alternative hypothesis, also composite, is
$$\{(m_1, m_2, \sigma): m_1 \neq m_2, \sigma > 0\}.$$

If we want to construct a test for the hypothesis H_0 at the significance level α we encounter a similar difficulty to that arising in the preceding example. The difference $\bar{x}-\bar{y}$, satisfactory in Example 3, now has a distribution depending on σ. It turns out, however, that we can proceed as in the preceding example, and construct a statistic with a distribution independent of the choice of a point from the set H_0. Indeed, we may take the statistic
$$u = \frac{\bar{x}-\bar{y}}{\sqrt{\frac{k+n}{kn} \cdot \frac{ns_1^2+ks_2^2}{k+n-2}}},$$
where
$$ns_1^2 = \sum_{j=1}^{n} (x_j-\bar{x})^2, \quad \bar{x} = \frac{1}{n}(x_1 + \ldots + x_n),$$
$$ks_2^2 = \sum_{j=1}^{k} (y_j-\bar{y})^2, \quad \bar{y} = \frac{1}{k}(y_1 + \ldots + y_k).$$

As proved in Theorem 2, the statistic u has Student's t distribution with $k+n-2$ degrees of freedom, independently of the choice of a point from H_0. Thus, we may select λ so that for the critical region of the form
$$W = \{(x_1, \ldots, x_n, y_1, \ldots, y_k): \lambda < |u|\}$$
we have
$$P\{W|(m_1, m_2, \sigma) \in H_0\} = \alpha.$$
In this manner we obtain the required test for the hypothesis H_0 at the significance level α.

We now give the theorem essential for the construction of the test in Example 5.

THEOREM 2. *If $\xi_1, \ldots, \xi_n, \eta_1, \ldots, \eta_k$ are independent random variables having identical normal distribution with the expectation m and standard deviation σ, then the statistic*

$$u = \frac{\bar{x} - \bar{y}}{\sqrt{\frac{k+n}{kn} \cdot \frac{ns_1^2 + ks_2^2}{k+n-2}}},$$

where

$$ns_1^2 = \sum_{j=1}^{n} (\xi_j - \bar{x})^2, \quad \bar{x} = \frac{1}{n}(\xi_1 + \ldots + \xi_n),$$

$$ks_2^2 = \sum_{j=1}^{k} (\eta_j - \bar{y})^2, \quad \bar{y} = \frac{1}{k}(\eta_1 + \ldots + \eta_k)$$

has Student's t distribution with $k+n-2$ degrees of freedom.

PROOF. By Theorem 1(a) $\bar{x}$ is independent of s_1^2 and $\bar{y}$ is independent of s_2^2. Moreover, $(\bar{x}, s_1^2)$ and $(\bar{y}, s_2^2)$ are independent bivariate random variables; it follows that the random variables $\bar{x}, \bar{y}, s_1^2, s_2^2$ are all independent. In particular, we have the independence of $\bar{x} - \bar{y}$ and $ns_1^2 + ks_2^2$. The difference $\bar{x} - \bar{y}$ has a normal distribution with expectation 0 and standard deviation $\sigma\sqrt{(k+n)/kn}$. By Theorem 1(b) the random variable ns_1^2/σ^2 has a chi-square distribution with $n-1$ degrees of freedom, while ks_2^2/σ^2 has a chi-square distribution with $k-1$ degrees of freedom, whence their sum $(ns_1^2 + ks_2^2)/\sigma^2$ has a chi-square distribution with $k+n-2$ degrees of freedom. Writing the formula defining u as

$$u = \frac{\bar{x} - \bar{y}}{\sigma\sqrt{\frac{k+n}{kn}}} \cdot \frac{1}{\sqrt{\frac{1}{k+n-2} \cdot \frac{ns_1^2 + ks_2^2}{\sigma^2}}}$$

we can easily see that u has the probability distribution described in Definition 2 of § 39, which completes the proof.

EXAMPLE 6. *Testing the hypothesis concerning variance in the normal distribution.* To determine the average dampness of seeds in a sack containing some fifty pounds of seed, we usually take a samples of a few ounces. In the laboratory this sample is proved out into several, usually three or four, bowls. By weighing the seeds before and after drying them in a special drying compartment, the average dampness is determined for each bowl separately. The average of these results is usually taken as the estimate of the average dampness of seeds in the sack. The procedure described above is to ensure a possibly exact determination of average dampness. Indeed, we may expect that, owing to various factors, such as inexact weighing, a somewhat uneven process of drying of seeds, and—

perhaps—variations in the dampness of seeds in the sack, the results of the measurements in different bowls may differ. The laboratory work is satisfactory if these deviations are not too large.

The central limit theorem of probability theory may be used as one of the arguments justifying the assumption that the measurements of dampness in different bowls may be treated as values of independent random variables having a normal distribution with an expectation equal to the actual average dampness of seeds in the population, and a standard deviation indicating the error of measurements; and sometimes called outright the *standard error of measurement*.

In situations such as the above we need to test the hypothesis $\sigma^2 \leqslant \sigma_0^2$ that the variance of the measurements does not exceed a given number σ_0^2.

Let $x_1, \ldots, x_n$ denote the results of measurements of dampness. As admissible hypotheses we take the assumptions asserting that these measurements have the probability density

$$f(x_1, \ldots, x_n; m, \sigma) = \prod_{i=1}^{n} \frac{1}{\sigma\sqrt{2\pi}} \exp\left\{-\frac{(x_i-m)^2}{2\sigma^2}\right\}$$

depending on two parameters, m and σ. The null hypothesis is composite:

$$H = \{(m, \sigma): -\infty < m < \infty, 0 < \sigma \leqslant \sigma_0\}$$

and the alternative hypothesis is also composite:

$$K = \{(m, \sigma): -\infty < m < \infty, \sigma_0 < \sigma < \infty\}.$$

By Theorem 1(b) we know that the statistic

$$\frac{ns^2}{\sigma^2} = \frac{1}{\sigma^2} \sum_{i=1}^{n} (x_i - \bar{x})^2$$

has a chi-square distribution with $n-1$ degrees of freedom, independently of the value of the parameter m. Thus, we may take the statistic ns^2 for the construction of the test.

Consider the test with the critical region

$$W = \{(x_1, \ldots, x_n): \lambda < ns^2\}$$

where λ is determined from the condition $P\{W|\sigma_0\} = \alpha$.

Since the statistic ns^2/σ^2 has the same distribution independently of the value of m, we have for $0 < \sigma < \sigma_0$:

$$P\{\lambda < ns^2|\sigma\} < P\{\lambda < ns^2|\sigma_0\} = \alpha$$

and for $\sigma_0 < \sigma < \infty$:

$$P\{\lambda < ns^2|\sigma\} > P\{\lambda < ns^2|\sigma_0\} = \alpha.$$

VIII. STATISTICAL INFERENCE

Thus, the number α is the upper bound of the probability of rejection of the null hypothesis H, the bound being taken with respect to all simple hypotheses contained in H. This upper bound will be called the *level of significance of the test* if the hypothesis is composite.

Thus, the test defined above is a test of the hypothesis H with the significance level α.

EXAMPLE 7. *Testing the hypothesis concerning the value of the regression coefficient.* Let us consider a population of elements characterized by two numerical characteristics with a joint normal distribution with expectations m_1, m_2 and the covariance matrix

$$\Lambda = \begin{bmatrix} \lambda_{11} & \lambda_{12} \\ \lambda_{21} & \lambda_{22} \end{bmatrix}.$$

Let $(x_1, y_1), \ldots, (x_n, y_n)$ be an n-element sample drawn from this population. The number

$$\beta = \lambda_{12}/\lambda_{11}$$

is the (see problem 1 of § 40) regression coefficient of the second characteristic on the first. We do not know the value of β, nor do we know any of the numbers $m_1, m_2, \lambda_{11}, \lambda_{12}, \lambda_{22}$. We want to test the hypothesis H that $\beta = \beta_0$.

We have here a five-parameter family of distributions in a $2n$-dimensional sample space of points $(x_1, y_1; \ldots; x_n, y_n)$. The parameters are $m_1, m_2, \lambda_{11}, \lambda_{12}$ and λ_{22}. The null hypothesis is defined by the condition

$$\beta \overset{\text{df}}{=} \frac{\lambda_{12}}{\lambda_{11}} = \beta_0.$$

The question arises of constructing a statistic which would have the same probability distribution for all distributions belonging to the null hypothesis. The statistic

(45.3) $$t = \frac{(b-\beta_0)s_1\sqrt{n-2}}{s_2\sqrt{1-r^2}}$$

where

$$s_1^2 = \frac{1}{n}\sum_{i=1}^{n}(x_i-\bar{x})^2, \quad \bar{x} = \frac{1}{n}(x_1+\ldots+x_n),$$

(45.4) $$s_2^2 = \frac{1}{n}\sum_{i=1}^{n}(y_i-\bar{y})^2, \quad \bar{y} = \frac{1}{n}(y_1+\ldots+y_n),$$

$$s_{xy} = \frac{1}{n}\sum_{i=1}^{n}(x_i-\bar{x})(y_i-\bar{y}), \quad b = \frac{s_{xy}}{s_1^2}, \quad r = \frac{s_{xy}}{s_1 s_2}$$

is found to have the required property. Here $\bar{x}$ and $\bar{y}$ are sample means, s_1^2 and s_2^2 are sample variances, r is the sample correlation coefficient, and b is the sample regression coefficient of the second characteristics on the first.

We have the following theorem:

THEOREM 3. *If $(x_1, y_1), \ldots, (x_n, y_n)$ are independent random variables having the same bivariate normal distribution with the vector (m_1, m_2) of expectations, then under the condition $\lambda_{12}/\lambda_{11} = \beta_0$ the statistic t defined by equality* (45.3) *has Student's t distribution with $n-2$ degrees of freedom.*

Thus, we can construct the test for the hypothesis $\beta = \beta_0$ at the significance level α, $0 < \alpha < 1$, by taking as the critical region the set

$$W = \{(x_1, y_1; \ldots; x_n, y_n): |t| > \lambda\},$$

where t is given by condition (45.3) and λ is chosen in such a way that the probability that the random variable having Student's distribution with $n-2$ degrees of freedom exceeds λ in absolute value equals α.

PROOF OF THEOREM 3. Write

(45.5) $$\eta_i = (y_i - m_2) - \beta(x_i - m_1).$$

Then η_i and x_i will have a joint normal distribution as linear combinations of the random variables x_i and y_i with a joint normal distribution. They will be uncorrelated (see problem 1(c) of § 40), whence also independent.

It follows from formula (45.5) that

(45.6) $$y_i - \bar{y} = \beta(x_i - \bar{x}) + (\eta_i - \bar{\eta})$$

where

$$\bar{\eta} = \frac{1}{n}(\eta_1 + \ldots + \eta_n).$$

Using this fact we may write, omitting the limits of summation from $i = 1$ to $i = n$:

$$b = \frac{s_{xy}}{s_1^2} = \frac{\frac{1}{n}\sum(x_i - \bar{x})(y_i - \bar{y})}{\frac{1}{n}\sum(x_i - \bar{x})^2} = \frac{\sum(x_i - \bar{x})[\beta(x_i - \bar{x}) + (\eta_i - \bar{\eta})]}{\sum(x_i - \bar{x})^2}$$

$$= \beta + \frac{\sum(x_i - \bar{x})(\eta_i - \bar{\eta})}{\sum(x_i - \bar{x})^2}.$$

In view of the fact that

$$\sum(x_i - \bar{x})\bar{\eta} = 0$$

VIII. STATISTICAL INFERENCE

we finally obtain

(45.7) $$b-\beta = \frac{\sum (x_i-\bar{x})\eta_i}{\sum (x_i-\bar{x})^2}.$$

Next, we have in view of equalities (45.5), (45.6) and (45.7):

(45.8)
$$t = \frac{(b-\beta)s_1\sqrt{n-2}}{s_2\sqrt{1-r^2}}$$

$$= \frac{\dfrac{\sum (x_i-\bar{x})\eta_i}{\sum (x_i-\bar{x})^2}\sqrt{\sum (x_i-\bar{x})^2}\sqrt{n-2}}{\sqrt{\sum (y_i-\bar{y})^2}\sqrt{1-\dfrac{[\sum (x_i-\bar{x})(y_i-\bar{y})]^2}{\sum (x_i-\bar{x})^2 \sum (y_i-\bar{y})^2}}}$$

$$= \frac{\sum \dfrac{x_i-\bar{x}}{\sqrt{\sum (x_i-\bar{x})^2}}\eta_i}{\sqrt{\dfrac{1}{n-2}\left[\sum (\eta_i-\bar{\eta})^2 - \left(\sum \dfrac{x_i-\bar{x}}{\sqrt{\sum (x_i-\bar{x})^2}}\eta_i\right)^2\right]}}.$$

As we have already noted, the random variables x_i and η_i are independent. It follows from the formula for t that for any fixed values $x_1, \ldots, x_n$ the random variable t has Student's t distribution with $n-2$ degrees of freedom. Indeed, when we consider the orthogonal transformation

(45.9)
$$\zeta_1 = c_{11}\eta_1 + \ldots + c_{1n}\eta_n,$$
$$\ldots\ldots\ldots\ldots\ldots\ldots\ldots\ldots\ldots\ldots$$
$$\zeta_n = c_{n1}\eta_1 + \ldots + c_{nn}\eta_n$$

with the first two rows given by the formulas

(45.10)
$$\zeta_1 = \sum \frac{x_i-\bar{x}}{\sqrt{\sum (x_i-\bar{x})^2}}\eta_i,$$
$$\zeta_2 = \frac{1}{n}(\eta_1 + \ldots + \eta_n)$$

we see that

(45.11) $$t = \frac{\zeta_1}{\sqrt{\dfrac{1}{n-2}(\zeta_2^2 + \ldots + \zeta_n^2)}}$$

where $\zeta_1, \zeta_2, \ldots, \zeta_n$ are independent random variables having normal distribution with expectation 0 and the same standard deviation. Since the conditional distribution of t given fixed values of the variables $x_1, \ldots, x_n$ is always the same, t must have Student's t distribution with $n-2$ degrees of freedom, which completes the proof.

EXAMPLE 8. *Testing the hypothesis concerning the value of the correlation coefficient.* Under the same assumptions and notations as in the preceding example, we shall consider the problem of testing the hypothesis that the correlation coefficient $\varrho = \lambda_{12}/\sqrt{\lambda_{11}\lambda_{22}}$ is equal to zero.

THEOREM 4. *If $(x_1, y_1), \ldots, (x_n, y_n)$ are independent random variables with the same non-degenerate bivariate normal distribution, and the random variables x_i and y_i are uncorrelated ($\varrho = 0$), then the statistic*

$$(45.12) \qquad t = \frac{r}{\sqrt{1-r^2}} \sqrt{n-2},$$

where the sample correlation coefficient r is defined by formula (45.5), *has Student's t distribution with $n-2$ degrees of freedom.*

It follows from this theorem that the test with the critical region

$$[W = \{(x_1, y_1; \ldots; x_n, y_n): |t| > \lambda\},$$

where t is defined by (45.12) and λ is a number chosen in such a way that the random variable having Student's t distribution with $n-2$ degrees of freedom exceeds λ with probability α, may be used as a test for the hypothesis $\varrho = 0$ at the significance level α.

Using Theorem 4, and transforming Student's t distribution according to relation (45.12), we may obtain the distribution of the sample correlation coefficient under the assumption $\varrho = 0$. These distributions are used for the construction of tables giving the critical values of sample correlation coefficients for testing the hypothesis of the independence of characteristics in the bivariate normal distribution.

It is more difficult to determine the distribution of the sample correlation coefficient under the assumption that the population coefficient ϱ has a value different from zero.

Fisher suggests in such cases the use of the transformation

$$(45.13) \qquad z = \frac{1}{2} \log \frac{1+r}{1-r}.$$

Even for small sample sizes, the statistic z has an approximately normal distribution with the standard deviation $1/\sqrt{n-3}$ and expectation

$$\frac{\varrho}{2(n-1)} + \frac{1}{2} \log \frac{1+\varrho}{1-\varrho},$$

VIII. STATISTICAL INFERENCE

where ϱ is the population correlation coefficient. In other words the statistic

(45.14) $$x = \frac{\sqrt{n-3}}{2}\left(\log\frac{1+r}{1-r} - \log\frac{1+\varrho}{1-\varrho} - \frac{\varrho}{n-1}\right)$$

has an approximately normal distribution with the mean 0 and variance 1. Thus, it may be used for testing the hypothesis that the population correlation coefficient has a certain specific value different from zero.

Problems

1. Suppose that $Y_1, \ldots, Y_n$ are independent random variables having normal distributions with a common variance σ^2 and expectations respectively equal $\beta x_1 + \mu, \ldots, \beta x_n + \mu$, where β and μ are parameters whose values are to be estimated, and $x_1, \ldots, x_n$ are known numbers.

The above assumptions describe the following situation: we know that the variable y is a linear function of the variable x, and we want to determine the coefficients of this linear function. We select n values $x_1, \ldots, x_n$ of the variable x and we determine the corresponding values of y with a random error normally distributed with expectation 0 and variance σ^2, whose value is not known. The values $Y_1, \ldots, Y_n$ are the results of our observations. We want to estimate the parameters β and μ of the linear function, and the variance σ^2.

Let

$$\bar{x} = \frac{1}{n}\sum_{i=1}^{n} x_i, \quad \bar{Y} = \frac{1}{n}\sum_{i=1}^{n} Y_i,$$

$$s_1^2 = s_{xx} = \frac{1}{n}\sum_{i=1}^{n}(x_i - \bar{x})^2, \quad s_2^2 = s_{yy} = \frac{1}{n}\sum_{i=1}^{n}(Y_i - \bar{Y})^2,$$

$$s_{xy} = \frac{1}{n}\sum (x_i - \bar{x})(Y_i - \bar{Y}),$$

$$r = \frac{s_{xy}}{s_1 s_2}, \quad b = \frac{s_{xy}}{s_{xx}}, \quad s^2 = s_{yy}(1 - r^2).$$

Prove that
(a) statistic b has a normal distribution with expectation β and variance σ^2/ns_1^2;
(b) statistic $m = \bar{Y} - b\bar{x}$ has a normal distribution with expectation μ and variance

$$\frac{\sigma^2}{n}\left(1 + \frac{\bar{x}^2}{s_1^2}\right);$$

(c) statistic s^2/σ^2 has a chi-square distribution with $n-2$ degrees of freedom;
(d) statistics m and s^2 are independent, hence for $\bar{x} = 0$ the statistic

$$\frac{m - \mu}{s_2\sqrt{1 - r^2}}\sqrt{n-2}$$

has Student's t distribution with $n-2$ degrees of freedom;
(e) statistics b and s^2 are independent, whence the statistic

$$\frac{(b - \beta)s_1}{s_2\sqrt{1 - r^2}}\sqrt{n-2}$$

has Student's distribution with $n-2$ degrees of freedom.

2. Under the assumptions of Example 7 prove that the statistic

$$(b-\beta)\frac{\sigma_1\sqrt{n-1}}{\sigma},$$

where σ_1 is the standard deviation of the characteristic x and σ is the standard deviation of the conditional distribution of the characteristic y for given x, has Student's t distribution with $n-1$ degrees of freedom.

3. Construct a test for the null hypothesis asserting that in two populations with a joint normal distribution of two numerical characteristics with the same residual variance of y with respect to x, the regression coefficients of y on x are identical. The size of the first sample is n_1 and the size of the second sample is n_2.

Hint. Show that the statistic

$$t = \frac{b'-b''}{\sqrt{\dfrac{n_1 s_2'^2(1-r'^2)+n_2 s_2''^2(1-r''^2)}{n_1+n_2-4}}\sqrt{\dfrac{1}{n_1 s_1'^2}+\dfrac{1}{n_2 s_1''^2}}}$$

(where b' and b'' are the sample regression coefficients of y on x, r' and r'' are the sample correlation coefficients, s_1', s_1'', s_2', s_2'' are the sample standard deviations of the characteristics x and y in the first and the second samples) has Student's t distribution with n_1+n_2-4 degrees of freedom.

4. Using the statistic defined by formula (45.14) construct a test, based on samples of equal sizes, for testing the hypothesis that in two populations with a non-degenerate bivariate normal distribution of two characteristics x and y, the correlation coefficients between x and y is the same.

5. Prove that if r is the sample correlation coefficient and s_1 is the sample standard deviation of the first characteristic computed from the same sample taken from a bivariate normal distribution, then r and s_1 are independent random variables.

§ 46. TESTS BASED ON LIMIT THEOREMS

The tests which we described in the preceding paragraph were constructed as follows: we selected a certain function of a sample, or statistic, suitable for measuring the effect which we were investigating, and then we found the distribution of this statistic. Having this distribution, we defined the critical region. It happens, however, that the problem of finding the probability distribution of a statistic for every n is a tedious numerical problem, while it is easy to find the limit distribution of that statistic for $n \to \infty$. This gives the possibility of constructing statistical tests, approximately at the required significance level, by treating the sample size which we already have as sufficiently large for the required approximation of the limit distribution. In other words, we treat the statistic as if it already had the limit distribution. To justify such a procedure, let us mention that the "exact" tests of the preceding paragraph were also approximate: they are "exact" only under the assumption that the elements of the

VIII. STATISTICAL INFERENCE

sample are independent, and the distribution in the population is normal, these assumptions, being as a rule, only approximately satisfied.

To construct tests in the manner outlined above one can use all the limit theorems which we discussed in § 38 and in particular in § 44. We shall not deal with such tests in detail. Instead, we shall present a few tests based on some new limit theorems which we have not discussed yet, and we shall formulate those theorems. We start from *Pearson's chi-square test*, called so because the statistic on which the test is based has in the limit the chi-square distribution, and the proof of this property being due to K. Pearson.

We have respectedly stressed how important it is to be able to assume that the distribution in a population is normal: we can then apply, say, Student's test and assign definite probabilities to the decisions obtained on the basis of this test. It would be valuable therefore to be able to test the hypothesis of normality of the distribution. We shall show to solve this problem by use of the chi-square test.

We start from a situation in which we want to test whether the distribution of the characteristic under investigation has the probability distribution function $F(x)$. Let $x_1, \ldots, x_n$ be the sample point representing n independent observations, and let $\xi_1, \ldots, \xi_n$ be independent random variables with distribution function $F(x)$. We can treat $x_1, \ldots, x_n$ as values of the random variables $\xi_1, \ldots, \xi_n$ with a distribution function $F(x)$. We still have to choose a statistic which will serve as a measure of deviation between the sample $x_1, \ldots, x_n$ and the hypothetical distribution function $F(x)$. Pearson suggested the following procedure: let us divide the real axis into $k+1$ disjoint parts, say by choosing k numbers $a_1 < a_2 < \ldots < a_k$, and considering the intervals

$$I_0 = \{x: x < a_1\},$$
$$I_j = \{x: a_j \leqslant x < a_{j+1}\}, \quad j = 1, 2, \ldots, k-1,$$
$$I_k = \{x: a_k \leqslant x\}.$$

Let p_j denote the probability that the random variable ξ_1 assumes a value from the interval I_j and suppose that the intervals I_j are chosen in such a way that all the probabilities p_j are positive. Then np_j equals the expected number of observations which should fall into the interval I_j if the sample size is n. Let n_j denote the number of observations which did really fall into interval I_j. Then the differences $n_j - np_j$ may serve as a measure of the consistency of our sample with our hypothesis. Pearson suggests using the statistic

$$\chi^2 = \frac{(n_0 - np_0)^2}{np_0} + \ldots + \frac{(n_k - np_k)^2}{np_k},$$

known as *Pearson's chi-square statistic,* and denoted traditionally by the symbol χ^2.

This suggestion is justified by the following theorem:

THEOREM 1 (Pearson). *If $\xi_1, \ldots, \xi_n$ are independent random variables with an identical probability distribution, $I_0, I_1, \ldots, I_n$ constitute a division of the real line into disjoint sets such that the probabilities $p_j = P\{\xi_i \in I_j\}$ are all positive, n_j denotes the random variable defined by the equality*

$$n_j = \operatorname{Card}\{i: \xi_i \in I_j, 1 \leqslant i \leqslant n\},$$

and

$$\chi^2 = \frac{(n_0 - np_0)^2}{np_0} + \ldots + \frac{(n_k - np_k)^2}{np_k},$$

then the distribution of the random variable χ^2 tends, as $n \to \infty$, to a chi-square distribution with k degrees of freedom.

Denote by λ the number satisfying, for a given α, the relation

$$1 - \alpha = K(\lambda),$$

where $K(x)$ denotes the probability distribution function of the chi-square distribution with k degrees of freedom. By Pearson's chi-square test for the null hypothesis asserting that $F(x)$ is the probability distribution function of the population we mean the test with the critical region

$$W = \{(x_1, \ldots, x_n): \lambda < \chi^2\}.$$

Some remarks concerning this test are in order. First, it can be asked what is the null hypothesis here. In constructing this test we made one very important arbitrary decision: we selected the intervals I_j. Once this choice has been made the statistic under consideration will have the same limit distribution as long as the probability of obtaining the observation from the set I_j is p_j. Thus, in situations where every distribution is *a priori* possible, the natural hypothesis is the class of all distributions for which $P(I_j) = p_j$. If we included some of the distributions of this class in the alternative hypothesis, then the probability of the rejection of the null hypothesis if it is true would be equal to the probability of its rejection if it is false (i.e. the true distribution belongs to the set of alternative hypotheses)—this situation is undesirable. The situation changes if we may exclude some of the distributions on *a priori* grounds. As an example we may take normal distributions of a continuous characteristic. Another example is provided by the situation where the characteristic may assume only a finite number of possible values, like the last significant digit in the measurement of, say, the width of the skull of a mouse. Another situation: the characteristic assumes only integer values, and we take Poisson distributions as admissible. In such situations, the null hypothesis may be treated as a simple hypothesis.

What can we say about the alternative hypothesis? In the first case we have to include in the set of alternatives all distributions such that for at least one index j we have $P(I_j) \neq p_j$. The fact that the null hypothesis may consist of arbitrary probability distributions, provided only that they satisfy a certain condition (under which the probability of a sample from the critical region will be already fixed), classifies this test as a non-parametric, unlike, say, Students's test, where the probability of a sample point from the critical region has been computed under the assumption of normality of the distribution. As a result, we could treat the null hypothesis as a condition imposed on the parameters of the normal distribution. This does not exclude, of course, the use of the chi-square test in a parametric situation, for instance, when we treat normal distributions as admissible hypotheses.

At the beginning of this paragraph we mentioned the possibility of testing the hypothesis of normality of distribution in the population. In the manner described above we may test the hypothesis that the distribution in the population is normal with given values of the parameters. Suppose that the distribution in a population is in fact normal, but we do not know the values of its parameters. If we wanted to test the hypothesis of normality by using the above test in a way involving the fixing of the values of the parameters of the distribution arbitrarily, we would run the risk of the parameters in the population being different and the sample turning out to be inconsistent with the distribution which we tried to fit to it, and in consequence we would reject the hypothesis of normality in spite of its being true. In other words, the probability of observing a sample point from the critical region would depend on which particular distribution assigned to the null hypothesis was actually true. Another way of using this test is to calculate first the expectation and variance from the sample; then, however, we run the risk of the sample fitting too well the distribution which we test, and we shall not discover the inconsistency even if it actually occurs. In other words, the question arises what will be the significance level of such a test. It turns out that the second approach can be justified, and the appropriate test may be constructed on the basis of the following limit theorem:

THEOREM 2. *If*

(a) $F(x; a_1, \ldots, a_m)$ *is a family of probability distribution functions depending on m parameters* $a_1, \ldots, a_m$;

(b) $I_1, \ldots, I_r$ *is a partition of the real line into r disjoint sets*;

(c) $p_j(a_1, \ldots, a_m) = P\{I_j | a_1, \ldots, a_m\}$ *is the probability that a random variable* with the probability distribution function $F(x; a_1, ..., a_m)$ assumes a value from the set I_j;

(d) in the non-degenerate cube $A = \{(a_1, ..., a_m): a'_1 < a_1 < a''_1, ..., a'_m < a_m < a''_m\}$ the following conditions hold:

(d$_1$) $p_j(a_1, ..., a_m) > c^2 > 0$ for all j;

(d$_2$) there exist continuous partial derivatives

$$\frac{\partial p_j}{\partial a_k}, \quad \frac{\partial^2 p_j}{\partial a_k \partial a_s},$$

(d$_3$) the matrix

$$B = \left[\frac{\partial p_j}{\partial a_k}\right]$$

where $j = 1, 2, ..., r$ and $k = 1, 2, ..., m$ is of the rank m;

(e) $\xi_1, \xi_2, ..., \xi_n$ are independent random variables with an identical distribution corresponding to a fixed point $(a_1^{(0)}, ..., a_m^{(0)})$ from the cube A;

(f) $n_1, ..., n_m$ are random variables defined by the equalities

$$n_j = \text{Card}\{i: \xi_i \in I_j, 1 \leq i \leq n\}, \quad j = 1, 2, ..., r;$$

(g) $a'_1 = a'_1(n_1, ..., n_m), ..., a'_r = a'_r(n_1, ..., n_m)$ is the only solution with respect to $a_1, ..., a_m$, convergent in probability to $(a_1^{(0)}, ..., a_m^{(0)})$, of the equations

$$\sum_{j=1}^{r} \frac{n_j}{np_j} \cdot \frac{\partial p_i}{\partial a_k} = 0, \quad k = 1, 2, ..., m$$

arising from comparing to zero the partial derivatives with respect to $a_1, ..., a_m$ of the function

$$\sum_{j=1}^{r} \frac{(n_j - np_j)^2}{np_j}$$

with the restriction that the partial derivatives are computed in such a way as if only the numerator depended on the parameters $a_1, ..., a_m$; moreover, in solving this system of equations we use an additional relation

$$\frac{\partial p_1}{\partial a_k} + ... + \frac{\partial p_r}{\partial a_k} = 0.$$

(h) $p'_1 = p'_1(a'_1, ..., a'_m), ..., p'_r = p'_r(a'_1, ..., a'_m)$,

then the statistic

(46.1) $$\chi^2 = \frac{(n_1 - np'_1)^2}{np'_1} + ... + \frac{(n_r - np'_r)^2}{np'_r}$$

has a probability distribution converging, as $n \to \infty$, to a chi-square distribution with $r-m-1$ degrees of freedom.

This theorem enables us to construct a test for the hypothesis that the distribution of the population investigated belongs to a certain m-parameter family of distributions: when we want to construct the test at the significance level α, we take as the critical region the set

$$W = \{(x_1, \ldots, x_n): \lambda < \chi^2\}$$

where χ^2 is the statistic defined by (46.1) and λ is the number satisfying the equation

$$1-\alpha = K(\lambda),$$

where $K(x)$ is the distribution function of a chi-square distribution with $r-m-1$ degrees of freedom.

We omit the proof of this important theorem, and restrict our considerations to illustrations of the application of this theorem in typical situations.

EXAMPLE 1. We know that the distribution of the characteristic studied under consideration has a standard deviation σ. We want to test the hypothesis that the distribution is normal. We have here only one parameter a_1, namely the expectation of the distribution. We take as a'_1 the sample mean $\bar{x} = (x_1 + \ldots + x_n)/n$ (in spite of the fact, that this is not, formally speaking, the solution of the system of equations described in Theorem 2); next, for the normal distribution with expectation $\bar{x}$ and variance σ^2 we compute the probabilities $p'_1, \ldots, p'_r$ and apply formula (46.1).

EXAMPLE 2. We want to test the hypothesis that the distribution is normal. We now have two parameters: a_1, the expectation of the distribution, and a_2, its standard deviation. We take as a'_1 the sample mean $\bar{x}$ and as a'_2 the sample standard deviation

$$s = \sqrt{\frac{(x_1-\bar{x})^2 + \ldots + (x_n-\bar{x})^2}{n}}$$

(here again, formally speaking, the sample mean and the sample standard deviation are only approximate solutions of the system of equations described in Theorem 2). We compute the probabilities $p'_1, \ldots, p'_r$ for the normal distribution with expectation $\bar{x}$ and variance s^2 and use them to compute the statistic χ^2 given by formula (46.1).

EXAMPLE 3. *The characteristic assumes integer values*. We want to test the hypothesis that the probability distribution of our characteristic is a Poisson distribution. As I_1 we take the first few integers, next, as $I_2, \ldots, I_{r-1}$ we take sets each consisting of one integer, and as I_r we take the rest. The only parameter is the expectation a_1 of the distribution. Again, the solution of the equation from Theorem 2 is rather complicated, but we may take as a_1' the sample mean, omitting the difficult numerical calculations. With the aid of tables we find the values necessary in formula (46.1), namely the values $p_1', \ldots, p_r'$ in Poisson distribution with expectation $\bar{x}$.

EXAMPLE 4. *Testing independence.* We can classify people according to the colour of their eyes and the colour of their hair. Are these two characteristics independent? In other words, the question is whether, knowing the colour of a person's hair, we can say more about that person's eyes than we could say if we did not have that information.

Using this example we shall discuss the application of the chi-square test to the investigation of the independence of traits. Let us imagine a population consisting of elements having two measurable characteristic each. Suppose that we fix classes for each of these characteristics, say r classes for the first characteristic and s classes for the second. Thus, each result of measurement will fall into one of the rs cells of this double classification. Let n denote the total number of observations, and let n_{ij} denote the number of observations which fall into the ith class according to the first, and the jth class according to the second characteristic. Next denote $n_{i.} = \sum_{j=1}^{s} n_{ij}$, $n_{.j} = \sum_{i=1}^{r} n_{ij}$. This information may be summarized in the form of the following table:

1st characteristic \ 2nd characteristic		class				total
		1	2	...	s	
class	1	n_{11}	n_{12}	...	n_{1s}	$n_{1.}$
	2	n_{21}	n_{22}	...	n_{2s}	$n_{2.}$
	.	.	.	...	$\vdots$	$\vdots$
	r	n_{r1}	n_{r2}	...	n_{rs}	$n_{r.}$
total		$n_{.1}$	$n_{.2}$	...	$n_{.s}$	n

If our characteristics were independent, then the probability distribution of these cells would be determined by giving the probabilities $p_1, \ldots, p_r$ that the first characteristic will assume a value from the first, second, ..., rth class, and analogous probabilities $q_1, \ldots, q_s$ for the second characteristic. Since we must

have
$$p_1 + \ldots + p_r = q_1 + \ldots + q_s = 1,$$
there are $r+s-2$ independent parameters determining the probability distribution of our double classification. If we take as these parameters the values $p_1, \ldots, p_{r-1}, q_1, \ldots, q_{s-1}$ and substitute $p_r = 1-p_1- \ldots -p_{r-1}$, $q_s = 1-q_1- \ldots -q_{s-1}$ then the equations described in point (g) of Theorem 2 will take the form

$$\frac{n_{i.}}{p_i} = \frac{n_{r.}}{p_r}, \quad i = 1, 2, \ldots, r-1,$$

$$\frac{n_{.j}}{q_j} = \frac{n_{.s}}{q_s}, \quad j = 1, 2, \ldots, s-1.$$

It follows that the solution is $p'_i = n_{i.}/n$, $i = 1, 2, \ldots, r$ and $q'_j = n_{.j}/n$, $j = 1, 2, \ldots, s$. Thus, formula (46.1) will take the form

$$\chi^2 = \sum_{i=1}^{r} \sum_{j=1}^{s} \frac{n \left(n_{ij} - \frac{n_{i.} n_{.j}}{n} \right)^2}{n_{i.} n_{.j}}.$$

By Theorem 2, the statistic χ^2 calculated above has in the limit, as $n \to \infty$, a chi-square distribution with $rs-(r+s-2)-1 = (r-1)(s-1)$ degrees of freedom.

The reader will verify that for $r = s = 2$ the above formula can be simplified to

$$\chi^2 = \frac{n(n_{11} n_{22} - n_{12} n_{21})^2}{n_{1.} n_{2.} n_{.1} n_{.2}}.$$

EXAMPLE 5. *Comparing two distributions.* In the preceding example we considered the case of investigating the independence of two characteristics. We treated observations as sampling elements from a population in which every element had two characteristics. Thus, the number of elements in each class with respect to every characteristic was a random variable. The situation is different when we have two populations characterized by one characteristic and we want to test the hypothesis that the characteristic has the same distribution in both populations. We select an n_1-element sample from the first population and an n_2-element sample from the second population. If we classify the elements selected into s classes, and denote by n_{ij} the number of elements of the ith population which fall into the jth class, we will be able to represent the results of our observations in the form of a two-row table similar to that of Example 4. The final formula will also be the same as in Example 4. However, the essential difference is that now the numbers $n_{1.}$ and $n_{2.}$ are not random but chosen in

class sample	1	2	...	s	total
1	n_{11}	n_{12}	...	n_{1s}	$n_1.$
2	n_{21}	n_{22}	...	n_{2s}	$n_2.$
total	$n_{.1}$	$n_{.2}$	...	$n_{.s}$	n

advance. Next, according to the null hypothesis, the distribution is the same in both populations, but we do not know this distribution. If the probability of obtaining an observation from the ith class is p_i, we shall have $s-1$ parameters of the form $p_1, \ldots, p_{s-1}$ and the relation $p_s = 1 - p_1 - \ldots - p_{s-1}$. According to Pearson's suggestion, the inconsistency of our samples with the hypothesis that the distribution is common to both populations may be measured by the statistic

$$\chi^2 = \sum_{i=1}^{2} \sum_{j=1}^{s} \frac{(n_{ij} - n_i p_i)^2}{n p_i}.$$

It can be proved that if we substitute $n_i./n$ for p_i, then the statistic obtained will have in the limit, as $n \to \infty$, a chi-square distribution with $s-1$ degrees of freedom. Thus, we may test our hypothesis using the same calculations as in Example 4, but this requires separate justification, independent of Theorem 2. What we said about two samples may be repeated for the case of r samples. Testing the hypothesis that all samples are drawn from populations with the same distribution of the characteristic is called *testing for homogeneity*.

At the beginning of this paragraph we discussed testing the hypothesis that the sample is drawn from a population having the given probability distribution with a probability distribution function $F(x)$. We used the chi-square test, and we stressed the disadvantage of this test, connected with a conventional choice of the classes I_j. Now, if the distribution function $F(x)$ of the characteristic is continuous, we can construct another test of this hypothesis, free of the above-mentioned disadvantage: namely, on the basis of the sample $x_1, \ldots, x_n$ we construct the empirical distribution function $S_n(x)$ and find the number

$$\sqrt{n}\,\delta_n = \sqrt{n} \sup_{-\infty < x < \infty} |S_n(x) - F(x)|.$$

If we take for λ the number satisfying the relation

$$1 - \alpha = K(\lambda)$$

where $K(x)$ is the Kolmogorov distribution defined in Theorem 4 of § 44, we define the test at the significance level α of the hypothesis that $F(x)$ is the distri-

VIII. STATISTICAL INFERENCE

bution of the population. As the critical region we take the set

$$W = \{(x_1, \ldots, x_n): \lambda < \sqrt{n}\,\delta_n\}.$$

This is the so-called *Kolmogorov test*.

The Kolmogorov test defined above leads to the verification, on the basis of one sample, of the hypothesis that the distribution in a population has a given probability distribution function $F(x)$. We can state an analogous problem: given two samples taken out of two populations without using the conventional partition into sets, test the hypothesis that the distributions in both populations are identical. The theorem of Smirnov, proved in 1939–1944, analogous to Theorem 4 of § 44, may serve as a basis for such verification:

THEOREM 3 (Smirnov). *Let* $\xi_1 = \xi_1(e), \ldots, \xi_{n_1} = \xi_{n_1}(e)$ *and* $\eta_1 = \eta_1(e), \ldots, \eta_{n_2} = \eta_{n_2}(e)$ *be independent random variables having the same probability distribution with a continuous probability distribution function* $F(x)$. *Let* $F_{n_1}(x)$ *and* $G_{n_2}(x)$ *be the empirical distribution function defined by formulas*

$$F_{n_1}(x) = \frac{1}{n_1}\,\mathrm{Card}\,\{i\colon \xi_i < x,\ i = 1, 2, \ldots, n_1\},$$

$$G_{n_2}(x) = \frac{1}{n_2}\,\mathrm{Card}\,\{i\colon \eta_i < x,\ i = 1, 2, \ldots, n_2\}.$$

Next, let

$$\delta_n' = \sup_{-\infty < x < \infty}\,(F_{n_1}(x) - G_{n_2}(x)),$$

$$\delta_n'' = \sup_{-\infty < x < \infty}\,|F_{n_1}(x) - G_{n_2}(x)|,$$

$$n = \frac{n_1 n_2}{n_1 + n_2}.$$

Then

$$\lim_{\substack{n_1 \to \infty \\ n_2 \to \infty}} P\{\sqrt{n}\,\delta_n' < x\} = \begin{cases} 1 - e^{-2x^2} & \text{for } x > 0, \\ 0 & \text{for } x < 0, \end{cases}$$

$$\lim_{\substack{n_1 \to \infty \\ n_2 \to \infty}} P\{\sqrt{n}\,\delta_n'' < x\} = K(x),$$

where

$$K(x) = \begin{cases} \sum_{k=-\infty}^{\infty} (-1)^k e^{-2k^2 x^2} & \text{for } x > 0, \\ 0 & \text{for } x < 0. \end{cases}$$

Using this theorem, we may take the statistic δ_n'' for testing the null hypothesis that in two populations the (continuous) distribution of the characteristic is the same, that is, the hypothesis that the distribution $F(x)$ in the first and the distribution $G(x)$ in the second population satisfy the relation

$$F(x) = G(x) \quad \text{for} \quad -\infty < x < \infty.$$

This test is sometimes called the *Kolmogorov–Smirnov test*.

Sometimes it can be assumed that the class of admissible distributions in a statistical problem has the following property: between distributions $F(x)$ and $G(x)$ of these populations we have the relation

$$F(x) \leqslant G(x) \quad \text{for all } x$$

or

$$F(x) \geqslant G(x) \quad \text{for all } x$$

with strict inequality for at least one value of x. This occurs, for instance, when we suspect that the admissible distributions arise from a given distribution by translation: there exists a distribution function $F(x)$ such that if $G(x)$ is one of the admissible distributions, then there exists a constant λ, called the *translation parameter*, such that for all x we have

$$G(x) = F(x-\lambda).$$

In such situations we may use the statistic δ'_n of Theorem 3 for testing the hypothesis that the distribution $F(x)$ of the first population (corresponding to the n_1-element sample) does not exceed the distribution $G(x)$ of the second population (corresponding to the n_2-element sample) against the alternative that $G(x)$ does not exceed $F(x)$ and is strictly smaller in a certain interval. Using the parameter λ, which we have called the translation parameter, we may write the null hypothesis in the form $\lambda \leqslant 0$, and the alternative hypothesis in the form $0 < \lambda$.

We have stressed the non-parametric character of Pearson's chi-square test. The Kolmogorov and Smirnov–Kolmogorov tests are of the same character. The latter test is used for testing the null hypothesis that the distributions in two populations coincide. Let us compare it with the Student's test interpreted as a test of the hypothesis of equality of two means in two populations which are known to have normal distributions and the same variances. There are two main differences between these tests.

First, the Kolmogorov–Smirnov test is much simpler in use, in the sense that the necessary arithmetical operations are much simpler. In fact, it requires only comparing the results of observations. In this respect, it belongs to the family of the so-called order tests, i.e. tests for which the value of the underlying statistic depends only on the relation "smaller than" between the results of observations. It does not require such of the operations appearing in computing Student's

VIII. STATISTICAL INFERENCE

statistic as addition, squaring, and taking square roots. Obviously, such operations constitute the main difficulty in practical applications of statistical tests. The wish to avoid the computational difficulties connected with the application of tests accounts for the interest of recent years, in order tests.

Second, Student's test is based on the assumption of normality of the underlaying distribution; only then the statistic t has the Student distribution. This assumption narrows the scope of applicability of this test. We often have to deal with non-normal distributions, e.g. asymmetric ones. Thus, we must have the tests for the basic hypotheses, such as equality of distributions, which would have certain definite probabilistic properties irrespectively of the distribution of the population. Thus, the generality of the assumptions under which the test retains its properties is the second reason for the interest in order tests.

Choosing from the many order tests, we shall describe here the Wilcoxon test, useful especially for testing the hypothesis concerning the translation parameter described above. The assumption under which we shall prove the properties of the test is the continuity of the distribution functions of characteristics. This assumption is needed for the following reason: the probability of observing two identical results in the sample should be zero. We make this assumption in all our subsequent reasonings.

The *Wilcoxon test* is defined as follows: let $x_1, \ldots, x_n$ and $y_1, \ldots, y_m$ be samples of sizes n and m chosen from two populations. Let us order both samples according to magnitude. Every pair of observations, of which one comes from the first population and the other from the second population and in which in the common ordering the observation from the second population precedes the observation from the first population will be called an *inversion*. We shall denote by U the total number of inversions.

In other words, if we order jointly both samples according to magnitude and in this ordering the smallest element of the sample $x_1, \ldots, x_n$ occupies the r_1th place, this gives $r_1 - 1$ inversions. Generally, if the kth in magnitude element of the sample $x_1, \ldots, x_n$ occupies the place r_k, this gives $r_k - k$ inversions. Thus, we can write

(46.2) $$U = r_1 + r_2 + \ldots + r_n - (1 + 2 + \ldots + n)$$
$$= r_1 + \ldots + r_n - \tfrac{1}{2}n(n-1).$$

The Wilcoxon test is based on the statistic U.

For small n and m it is easy to find directly the distribution of the statistic U under the assumption that the distribution in both populations are identical. Neglecting, as irrelevant for the value of U, the order of elements of the samples $x_1, \ldots, x_n$ and $y_1, \ldots, y_m$ separately, we may reduce the problem by denoting by x the places occupied in the common ordering by elements of the sample $x_1, \ldots, x_n$ and by y the places occupied by elements of the sample $y_1, \ldots, y_m$. Then the problem is reduced to the problem of a random ordering of n letters x and m letters y. There are $\binom{n+m}{n}$ such orderings, and they all are equally probable. The calculation of the number of inversions does not change: one has to assign to each letter x the number of letters y preceding it, and add all these numbers together. Obviously, this is equivalent to formula (46.2).

For example, if $m = n = 2$, we have the following orderings and numbers of inversions:

ordering	number of inversions U
xxyy	0
xyxy	1
xyyx	2
yxxy	2
yxyx	3
yyxx	4

We easily find the expectation and variance of the statistic U under the assumption of equal distributions of the populations. We have

THEOREM 4. *If $x_1, \ldots, x_n, y_1, \ldots, y_m$ are $n+m$ independent random variables with the same continuous distribution, and U is the total number of inversions computed according to formula (46.2), then the expectation of U is given by the formula*

(46.3) $$EU = \tfrac{1}{2}nm$$

and the variance of U is

(46.4) $$D^2 U = \tfrac{1}{12}nm(n+m+1).$$

PROOF. For given i, j ($1 \leq i \leq n$, $1 \leq j \leq m$) we define

$$z_{ij} = \begin{cases} 1 & \text{if } y_j < x_i, \\ 0 & \text{if } x_i < y_j. \end{cases}$$

Thus, z_{ij} is the random variable which indicates whether x_i and y_j give an inversion or not. According to the definition of U we may write

$$U = \sum_{i=1}^{n} \sum_{j=1}^{m} z_{ij}.$$

VIII. STATISTICAL INFERENCE

Our problem now consists in computing the expectation and variance of the sum of the random variables z_{ij}.

First, under our assumptions all random variables z_{ij} have the same distribution: they assume value 1 with probability 1/2 and value 0 also with probability 1/2. Thus, they have expectation 1/2, and since there are nm of them, the expectation of U is $\frac{1}{2}nm$, which proves formula (46.3).

Next, we have

$$D^2 U = E(U-EU)^2 = E\left(\sum_{i=1}^{n}\sum_{j=1}^{m}(z_{ij}-\tfrac{1}{2})\right)^2$$

$$= E\sum_{i=1}^{n}\sum_{j=1}^{m}\sum_{r=1}^{n}\sum_{s=1}^{m}(z_{ij}-\tfrac{1}{2})(z_{rs}-\tfrac{1}{2})$$

$$= \sum_{i=1}^{n}\sum_{j=1}^{m}\sum_{r=1}^{n}\sum_{s=1}^{m}E(z_{ij}-\tfrac{1}{2})(z_{rs}-\tfrac{1}{2}).$$

If $i \neq r$ and $j \neq s$, then the random variables z_{ij} and z_{rs} are independent, whence

$$E(z_{ij}-\tfrac{1}{2})(z_{rs}-\tfrac{1}{2}) = 0.$$

There are $n(n-1)m(m-1)$ of such systems (i, j, r, s).

Similarly, if $i = r$ and $j \neq s$ or if $i \neq r$ and $j = s$, then

$$E(z_{ij}-\tfrac{1}{2})(z_{rs}-\tfrac{1}{2}) = \tfrac{1}{12},$$

since, as we can easily see, there are then 6 equiprobable possibilities, two of them giving value $-1/4$ and four value $1/4$ for the product in question. There are $nm(m-1)+n(n-1)m$ of such systems.

Finally, if $i = r$ and $j = s$ then

$$E(z_{ij}-\tfrac{1}{2})(z_{rs}-\tfrac{1}{2}) = \tfrac{1}{4}$$

since the expectation of this product then equals simply the variance of the random variable z_{ij}. There are nm such systems.

Combining all these possibilities, we obtain

$$D^2 U = \tfrac{1}{12}nm(m-1)+n(n-1)m+\tfrac{1}{4}nm = \tfrac{1}{12}nm(n+m+1),$$

which completes the proof of Theorem 4.

Mann and Whitney, who proved Theorem 4, have also proved the following theorem*

THEOREM 5. *Under the assumptions of Theorem 4, if $n \to \infty$ and $m \to \infty$, then*

---

* See B. L. van der Waerden, *Mathematische Statistik*, Berlin–Göttingen–Heidelberg, 1957, Chapter 12.

the statistic U is asymptotically normal with the expectation and variance given by (46.3) and (46.4).

This theorem was complemented by van der Waerden, who proved

THEOREM 6. *Under the assumptions of Theorem 4, for fixed n and for $m \to \infty$ the statistic U has asymptotically a distribution equal to the distribution of the sum of n independent random variables uniformly distributed in the interval $(0, m)$.*

The last two theorems and the tables for small n and m show that for $\min(n, m) \geqslant 4$ and $m+n \geqslant 20$ the normal approximation to the distribution of the statistic U is quite sufficient.

Problems

1. Find the probability density and probability distribution function of the sum of (a) two and (b) three independent random variables each having a distribution uniform in $(0, 1)$.

2. *Galton's order test.* Let $x_1, \ldots, x_n, y_1, \ldots, y_n$ be independent random variables with the same continuous probability distribution. Let $x_1^* < x_2^* < \ldots < x_n^*$ be the sample $x_1, \ldots, x_n$ ordered according to magnitude, and let $y_1^* < y_2^* < \ldots < y_n^*$ be the sample $y_1, \ldots, y_n$ ordered according to magnitude. Denote by t the random variable which tells for how many indices i the inequality $x_i < y_i$ is satisfied. In other words,
$$t = \text{Card}\{i: x_i < y_i\}.$$
Prove that t has uniform distribution, i.e.
$$P\{t = k\} = 1/(n+1) \text{ for } k = 0, 1, \ldots, n.$$

Galton suggested the use of the statistic t defined above for verifying the hypothesis that in two populations the distributions of the characteristic are the same.

Hint. Each of the $\binom{2n}{n}$ sets of places which may be occupied by the elements of the sample $x_1, \ldots, x_n$ among the $2n$ successive places has the same probability. Prove by induction on n, that for every $k = 1, 2, \ldots, n$ the number of systems for which the statistic t assumes the value k is the same as the number of systems for which this statistic assumes the value $k-1$.

§ 47. MOST POWERFUL AND UNIFORMLY MOST POWERFUL TESTS

In §§ 45 and 46 we discussed several statistical tests based on a number of independent observations that was fixed in advance. Now, we shall discuss the construction of such tests in more detail; we consider several desirable properties of the tests and show that some of the tests discussed in the preceding paragraphs possess these properties.

In our considerations we shall make essential use of the information about the probability distribution admissible in the given statistical problem (see the remarks at the beginning of § 45).

When testing hypotheses two types of errors are possible: *errors of the first kind* consisting in rejecting the null hypothesis when it is true, and *errors of the second kind* consisting in accepting the null hypothesis when the alternative hypothesis is true. Obviously, it is desirable that the probabilities of both kinds of errors are small. Unfortunately, one cannot decrease the probabilities of both kinds of errors simultaneously: when we increase the critical set, we decrease the probability of error of the second kind, but we also increase the probability of error of the first kind. One of the ways of reaching a compromise between the probabilities of errors of the first and the second kinds is to fix the probability of error of the first kind, and minimize the probability of error of the second kind under this condition. The fixed probability of error of the first kind is called the *level of significance of the test*; a test with a given significance level and the minimal probability of error of the second kind is called the *most powerful*. What we said above is simple and clear when the null and the alternative hypotheses are both simple, but becomes complicated when we consider composite hypotheses. From the mathematical point of view, the problem of finding the most powerful test with a given significance level reduces to the problem of finding the extremum under a side condition. We shall now discuss this problem.

Let us start with listing formal definitions concerning the construction of tests based on a fixed finite number of observations in cases of parametric tests (see § 45).

We are given the space X of sample points x. As a rule, we consider X to be a finite-dimensional Euclidean space, and we consider a family $P_\theta(A)$ of probability distributions on a fixed σ-field S of subsets of X. These probability distributions depend on a certain parameter θ, equal to a real number, or a vector; we shall denote the space of all parameters by Ω. Whenever convenient, we shall speak of the probability distribution $P_\theta(A)$ in X as of the probability distribution of a system of random variables $\xi_1, ..., \xi_n$ representing the results of observations. We shall distinguish a certain class of distributions $P_\theta(A)$ or—which amounts to the same—a certain set $H \subset \Omega$ which will be called the *null hypothesis*. The null hypothesis is *simple* or *composite* according to whether H contains one point or more than one point. Next, we shall distinguish a certain set $K \subset \Omega$ disjoint with H and we shall call it the *alternative hypothesis*. Again the alternative hypothesis is *simple* or *composite* according to whether K consists of one or more than one point. By the *test of the hypothesis H against the alternative K* we mean a function $\Phi(x)$ which assigns to the sample points x one of the two decisions: "the hypothesis H is false" and "the hypothesis H is true". Since this function is uniquely determined by indicating the set W of those points which correspond to the first of these decisions, in subsequent formulations we shall identify the test with the set W, to be called the *critical region of the* *test*. If the null hypothesis is simple, $H = \{\theta_0\}$, then the number $\alpha = P_{\theta_0}(W)$ is called the *significance level of the test*. If the hypothesis H is composite, by the *significance level of test W* we mean the number

$$\alpha = \sup_{\theta \in H} P_\theta(W).$$

The probability $P_\theta(W)$ regarded for a fixed W as a function of θ on the set $H \cup K$ is called the *power of the test W*, and the value of this function for a given $\theta \in K$ is called the *power of the test W against the alternative* θ. To stress the fact that we are concerned with the power of the test W we shall sometimes write $M(\theta, W)$ instead of $P_\theta(W)$.

Besides the tests described above it is convenient to consider also the so-called randomized tests. By a *randomized test*, or *randomized critical function*, we mean a function which to every sample point x assigns the probability $\Phi(x)$ of the decision "the hypothesis H is false" if we observe the sample point x. This is to be interpreted as follows: after observing the point x we organize a lottery, in which with the probability $\Phi(x)$ we choose the decision "the hypothesis H is false" and with the probability $1-\Phi(x)$ we choose the decision "the hypothesis H is true". Randomized tests are necessary in cases where the distribution $P_\theta(A)$ is discrete: without randomization it might be impossible to construct a test with a given significance level. Obviously, a randomized test in which the critical function assumes only the values 0 and 1 is equivalent to an ordinary test with the critical region $W = \{x : \Phi(x) = 1\}$.

The use of randomized test may appear pointless. Formally, however, it enriches the chances of the statistician. Thus, one should find out whether such enrichment does in fact improve the statistician's situation. In the sequel we shall discuss randomized tests. It will be found that, as a rule, the optimal randomized test reduces in fact to a non-randomized test, except for a small border set, where randomization may be necessary to obtain the required significance level of the test.

By the *power of a randomized test* Φ we shall mean the probability $M(\theta, \Phi)$ of obtaining the decision "the hypothesis H is false" treated as a function of the parameter θ on the set $H \cup K$. Under this definition of the power of a randomized test, we may retain without formal changes the definitions of the significance level of a test and of the most powerful test.

In the sequel, in order to be able to obtain formulations pertaining both to discrete distributions (such as binomial, geometric or Poisson distribution) and to continuous distributions (such as normal, or gamma distribution), we introduce the following convention: we introduce an auxiliary measure $\mu(A)$ in the sample space, i.e. in the n-dimensional Euclidean space. For $\mu(A)$ we take

VIII. STATISTICAL INFERENCE

either an n-dimensional Lebesgue measure, or a "counting" measure, for which $\mu(A)$ equals the number of points with integer coordinates in A. Substituting one of these measures for $\mu(A)$ as required in the context, we shall be able to assume that all distributions $P_\theta(A)$ appearing in the hypotheses in question have probability densities $f(x; \theta)$ with respect to the measure $\mu(A)$ fixed in the given problem.

Consider the problem of testing a simple hypothesis H against a simple alternative K. It seems natural to accept as the best test of H against the alternative K at a given significance levels α the most powerful test, i.e., a test for which the probability of rejecting H if K is true is the greatest.

We now give the theorem describing the structure of the most powerful randomized test of a simple hypothesis against a simple alternative.

THEOREM 1 (Fundamental lemma of Neyman–Pearson). *Let H be a simple hypothesis asserting that $f_0(x)$ is the density of the probability distribution in X taken with respect to a measure $\mu(A)$ and let K be a simple alternative asserting that $f_1(x)$ is the probability density in X, taken with respect to a measure $\mu(A)$. Let us fix the significance level α. Then*

(a) *there exists a randomized test $\Phi_0(x)$ of the hypothesis H against the alternative K on the significance level α such that for some k*

$$\Phi_0(x) = \begin{cases} 1 & \text{if} \quad f_1(x) > kf_0(x), \\ 0 & \text{if} \quad f_1(x) < kf_0(x); \end{cases}$$

(b) *the test described in (a) is the most powerful test of the hypothesis H against the alternative K;*

(c) *if Φ is the most powerful, it must be of the form described in (a);*

(d) *if $0 < \alpha < 1$, then the probability β of rejecting the null hypothesis when the alternative is true exceeds α, provided only that the densities $f_0(x)$ and $f_1(x)$ do not coincide almost everywhere with respect to the measure μ.*

PROOF. Consider the function

$$a(c) = \int_{W(c)} f_0(x) d\mu(x)$$

where

$$W(c) = \{x: f_1(x) \geq cf_0(x)\}.$$

The above formulas imply that $a(c)$ equals the probability that the random variable $f_1(x)/f_0(x)$, whose probability is determined by assuming a distribution with density $f_0(x)$ in the space X, assumes a value not smaller than c. Thus, $1-a(c)$ is the probability distribution function of this random variable, and

as such it is non-decreasing and continuous from the left. Let us take as k the quantile of the order $1-\alpha$ of the distribution function $1-a(c)$, i.e., a number k such that
$$1-a(k) \leqslant 1-\alpha \leqslant 1-a(k+0)$$
where $a(k+0)$ denotes the right-hand side limit of the function $a(c)$ at the point $c = k$.

Several cases are possible. Thus, the quantile k may be non-unique if the distribution function $1-a(c)$ has a period of constancy. This means that the random variable $f_1(x)/f_0(x)$ assumes values from the interior of this interval with probability 0. In this case we take as k the largest quantile of the order $1-\alpha$, which is possible in view of the continuity from the left of the function $1-a(c)$. It may happen that for such a quantile k of the order $1-\alpha$ the probability distribution function $1-a(c)$ has a jump. This means that the random variable $f_1(x)/f_0(x)$ assumes the value k with a positive probability equal to $1-a(k+0)-(1-a(k)) = a(k)-a(k+0)$.

At any rate, it is clear that if we put
$$\Phi_0(x) = \begin{cases} 1 & \text{if } f_1(x) > kf_0(x), \\ \gamma & \text{if } f_1(x) = kf_0(x), \\ 0 & \text{if } f_1(x) < kf_0(x), \end{cases}$$
where γ is the root of the equation
$$1-a(k)+\gamma(a(k)-a(k+0)) = 1-\alpha,$$
if $a(k)-a(k+0) > 0$ and $\gamma = 0$ if $a(k)-a(k+0) = 0$, then $\Phi_0(x)$ is a test of the hypothesis H at the significance level α. This proves assertion (a).

Let $\Phi_0(x)$ be the test described in (a), and let $\Phi(x)$ be an arbitrary test of the hypothesis H at the significance level α, i.e., such that
$$\int_X \Phi(x)f_0(x)\,d\mu(x) = \alpha.$$
Let
$$S^{(>)} = \{x\colon \Phi_0(x) > \Phi(x)\},$$
$$S^{(<)} = \{x\colon \Phi_0(x) < \Phi(x)\}.$$
For $x \in S^{(>)}$ we have $\Phi_0(x) > 0$, whence $f_1(x) \geqslant kf_0(x)$. For $x \in S^{(<)}$ we have $\Phi_0(x) < 1$, whence $f_1(x) \leqslant kf_0(x)$. It follows that
$$\int_X (\Phi_0(x)-\Phi(x))(f_1(x)-kf_0(x))\,d\mu(x)$$
$$= \int_{S^{(>)} \cup S^{(<)}} (\Phi_0(x)-\Phi(x))(f_1(x)-kf_0(x))\,d\mu(x) \geqslant 0$$

VIII. STATISTICAL INFERENCE

and the inequality is strict if the set $(S^{(>)} \cup S^{(<)}) \cap \{x: f_1(x) \neq kf_0(x)\}$ has a positive measure μ. Since both tests $\Phi_0(x)$ and $\Phi(x)$ are, by assumption, tests on the significance level α, that is,

$$\int_X \Phi_0(x) f_0(x) \, d\mu(x) = \int_X \Phi(x) f_0(x) \, d\mu(x),$$

we obtain the inequality

$$\int_X \Phi_0(x) f_1(x) \, d\mu(x) \geq \int_X \Phi(x) f_1(x) \, d\mu(x),$$

strict if the tests $\Phi_0(x)$ and $\Phi(x)$ differ on a set of positive measure of points x such that $f_1(x) \neq kf_0(x)$. This proves assertions (b) and (c).

The test $\Phi(x) \equiv \alpha$, which for every x rejects H with probability α, is obviously a test at the significance level α. For this test we have the relation $\beta = \alpha$. In view of (c), this test is most powerful only if we have $f_1(x) = kf_0(x)$ almost everywhere with respect to the measure μ, i.e., if $\mu\{x: f_1(x) \neq kf_0(x)\} = 0$. We then have $f_1(x) = f_0(x)$ almost everywhere with respect to μ. This proves assertion (d), and completes the proof of the theorem.

Theorem 1 gives the method of constructing the most powerful test for a simple hypothesis against a simple alternative. We shall now formulate some corollaries which give the explicit form of the most powerful test for a few important probability distributions. In this context it will be convenient to speak of the probability distributions $P_\theta(A)$ in X as of the probability distributions of systems of random variables $\xi_1, \ldots, \xi_n$ representing the results of observations.

COROLLARY 1. INDEPENDENT TRIALS. BINOMIAL DISTRIBUTION. *Let $\xi_1, \ldots, \xi_n$ be independent random variables with the same probability distribution. Let H be the null hypothesis, and let*

$$P\{\xi_i = 1|H\} = p_0, \quad P\{\xi_i = 0|H\} = q_0 = 1 - p_0, \quad i = 1, \ldots, n.$$

Let K be the alternative hypothesis, and let

$$P\{\xi_i = 1|K\} = p_1, \quad P\{\xi_i = 0|K\} = q_1 = 1 - p_1, \quad i = 1, \ldots, n,$$

where $0 \leq p_0 < p_1 \leq 1$.

Given α, the most powerful test of the hypothesis H against the alternative K at the significance level α is given by the relations

(47.1)
$$\Phi(m) = \begin{cases} 1 & \text{for } m > c, \\ \gamma & \text{for } m = c, \\ 0 & \text{for } m < c, \end{cases}$$

where $m = \xi_1 + \ldots + \xi_n$ and integer c and γ from the interval $0 \leqslant \gamma < 1$ are determined from the equation

(47.2) $$\gamma \binom{n}{c-1} p_0^{c-1} q_0^{n-c+1} + \sum_{i=c}^{n} p_0^i q_0^{n-i} = \alpha.$$

The test defined by (47.1) is the most powerful test on the significance level α of the hypothesis H' asserting that m has the binomial distribution

$$P\{m = i|H'\} = \binom{n}{i} p_1^i (1-p_0)^{n-i}, \quad i = 0, 1, \ldots, n,$$

against the alternative K' asserting that m has the binomial distribution

$$P\{m = i|K'\} = \binom{n}{i} p_1^i (1-p_1)^{n-i}, \quad i = 0, 1, \ldots, n.$$

PROOF. The role of X is played here by the class of all sequences $\{i_1, \ldots, i_n\}$ consisting of ones and zeros. We can write

(47.3)
$$f_0(i_1, \ldots, i_n) = p_0^m q_0^{n-m},$$
$$f_1(i_1, \ldots, i_n) = p_1^m q_1^{n-m},$$

where m is the number of ones in the sequence $\{i_1, \ldots, i_n\}$.

According to Theorem 1, we obtain the most powerful test of the hypothesis H against the alternative K on the significance level α if we find a k such that the test

$$\Phi(i_1, \ldots, i_n) = \begin{cases} 1 & \text{if } f_1(i_1, \ldots, i_n) > kf_0(i_1, \ldots, i_n), \\ \gamma & \text{if } f_1(i_1, \ldots, i_n) = kf_0(i_1, \ldots, i_n), \\ 0 & \text{if } f_1(i_1, \ldots, i_n) < kf_0(i_1, \ldots, i_n), \end{cases}$$

where $0 \leqslant \gamma < 1$ will have the significance level α.

In view of equalities (47.3) the above definition of the test Φ is, for a certain c, equivalent to the following definition:

$$\Phi(i_1, \ldots, i_n) = \begin{cases} 1 & \text{if } m > c, \\ \gamma & \text{if } m = c, \\ 0 & \text{if } m < c. \end{cases}$$

It remains to choose c and γ so as to obtain a test at the significance level α. Since under the hypothesis H the random variable m has the binomial distribution

$$P\{m = i|H\} = \binom{n}{i} p_0^i q_0^{n-i}, \quad i = 0, 1, \ldots, n,$$

it suffices to choose integer c and γ such that $0 \leqslant \gamma < 1$ from equation (47.2). This proves that the test (47.1) is the most powerful test of the hypothesis H against the alternative K at the significance level α.

VIII. STATISTICAL INFERENCE

In the case of the hypotheses H' and K' the role of X is assumed by the set $\{0, 1, \ldots, n\}$ and we have

$$f_0(i) = \binom{n}{i} p_0^i q_0^{n-i}, \quad f_1(i) = \binom{n}{i} p_1^i q_1^{n-i}.$$

The relations

$$f_1(m) > kf_0(m), \quad f_1(m) = kf_0(m), \quad f_1(m) < kf_0(m),$$

required by Theorem 1 for the most powerful test, reduce, as before, to the relations

$$m > c, \quad m = c, \quad m < c,$$

which completes the proof of Corollary 1.

Note that the above test will not change if we change the alternative hypothesis, provided that the inequality $p_0 < p_1$ is preserved. Thus, this test is most powerful against any simple alternative $K: p = p_1$ where $p_0 < p_1$. If the alternative hypothesis is composite, one may expect that the most powerful test of the hypothesis H against a simple hypothesis contained in K may depend on this hypothesis. In our case, for a composite hypothesis K asserting that p lies in the interval $p_0 < p_1 \leqslant 1$ there is no such dependence. Therefore, the test which we have found is called the *uniformly most powerful test of composite alternative hypothesis* $K = \{p_1: p_0 < p_1 \leqslant 1\}$.

Our original problem concerning the hypotheses H and K and the system of variables $\xi_1, \ldots, \xi_n$ reduced to that of testing the hypotheses H' and K' concerning the random variable m. This reduction was possible because the statistic m equal to the number of ones in the sequence $i_1, \ldots, i_n$ is sufficient.

The intuitive background of the concept of sufficiency is the following: roughly speaking, the problem facing the statistician lies in the best possible identification of the value of the parameter θ on the basis of the observation x, provided we know that the random experiment is described by one of the probability distributions on the space X, these probability distributions having the densities $f(x; \theta)$ with an unknown value of parameter θ. One can ask, what values of the observation x should be assigned to the same decision about the parameter θ. The set A_t of observations x to which the statistician assigns the same decision about the parameter θ, or: the orbit $A_t = \{x: t(x) = t\}$ of the statistic $t(x)$ used by the statistician, should have the following property: the conditional distribution on this orbit should be independent of the parameter θ. Indeed, if such is the case, all the observations $x \in A_t$ yield the same information about the parameter θ, and their further differentiation yields nothing.

We shall present the above considerations in the form of the following definition:

DEFINITION. The statistic $t(x)$ defined for $x \in X$ is *sufficient* for the family of probability distributions $P_\theta(A)$ defined on X for $\theta \in \Omega$ or, shortly, is sufficient for the parameter θ, if for every orbit $A_t = \{x : t(x) = t\}$ the conditional probability distribution on this orbit, corresponding to the distribution $P_\theta(A)$ is independent of θ.

Naturally, it is desirable for a sufficient statistic to have as few orbits as possible, since this leads to a considerable simplification. The sufficient statistic $t(x)$ is called *minimal* if none of its orbits is a proper subset of an orbit of another sufficient statistic.

In independent trials discussed in Corollary 1, the role of X was played by the class of all possible sequences $\{i_1, \ldots, i_n\}$ of ones and zeros. The orbits of the statistic m (equal to the number of ones in such sequences) are sets of sequences with the same number of ones. For $m = k$, the number of such sequences equals $\binom{n}{k}$, and each of these sequences has the same probability $p^k q^{n-k}$. Thus, the conditional probability on such an orbit is the same for every value of the parameter p, namely uniform.

The following theorem can be proved.

THEOREM 2 (on factorization). *A statistic $t(x)$ is sufficient for the family of probability distributions with densities $f(x; \theta)$, $\theta \in \Omega$, if and only if the density $f(x; \theta)$ can be written in the form*

$$f(x; \theta) = h(x) g(t(x), \theta),$$

where $h(x)$ is a function independent of θ and the function g depends on x only through the statistic $t(x)$.

As an example, consider the probability density of a system of n independent random variables $\xi_1, \ldots, \xi_n$ having a normal distribution with a known variance σ^2 and an unknown expectation θ. We have

$$f(x; \theta) = \prod_{i=1}^{n} \frac{1}{\sigma\sqrt{2\pi}} \exp\left\{-\frac{(x_i - \theta)^2}{2\sigma^2}\right\}$$

$$= \underbrace{\frac{1}{\sigma^n(\sqrt{2\pi})^n} \exp\left\{-\frac{1}{2\sigma^2} \sum_{i=1}^{n} x_i^2\right\}}_{h(x)} \underbrace{\exp\left\{-\frac{1}{2\sigma^2}(-2n\theta\bar{x} + n\theta^2)\right\}}_{g(t(x), \theta)},$$

which shows that the sample mean $\bar{x} = \dfrac{x_1 + \ldots + x_n}{n}$ is a sufficient statistic for the parameter θ.

VIII. STATISTICAL INFERENCE

COROLLARY 2. POISSON DISTRIBUTION. *Let H be the null hypothesis that the random variable ξ has a Poisson distribution*

(47.4) $$P\{\xi = i | H\} = \frac{\lambda_0^i}{i!} e^{-\lambda_0}, \quad i = 0, 1, 2, \ldots$$

and let K be the alternative asserting that ξ has a Poisson distribution

(47.5) $$P\{\xi = i | K\} = \frac{\lambda_1^i}{i!} e^{-\lambda_1}, \quad i = 0, 1, 2, \ldots,$$

where $0 < \lambda_0 < \lambda_1$.

Then for a given α from the interval $0 < \alpha < 1$ the test defined as

(47.6) $$\Phi(i) = \begin{cases} 1 & \text{if } i > c, \\ \gamma & \text{if } i = c, \\ 0 & \text{if } i < c, \end{cases}$$

where integer c and γ satisfying the inequality $0 \leqslant \gamma < 1$ are chosen from the equation

(47.7) $$\gamma \frac{\lambda_0^c}{c!} e^{-\lambda_0} + \sum_{i=c+1}^{\infty} \frac{\lambda_0^i}{i!} e^{-\lambda_0} = \alpha,$$

is the most powerful test of the hypothesis H against the alternative K on the significance level α.

PROOF. According to Theorem 1 we have to determine k and γ for which the test

$$\Phi(i) = \begin{cases} 1 & \text{if } f_1(i) > kf_0(i), \\ \gamma & \text{if } f_1(i) = kf_0(i), \\ 0 & \text{if } f_1(i) < kf_0(i), \end{cases}$$

where

$$f_0(i) = \frac{\lambda_0^i}{i!} e^{-\lambda_0}, \quad f_1(i) = \frac{\lambda_1^i}{i!} e^{-\lambda_1}, \quad i = 0, 1, 2, \ldots$$

is a test on the significance level α. The ratio

$$\frac{f_1(i)}{f_0(i)} = \left(\frac{\lambda_1}{\lambda_0}\right)^i \exp[-(\lambda_1 - \lambda_0)]$$

is an increasing function of the variable i. This proves that the most powerful test is of the form (47.6) and by the choice of c and γ satisfying (47.7) we obtain the test on the required level of significance.

Since the test will not change if we change the alternative hypothesis, provided only that $\lambda_0 < \lambda_1$, the above test is uniformly most powerful against the composite alternative hypothesis $K = \{\lambda : \lambda_0 < \lambda\}$.

COROLLARY 3. EXPECTATION IN NORMAL DISTRIBUTION. *Let H be the null hypothesis asserting that independent random variables $\xi_1, \ldots, \xi_n$ have a normal distribution $N(\mu_0, \sigma)$ and let K be the alternative asserting that these random variables have a normal distribution $N(\mu_1, \sigma)$, where $\mu_0 < \mu_1$ and σ is known.*

For any given α from the interval $0 < \alpha < 1$ the most powerful test of the hypothesis H against the alternative K is given by the relation

(47.8) $$\Phi(\bar{x}) = \begin{cases} 1 & \text{if } \bar{x} > c, \\ 0 & \text{if } \bar{x} < c, \end{cases}$$

where $\bar{x} = \dfrac{1}{n}(x_1 + \ldots + x_n)$, $c = \mu_0 + \dfrac{\sigma \lambda_\alpha}{\sqrt{n}}$ and λ_α is determined from the condition

(47.9) $$\int_{\lambda_\alpha}^{\infty} \frac{1}{\sqrt{2\pi}} e^{-u^2/2} du = \alpha.$$

PROOF. The joint density of the random variables $\xi_1, \ldots, \xi_n$ under the hypothesis H is

$$f_0(x_1, \ldots, x_n) = \prod_{i=1}^{n} \frac{1}{\sigma \sqrt{2\pi}} \exp\left\{-\frac{1}{2\sigma^2}(x_i - \mu_0)^2\right\},$$

and under the alternative K it is

$$f_1(x_1, \ldots, x_n) = \prod_{i=1}^{n} \frac{1}{\sigma \sqrt{2\pi}} \exp\left\{-\frac{1}{2\sigma^2}(x_i - \mu_1)^2\right\}.$$

Next, we have

$$\frac{f_1(x_1, \ldots, x_n)}{f_0(x_1, \ldots, x_n)} = \exp\left\{-\frac{1}{2\sigma^2} \sum_{i=1}^{n}[(x_i - \mu_1)^2 - (x_i - \mu_0)^2]\right\}$$

$$= \exp\left\{-\frac{1}{2\sigma^2} \sum_{i=1}^{n}[-2(\mu_1 - \mu_0)x_i + \mu_1^2 - \mu_0^2]\right\}$$

$$= \exp\left\{-\frac{1}{2\sigma^2}\left[-2(\mu_1 - \mu_0)\sum_{i=1}^{n} x_i + n(\mu_1^2 - \mu_0^2)\right]\right\}$$

$$= \exp\left\{\frac{\mu_1 - \mu_0}{\sigma^2} \sum_{i=1}^{n} x_i - \frac{1}{2\sigma^2} n(\mu_1^2 - \mu_0^2)\right\}$$

$$= \exp\left\{\frac{\mu_1 - \mu_0}{\sigma^2} n\bar{x} - \frac{1}{2\sigma^2} n(\mu_1^2 - \mu_0^2)\right\}.$$

Since $\mu_0 < \mu_1$, the ratio under consideration is an increasing function of the average $\bar{x} = \frac{1}{n}(x_1 + \ldots + x_n)$. Thus, by Theorem 1, the most powerful test is of the form (47.8) and (47.9) gives the required level of significance.

As before, this test will not change if we modify the alternative hypothesis, provided only that $\mu_0 < \mu_1$. Thus, the above test is uniformly the most powerful test of the hypothesis H against the composite alternative $K = \{\mu: \mu_0 < \mu\}$.

COROLLARY 4. VARIANCE IN NORMAL DISTRIBUTION WITH KNOWN EXPECTATION. *Let H be the hypothesis asserting that independent random variables $\xi_1, \ldots, \xi_n$ have identical normal distribution $N(\mu, \sigma_0)$, and let K be the hypothesis that these random variables have a normal distribution $N(\mu, \sigma_1)$ where $\sigma_0 < \sigma_1$.*

Then, for given α from the interval $0 < \alpha < 1$, the most powerful test of H against the alternative K on the significance level α is

(47.10)
$$\Phi(s^2) = \begin{cases} 1 & \text{if } s^2 > c, \\ 0 & \text{if } s^2 < c, \end{cases}$$

where $s^2 = \frac{1}{n}\sum_{i=1}^{n}(\xi_i - \mu)^2$, $c = \frac{\sigma_0^2 \lambda_\alpha}{n}$ and λ_α is determined from the relation

(47.11)
$$P\{\lambda_\alpha < \chi_n^2\} = \alpha,$$

where χ_n^2 denotes a random variable with a chi-square distribution with n degrees of freedom.

PROOF. Under the hypothesis H the joint probability density of the random variables $\xi_1, \ldots, \xi_n$ is

$$f_0(x_1, \ldots, x_n) = \prod_{i=1}^{n} \frac{1}{\sigma_0 \sqrt{2\pi}} \exp\left\{-\frac{(x_i - \mu)^2}{2\sigma_0^2}\right\},$$

while under the hypothesis K is equals to

$$f_1(x_1, \ldots, x_n) = \prod_{i=1}^{n} \frac{1}{\sigma_1 \sqrt{2\pi}} \exp\left\{-\frac{(x_i - \mu)^2}{2\sigma_1^2}\right\}.$$

For the ratio of these two densities we have

$$\frac{f_1(x_1, \ldots, x_n)}{f_0(x_1, \ldots, x_n)} = \exp\left\{\frac{1}{2}\left(\frac{1}{\sigma_0^2} - \frac{1}{\sigma_1^2}\right)\sum_{i=1}^{n}(x_i - \mu)^2\right\}.$$

Since $\sigma_0 < \sigma_1$, this ratio is an increasing function of the sum $\sum_{i=1}^{n}(x_i - \mu)^2$. Thus, in view of Theorem 1, the most powerful test is of the form (47.10), and if we determine c according to condition (47.11) we obtain the test with the significance level α.

Once more we see that the test which we have found is most powerful against the composite alternative $K = \{\sigma: \sigma_0 < \sigma\}$.

In the above constructions of the most powerful tests we used the fact that the ratio of the densities corresponding to two different values of the parameter was an increasing function of the sample point (if X was equal to the real line or its subset), or it was an increasing function of a certain function $T(x)$ of the sample point. We shall discuss in more detail the use of this monotonicity in constructing the most powerful tests. Since this property appears, as we have seen, for different distributions, we shall consider the general case in order to avoid repeating the reasoning for different particular cases. It will be found that the tests which we have just discussed are not only uniformly most powerful tests of a simple null hypothesis against one-sided composite alternatives, but have also some optimal properties as tests of certain composite null hypotheses.

Consider a family of probability distributions $P_\theta(A)$ on the space X of sample points x, with densities $f_\theta(x)$ (with respect to a fixed measure μ), depending on the real parameter θ from the set Ω equal to the real line, a half-line or an interval. To simplify the formulations, we assume tacitly that all the values of parameters appearing in the considerations below belong to the set Ω. We shall say that our family of distribution has a *monotone likelihood ratio* with respect to the function $T(x)$ defined on X, if for $\theta' < \theta''$ the probability distributions are distinct, and the ratio $f_{\theta''}(x)/f_{\theta'}(x)$ is a non-decreasing function of $T(x)$; in other words, this ratio satisfies the inequality $f_{\theta''}(x')/f_{\theta'}(x') \leqslant f_{\theta''}(x'')/f_{\theta'}(x'')$ provided that $T(x') \leqslant T(x'')$. If there exists a function $T(x)$ with respect to which the family of distributions has a monotone likelihood ratio, we shall simply say that this family has a monotone likelihood ratio.

We now formulate the theorem on uniformly most powerful tests of composite null hypotheses against composite alternative hypotheses in the case of families with monotone likelihood ratio.

THEOREM 3. *Let $P_\theta(A)$ be a family of probability distributions on X depending on a real parameter θ, $\theta \in \Omega$, with a monotone likelihood ratio with respect to a function $T(x)$. Let H be the null hypothesis asserting that $\theta \leqslant \theta_0$, and let K be the alternative hypothesis asserting that $\theta_0 < \theta$. Then for a given α from the interval $0 < \alpha < 1$, the test Φ_0 of the form*

$$\Phi_0(x) = \begin{cases} 1 & \text{if} \quad T(x) > c, \\ \gamma & \text{if} \quad T(x) = c, \\ 0 & \text{if} \quad T(x) < c, \end{cases}$$

VIII. STATISTICAL INFERENCE

where the constants c and γ are determined from the condition

$$M(\Phi_0, \theta_0) \stackrel{\mathrm{df}}{=} \int_X \Phi_0(x) f_{\theta_0}(x) d\mu(x) = \alpha,$$

satisfies the following properties:

(a) Φ_0 is a test of the hypothesis H against the alternative K on the significance level α;

(b) Φ_0 is uniformly most powerful;

(c) its power

$$M(\Phi_0, \theta) = \int_X \Phi_0(x) f_\theta(x) d\mu(x)$$

is an increasing function of the parameter θ;

(d) among the tests of the hypothesis H which satisfy the condition

$$M(\Phi, \theta_0) = \alpha$$

the test Φ_0 described in (a) yields the minimum value of power $M(\Phi, \theta)$ for each $\theta \in H$ (it minimizes in H the probability of error of the first kind).

PROOF. Note first that since the distributions P_θ have monotone likelihood ratio with respect to $T(x)$, then for every $\theta' < \theta''$ the inequality

$$\frac{f_{\theta''}(x)}{f_{\theta'}(x)} > k$$

is equivalent to the inequality

$$T(x) > c$$

for a certain c.

Consider now the null hypothesis $H_0 = \{\theta: \theta = \theta_0\}$ and the simple alternative $\theta = \theta_1$ with $\theta_0 < \theta_1$. By Theorem 1, there exists a most powerful test of the hypothesis H against the alternative $\theta = \theta_1$ at the significance level α given in (a); let us denote that test by Φ_0.

Suppose now that $\theta' < \theta''$. By the same Theorem 1 and by the monotonicity of the likelihood ratio, the test Φ_0 is also most powerful for the hypothesis $\theta = \theta'$ against the alternative $\theta = \theta''$ on the significance level $M(\Phi_0, \theta')$. By assertion (d) of Theorem 1, we have $M(\Phi_0, \theta') < M(\Phi_0, \theta'')$ which implies assertions (a) and (c).

Let us fix $\theta_1 > \theta_0$. The test Φ_0 defined above gives the maximum power $M(\Phi, \theta_1)$ at the point $\theta = \theta_1$ in the class of all tests satisfying the condition $M(\Phi, \theta_0) = \alpha$, and the class of these tests contains the class of tests of the hypothesis $H = \{\theta: \theta < \theta_0\}$ on the significance level α. The tests satisfying the stronger condition $\sup_{\theta \in H} M(\Phi, \theta) = \alpha$ belong to the latter class. Since the test Φ_0 gives the maximum power $M(\Phi, \theta)$ in a wider class of tests, it has the same property in a smaller class; thus, it is most powerful for the hypothesis H against the alternative $\theta = \theta_1$. Since the constants c and γ defining the test Φ_0 do not depend on θ_1, the test Φ_0 is uniformly most powerful. This proves assertion (b).

The proof of (d) is analogous to the proof of Theorem 1, with the inequalities reversed, and we shall omit it.

COROLLARY 5. *The test found in Corollary 1 is uniformly most powerful for the hypothesis* $H = \{p: 0 \leqslant p \leqslant p_0\}$ *against the alternative* $K = \{p: p_0 < p \leqslant 1\}$.

COROLLARY 6. *The test found in Corollary 2 is uniformly most powerful for the hypothesis* $H = \{\lambda: 0 < \lambda \leqslant \lambda_0\}$ *against the alternative* $K = \{\lambda: \lambda_0 < \lambda < \infty\}$.

COROLLARY 7. *The test found in Corollary 3 is uniformly most powerful for the hypothesis* $H = \{\mu: -\infty < \mu \leqslant \mu_0\}$ *against the alternative* $K = \{\mu: \mu_0 < \mu < \infty\}$.

COROLLARY 8. *The test found in Corollary 4 is uniformly most powerful for the hypothesis* $H = \{\sigma: 0 < \sigma \leqslant \sigma_0\}$ *against the alternative* $K = \{\sigma: \sigma_0 < \sigma < \infty\}$.

Problems

1. Find the uniformly most powerful tests of the hypothesis K against the alternative H taking as K and H the hypotheses appearing in Corollaries 1, 2, 3 and 4.

2. Suppose that for real x and θ the functions

$$f_\theta(x) = C(\theta) e^{Q(\theta) T(x)}$$

where $Q(\theta)$ is an increasing function of the parameter θ, are probability densities. Suppose that $\theta_1 < \theta_2$ and consider the null hypothesis $H = \{\theta: \theta \leqslant \theta_1 \text{ or } \theta_2 \leqslant \theta\}$ and the alternative $K = \{\theta: \theta_1 < \theta < \theta_2\}$. Find the uniformly most powerful test of the hypothesis H against the alternative K at the given level of significance α from the interval $0 < \alpha < 1$.

3. Suppose that in a certain town there are N cars with licence plates numbered $1, 2, \ldots, N$. We observed the numbers $x_1, \ldots, x_n$ on the licence plates of n successive cars we encounter. Assume that the probability of encountering each of the cars is the same, and we encounter them independently. Prove that the maximum of the observed numbers is a sufficient statistic for the number N of cars in the town.

§ 48. PROBLEM OF DISCRIMINATION

At the beginning of § 45 we established the place of testing hypotheses among the variety of statistical problems: the object of testing hypotheses was to decide on the basis of observations which of the two classes of distributions admissible in the given problem contains the actual distribution of the population, or the actual distribution of the results of the experiment which is being conducted.

VIII. STATISTICAL INFERENCE

The problem of testing hypotheses, however, had one special feature: out of the two classes of distributions under consideration one was distinguished as the null hypothesis, and the side condition concerning the construction of the test, namely the specification of significance level, referred to the null hypothesis.

In certain statistical problems, where the object of considerations is basically the same as in testing hypotheses, i.e., we want to decide on the basis of observations which of the two hypotheses is true, such distinguishing of one hypothesis may appear unnatural. As an example we may use the situation in which on the basis of the morphological characteristics of a fossil skull we want to determine the sex of the person to whom the skull belonged, or another situation in which on the basis of the appearance of the received signal we want to ascertain whether the expected signal was transmitted or not.

Problems of the above type, whose practical aspect is well illustrated by the example of the skulls and in which we have two competing simple hypotheses and the distinguishing of the null hypothesis, typical in testing hypotheses, is unjustified, are known under the name of *problems of discrimination*. The difference between these problems and the problem of testing hypotheses consists in a different formulation of the criterion of optimality: we usually want to minimize the probability of erroneous classification.

We shall discuss the minimax approach and Bayes' approach to such problems. Using the example of the skulls, we shall speak of populations and of classifying the observed element to one of two or more populations on the basis of the observation of the characteristic x of that element. Apart from the terminology, our description is equally suitable for the case of deciding which of the two or more hypothese is true on the basis of the observation though then such suggestive terms as "population" or "element sampled from the population" seem to be less appropriate. We shall speak of a numerical characteristic x, but our considerations will concern equally well the probability distributions in an arbitrary Euclidean space. We shall discuss several possible cases.

CASE 1. Two populations; the prior distribution known. Let $f_1(x)$ and $f_2(x)$ be two probability densities of a numerical characteristic x in two populations. Next, let p_1 be the prior probability that the element considered comes from the first population, and let $p_2 = 1-p_1$ be the prior probability that it comes from the second population. Let L_1 denote the loss resulting from classifying the element of the first population as an element of the second population, and let L_2 denote the loss of classifying an element of the second population as an element of the first population. When the classification is correct, we incur no loss. Our problem consists in determining the discrimination function $\Phi(x)$, defined on the set of values of x by specifying two sets A_1 and A_2, whose union equals the whole real line, where A_1 is the set of those points x for which we decide that the element in question comes from the first population, and A_2 is the set of those points x for which we decide that it comes from the second population. We want to choose sets A_1 and A_2 so as to minimize the risk, i.e. the expected loss resulting from the decision procedure defined by the sets A_1 and A_2.

Let us determine the risk r for a given pair of sets A_1, A_2. With probability p_1 the element comes from the first population. We suffer the loss L_1 if the result of the observation belongs to the set A_2; the probability of this event is $\int_{A_2} f_1(x)dx$. In other case the loss is 0. Thus, the expected loss, provided that the element comes from the first population, is $L_1 \int_{A_2} f_1(x)dx$. Finally, taking into account both possibilities and their probabilities, we obtain

(48.1) $$r = p_1 L_1 \int_{A_2} f_1(x)dx + p_2 L_2 \int_{A_1} f_2(x)dx.$$

We see that the risk equals the integral extended over the whole real line, the integrand being equal either to $p_1 L_1 f_1(x)$ or to $p_2 L_2 f_2(x)$ depending on whether x is in A_2 or A_1. Thus, in order to minimize the risk, one has to choose sets A_1 and A_2 in such a way as to minimize the integrand. This leads to the following definition of the optimal sets A_1 and A_2:

(48.2)
$$A_1 = \{x\colon p_2 L_2 f_2(x) \leqslant p_1 L_1 f_1(x)\},$$
$$A_2 = \{x\colon p_1 L_1 f_1(x) < p_2 L_2 f_2(x)\}.$$

The points for which we have $p_1 L_1 f_1(x) = p_2 L_2 f_2(x)$ have been assigned to A_1, although they could equally well be assigned to A_2 without changing the risk. In applications, the probability of such equality is usually zero.

Thus, formulas (48.2) define the optimal discrimination procedure if the prior distribution is known.

Note that the above procedure of discrimination is in the case of $L_1 = L_2$ equivalent to the procedure which consists in assigning elements to the second population if and only if the posterior probability of their belonging to the second population exceeds $1/2$. According to Bayes' theorem (see § 16) the posterior probabilities that the element belongs to the first and the second populations, x being the observed value of the characteristic, are

$$\frac{p_1 f_1(x)}{p_1 f_1(x) + p_2 f_2(x)}, \quad \frac{p_2 f_2(x)}{p_1 f_1(x) + p_2 f_2(x)},$$

and the condition that x belongs to A_2 takes the form

$$p_1 f_1(x) < p_2 f_2(x).$$

In view of this connection with Bayes' theorem and the prior distribution, the discrimination procedure (48.2) is called the *Bayesian discrimination function*.

CASE 2. Two populations; the prior distribution unknown. We already know the best procedure of discrimination when the prior probabilities of the first and the second populations are known. It does happen that these probabilities are not known. While for an ordinary cemetery, we may reasonably assume that $p_1 = p_2 = 1/2$, frequently these probabilities are unknown, even in the case of skulls: in the case of prehistoric cemeteries, we do not always know whether what we are observing is an ordinary cemetery close to a human settlement, or a battle-field, where we should expect a large majority of male skulls. Also, it would be difficult to speak without reservations about a prior distribution in the case of the already mentioned receiving installation.

Now, we find that it is possible avoid discussions connected with the prior distribution by indicating a set A_1 such that the risk does not depend on the value of p_1. Moreover, we can find a set A_1 which has the above property, and which constitutes, moreover, the optimal set against a certain prior distribution.

Indeed, using the equality $p_2 = 1-p_1$, we modify the formula (48.1) as follows:

$$(48.3) \qquad r = \int_{A_1} L_2 f_2(x) dx + p_1 \left\{ \int_{A_2} L_1 f_1(x) dx - \int_{A_1} L_2 f_2(x) dx \right\}.$$

If there is no set of points x of positive measure on which the densities $f_1(x)$ and $f_2(x)$ would be positive and proportional (in applications this condition is usually satisfied), there exists a C such that for the sets A_1 and A_2 of the form

$$(48.4) \qquad \begin{aligned} A_1 &= \{x: L_2 f_2(x) \leqslant C L_1 f_1(x)\}, \\ A_2 &= \{x: C L_1 f_1(x) < L_2 f_2(x)\} \end{aligned}$$

the coefficient at p_1 in formula (48.3) vanishes. For the discrimination function obtained in this way, the risk clearly does not depend on the prior distribution. Consequently, one cannot increase the risk by the choice of a prior distribution. On the other hand, by comparison with formulas (48.2) we see that discrimination function (48.4) is a Bayesian discrimination function corresponding to the prior distribution with $p_1 = C/(1+C)$. Thus, there is no discrimination function which would give a smaller risk than that given by (48.4) for all prior distributions. If we evaluate discrimination functions by their maximum risk, the maximum being taken with respect to all prior distributions, then this maximum is smallest for the discrimination function (48.4). In this sense (48.4) is the *minimax discrimination function*.

CASE 3. Generalization to an arbitrary finite number of populations.

If n is the number of populations, $f_i(x)$ is the probability density of the characteristic x in the ith population, A_i is the set of points x for which we decide that the element considered belongs to the ith population, L_{ij} is the loss due to classifying an element of the ith population as an element of the jth population, and p_i is the prior probability that the element considered belongs to the ith population, and if the densities $f_i(x)$ are such that their linear combinations with non-equal coefficients may vanish at most on a set of points x of measure zero, then

(a) the Bayesian discrimination function corresponding to the prior distribution $p_1, \ldots, p_n$ is determined by sets A_j defined as

$$(48.5) \qquad A_j = \left\{ x : \sum_{i=1}^{n} p_i L_{ij} f_i(x) = \min_{1 \leqslant k \leqslant n} \sum_{i=1}^{n} p_i L_{ik} f_i(x) \right\},$$

(b) if $C_1 = 1, C_2, \ldots, C_n$ are non-negative constants such that for the sets

$$(48.6) \quad A_j = \left\{ x : \sum_{i=1}^{n} C_i L_{ij} f_i(x) = \min_{1 \leqslant k \leqslant n} \sum_{i=1}^{n} C_i L_{ik} f_i(x), \ k = 1, \ldots, n \right\},$$

$j = 1, 2, \ldots, n$, we have

$$(48.7) \qquad \sum_{i=1}^{n} \int_{A_j} L_{ij} f_i(x) dx = \sum_{j=1}^{n} \int_{A_j} L_{nj} f_n(x) dx, \quad i = 1, \ldots, n-1,$$

then (48.6) gives the minimax discrimination function.

Note that the sets A_j defined by (48.5) and (48.6) are not disjoint. In view of the assumptions about the densities $f_i(x)$, the intersections of these sets must be of measure zero. Thus, the risks and probabilities computed would not change if we made these sets disjoint in an arbitrary manner. We could easily avoid this lack of uniqueness, and also avoid the inessential assumption concerning linear combinations of densities $f_i(x)$, if we considered randomized discrimination functions, just as we considered randomized decisions before.

We now proceed to the proof.

PROOF. To prove (a) it suffices to express the risk by a formula analogous to (48.1):

$$(48.8) \qquad r = \sum_{i=1}^{n} p_i \sum_{j=1}^{n} L_{ij} \int_{A_j} f_i(x) dx = \sum_{i=1}^{n} \sum_{j=1}^{n} p_i L_{ij} \int_{A_j} f_i(x) dx$$

$$= \sum_{j=1}^{n} \sum_{i=1}^{n} p_i L_{ij} \int_{A_j} f_i(x) dx = \sum_{j=1}^{n} \int_{A_j} \sum_{i=1}^{n} p_i L_{ij} f_i(x) dx.$$

VIII. STATISTICAL INFERENCE

Since the risk can be expressed as an integral extended over the whole real line, where the integrand is equal at every point x to one of the functions $\sum_{i=1}^{n} p_i L_{ij} f_i(x)$, $j = 1, \ldots, n$, it follows that (48.6) gives the best discrimination procedure if the probabilities $p_1, \ldots, p_n$ are known. This proves (a).

Note that if $L_{ij} = 0$ for $i = j$ and $L_{ij} = 1$ for $i \neq j$, then x belongs to A_j if and only if for this point x the posterior probability that the element considered belongs to the jth population is not smaller than the posterior probability that it belongs to any other population.

We now pass to the proof of assertion (b). Using the equality $p_n = 1 - p_1 - \ldots - p_{n-1}$, we transform (48.8) into the form

$$(48.9) \qquad r = \sum_{j=1}^{n} \int_{A_j} L_{nj} f_n(x) \, dx + \sum_{i=1}^{n-1} p_i \left\{ \sum_{j=1}^{n} \int_{A_j} L_{ij} f_i(x) \, dx - \sum_{j=1}^{n} \int_{A_j} L_{nj} f_n(x) \, dx \right\}.$$

By a proper choice of non-negative constants $C_1 = 1$, $C_2, \ldots, C_n$, and, consequently, the sets A_j given by (48.6), we can obtain (48.9) with the coefficients at p_i on the right-hand side vanishing. This gives therefore a discrimination function whose risk does not depend on the prior distribution, this discrimination function being, at the same time, the best procedure against the prior distribution with $p_i = C_i/(C_1 + \ldots + C_n)$. This shows that the discriminating function in question is minimax.

Problems

1. We know that in the first population the characteristic x has a normal distribution with mean 0 and standard deviation 1, and in the second population the characteristic x has a normal distribution with mean 1 and standard deviation 2. Construct the minimax discrimination function assuming that the loss resulting from the wrong classification equals 1.

A n s w e r. The set of points x for which we are to decide that the element belongs to the first population is of the form $\{x: -\frac{1}{3} - C < x < -\frac{1}{3} + C\}$, $C \approx 1.04$; the probability of the wrong classification equals 0.324 (see Fig. 2).

2. The pair (x, y) has in the first population a normal distribution with the covariance matrix

(a) $\quad \Lambda = \begin{bmatrix} 1 & 0 \\ 0 & 4 \end{bmatrix}$, \qquad (b) $\quad \Lambda = \begin{bmatrix} 1 & 1 \\ 1 & 4 \end{bmatrix}$

and the vector of expectations $(0, 0)$, and in the second population it has a normal distribution with the same covariance matrix and the vector of expectations $(1, 1)$. Construct the minimax discrimination function and compute the probability of the wrong classification.

A n s w e r. We should assign the element to the first population (a) if $4x + y < 5/2$, (b) if $x < 1/2$; the probability of the wrong classification equals approximately (a) 0.131, (b) 0.309.

3. The system of n characteristics $(x_1, \ldots, x_n)$ has in two populations a joint normal distribution with the same covariance matrix

$$\Lambda = [\lambda_{ii}]$$

and vectors of expectations $(m_1, \ldots, m_n)$ and $(m_1+z_1, \ldots, m_n+z_n)$. Prove that for every Bayesian discrimination function $\delta(x_1, \ldots, x_n)$ the set of points $(x_1, \ldots, x_n)$ for which one decides that the element considered belongs to the first population is of the form

$$\left\{(x_1, \ldots, x_n): \sum_{i=1}^{n} \lambda_i x_i > C\right\},$$

where C is a constant to be determined, and

$$\lambda_i = \sum_{j=1}^{n} \frac{\Lambda_{ij}}{|\Lambda|} z_j$$

(here Λ_{ij} is the algebraic complement of the element λ_{ij} in the covariance matrix, and $|\Lambda|$ is the determinant of the covariance matrix; in other words, $[\Lambda_{ij}/|\Lambda|]$ is the inverse of the covariance matrix).

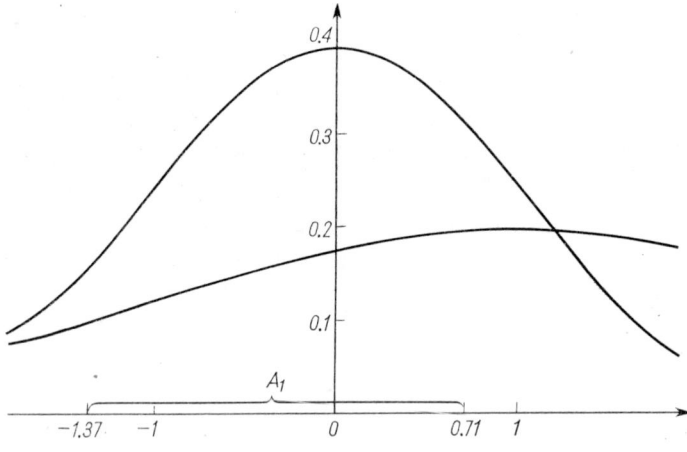

Fig. 2

4. In an n dimensional Euclidean space of points $(x_1, \ldots, x_n)$ consider two probability distributions with the same covariance matrix

$$\Lambda = [\lambda_{ij}]$$

and vectors of expectations $(0, \ldots, 0)$ and $(z_1, \ldots, z_n)$.

Consider the linear combinations

$$\varkappa = \sum_{i=1}^{n} \lambda_i x_i$$

and the ratio of the square of the difference of expectations of such linear combinations in the first and the second population to the variance (equal in both populations). Show that (up to the proportionality constant) this ratio has its maximum for the linear combination with λ defined in problem 3.

Prove this theorem for $n = 2$.

§ 49. PROBLEM OF ESTIMATION. BAYESIAN ESTIMATES

In Sections 45–47 we dealt with statistical tests. The problems discussed there concerned situations in which we restrict ourselves to making one of two decisions on the basis of observations. Thus, in comparing two varieties of tomatoes, discussed in the examples of § 45, we only wanted to know whether the new variety is more fertile than the old one, or not. The problem would be different if we wanted to assess on the basis of observations, as accurately as possible, how much better is the new variety. Problems in which we want to assess, or estimate, certain parameters of the distribution in the population are called *estimation problems*. We shall discuss them in this section.

In these problems the object is to construct a function which assigns numerical values to observations, these values being as close as possible to the estimated parameter. Such a function is called a *statistic*. In cases where we deal with statistics suitable for estimating certain parameters of distribution in the population, we shall call such statistics *estimators*, usually indicating to which particular parameter they refer.

In the preceding parts of the book, in particular in §§ 33 and 44, we presented that information concerning estimators which is contained in the laws of large numbers and the limit theorems of probability theory. These theorems formalize the intuitive concept according to which samples constitute miniatures of the population. Generally, these theorems assert that as an estimator of a parameter of the population we may take an analogous parameter from the sample, i.e., a parameter calculated for the empirical distribution function.

The weak laws of large numbers assert that as the sample size increases these estimators tend in probability to the estimated parameters. Estimators with this property are called *consistent*. Thus, according to the laws of large numbers, the mean, the variance and the quantiles calculated from the empirical distribution function are consistent estimators of the expectation, variance and quantiles of the population respectively. The consistency of an estimator is one of its desirable properties.

At the beginning of §45, we pointed out the special character of the laws of large numbers and limit theorems: they assert that for a given distribution of the population certain statistics computed from the sample of this population will have probability distributions with certain specified properties. However, the problem facing statisticians is, in a sense, inverse: one is interested not so much in finding the distribution of the statistic when the distribution of the population is known as in deciding what can be said about the distribution of the population when we know the result of observation.

The key to the solution of such an inverse problem is contained in Bayes' theorem mentioned in § 16. Here we give a more general and abstract version of this theorem.

THEOREM 1. *Suppose that X is a space of points x representing the possible results of observations, S is a σ-field of subsets of X, Ω is the space of parameters θ determining the probability distributions $P_\theta(A)$ on S, and $f_\theta(x) = f(x|\theta)$ is the density of the distribution $P_\theta(A)$ with respect to a fixed measure μ on S. Next, let $g(\theta)$ be the density of a probability distribution on a σ-field T of subsets of Ω, taken with respect to a measure ν on T. Suppose that $f(x|\theta)$, as a function of x and θ, is measurable with respect to the σ-field $S \times T$. Then $g(\theta)f(x|\theta)$ is the density (with respect to measure $\mu \times \nu$) of a probability distribution in space $X \times \Omega$, this probability distribution being defined on $S \times T$, $g(\theta)$ is the marginal distribution on Ω of this joint probability distribution, and $f(x|\theta)$ is the density of a conditional distribution on X for a given $\theta \in \Omega$.*

Next, the density of the conditional distribution on Ω for a fixed x is given, perhaps except on a set of points x of measure μ equal to zero, by the formula

$$(49.1) \qquad g(\theta|x) = \frac{g(\theta)f(x|\theta)}{\int_\Omega g(\theta)f(x|\theta)\,d\nu(\theta)}.$$

The probability distribution on Ω with density $g(\theta)$ is called the *prior distribution* of the parameter θ, and the distribution on Ω with density $g(\theta|x)$ is called the *posterior distribution* of the parameter θ. The posterior distribution gives all the information about θ contained in the observation x. Thus, (49.1) provides a foundation for inference about θ on the basis of the observation x. If θ is a real number, and we want to estimate it so as to minimize the average square error, we ought to take as an estimator of θ, at the point x, the expectation of the conditional distribution of θ:

$$(49.2) \qquad t(x) = \int_\Omega \theta g(\theta|x)\,d\nu(\theta).$$

If we want to minimize the expected absolute value of the error, the best estimator of θ proves to be given by the median of the distribution $g(\theta|x)$.

If θ assumes only a finite or countable number of values, and we want to guess the actual value of the parameter θ with maximum probability of a correct guess, we ought to take as an estimator of θ corresponding to the observation x the *mode* of distribution with density $g(\theta|x)$, i.e., a value of θ which is the most probable for this distribution.

The estimators of the parameter θ obtained in the above manner from formula (49.1) are called *Bayesian estimators*.

However, in order to be able to use formula (49.1) one has to have certain

VIII. STATISTICAL INFERENCE

information. First, one has to know the densities $f(x|\theta)$ of the conditional distributions of the results of observations for a given value of the parameter. Secondly, it must be possible to treat the value of the parameter as a random variable. Thirdly, one has to know the probability density $g(\theta)$ of the distribution of this random variable. The easiest problem is to define the densities $f(x|\theta)$; they are usually controlled experimentally, and may be regarded as known.

Sometimes it happens that also the second and third conditions are satisfied. For instance, if we produce water meters, under fixed conditions, the parameter θ is the systematic error of their readings, and we want to determine on the basis of control measurements the magnitude of this systematic error for each meter produced. Then, on the basis of observations one can determine the prior distribution of these errors. If we know, in addition, the accuracy of the control measurements, we can apply formula (49.1) and base on it the procedure of accepting the items produced.

It happens sometimes that we suppose the parameter θ to be a random variable, but we do not know its prior distribution. As an example consider the problem of acceptance of a lot of merchandise coming from different producers. If there are many producers, if they come from different towns, and the quality of the merchandise is determined by various random factors, one can assume that the parameter characterizing the quality of the whole lot has a certain prior distribution in the sense that the lot in question may be regarded as being drawn at random from a large number of lots presented for acceptance. However, we do not know this prior distribution.

In most cases, however, one cannot treat the parameter as a random variable. It does not help to say that we deal here with a one-point prior distribution, i.e. that the value of the parameter is, with probability one, such as it really is. Formula (49.1) would tell us that we should give this value as an estimator, independently of the value of the observation, and since we do not know this value, we are forced to guess. In such cases it has been suggested that the use of the so-called *Bayes rule* may provide the solution: when we do not have any information about the prior distribution, this rule tell us to assume that this distribution is uniform. The so-called *retrospective plans* in quality control are based on this rule. They are based on the assumption that the defectiveness of the lot, i.e. the fraction of defective items in the lot, has a prior uniform distribution in the interval (0, 1). This rule loses its intuitive appeal when the parameter can assume values from an unbounded set. Moreover, the Bayes rule has been violently criticized, the main issue being its complete arbitrariness: why should the deffectiveness have the uniform distribution, and not, say, its sixth power? Why the uniform, and not some other distribution?

As a result of this criticism several methods of statistical inference, in particular the construction of estimators, have been suggested; these methods do not use Bayes' theorem, and do not rely on the prior distribution. In place of this theorem they introduce the postulate of unprejudiced, or symmetric, treatment of all hypotheses.

Problems

1. The systematic error of some water meters has a prior normal distribution with expectation 0 and standard deviation σ_0. The result of a control measurement in a meter having a systematic error x has the normal distribution with expectation x and standard deviation σ_1. Find the posterior distribution of the systematic error of the meter (a) based on a single control measurement, and (b) based on two independent control measurements.

A n s w e r. Normal distribution with the expectation

$$\text{(a) } m_1\sigma_0^2/(\sigma_0^2+\sigma_1^2), \quad \text{(b) } 2m_1\sigma_0^2/(2\sigma_0^2+\sigma_1^2)$$

and standard deviation

$$\text{(a) } \sigma_1/\sqrt{1+(\sigma_1/\sigma_0)^2}, \quad \text{(b) } \sigma_1/\sqrt{1+(\sigma_1/\sigma_0\sqrt{2})^2}$$

(here m_1 denotes the arithmetic mean of the control measurements).

§ 50. INTERVAL ESTIMATION. CONFIDENCE INTERVALS

Among the methods of estimation which do not utilize Bayes' theorem the most important is the method of confidence intervals introduced by J. Spława-Neyman. Instead of asking, for a fixed value of observation and a fixed interval, about the posterior probability that the parameter belongs to this interval (which cannot be answered without the knowledge of the prior distribution), Neyman suggests that we should assign intervals to the results of observations in such a way as to have, independently of the actual distribution of the population, a guaranteed fixed probability that the interval constructed on the basis of observations will contain the unknown value of the parameter. If the intervals obtained in this way are sufficiently short, then, from the practical point of view, the information that the unknown value of the parameter is contained within such an interval may prove sufficiently accurate for making further attempts at determining the value of the parameter unnecessary.

We shall explain this construction by means of a few examples.

EXAMPLE 1. Suppose that we know that the distribution in the population is normal with a known variance σ^2, and we want to determine the unknown value of the expectation m of this distribution. Now, it is known that the sample mean $\bar{x} = (x_1 + \ldots + x_n)/n$ has a normal distribution with expectation m and variance σ^2/n. In tables of the normal distribution we may find a number λ

VIII. STATISTICAL INFERENCE

such that the inequality

$$(50.1) \qquad m - \frac{\lambda\sigma}{\sqrt{n}} < \bar{x} < m + \frac{\lambda\sigma}{\sqrt{n}}$$

has a fixed probability α, say $\alpha = 0.95$. The interval (50.1) has fixed but unknown end-points and contains the sample mean with a fixed probability, which we will call the *confidence level*. The inequalities appearing in (50.1) may be written in the following equivalent form:

$$(50.2) \qquad \bar{x} - \frac{\lambda\sigma}{\sqrt{n}} < m < \bar{x} + \frac{\lambda\sigma}{\sqrt{n}}.$$

Now the first and third terms of this inequality determine an interval with known end-points, and the position of this interval depends on the results of observations. Moreover, this random interval of length $2\lambda\sigma/\sqrt{n}$ is such that it covers the unknown true value of the parameter with a fixed probability α.

If we claim that the expectation m is contained in the interval with the end-points $\bar{x} - \lambda\sigma/\sqrt{n}$ and $\bar{x} + \lambda\sigma/\sqrt{n}$, then the probability of our claim to be consistent with the reality will be equal to α. The desired independence of the probability of the truth of our claim from the prior distribution has been obtained at the cost of referring the probability α to the procedure—each time the interval (50.2) may have a different position—instead of referring it to a fixed interval, as is the case when we apply Bayes' theorem.

EXAMPLE 2. When both the expectation m and the standard deviation σ are unknown, and we know only that the distribution is normal, we can construct confidence intervals for the expectation with the given confidence level using the Student's t statistic. We know (see Theorem 1 of § 45) that the statistic

$$t = \frac{\bar{x} - m}{s}\sqrt{n-1},$$

where

$$\bar{x} = \frac{1}{n}\sum_{i=1}^{n} x_i, \qquad s^2 = \frac{1}{n}\sum_{i=1}^{n}(x_i - \bar{x})^2$$

computed from an n-element sample $x_1, \ldots, x_n$ has Student's t distribution with $n-1$ degrees of freedom. Thus, we can find in the tables of Student's distribution a number t_α such that the inequalities

$$(50.3) \qquad -t_\alpha < \frac{\bar{x}-m}{s}\sqrt{n-1} < t_\alpha$$

will be satisfied with probability α. Multiplying them by $s/\sqrt{n-1}$ and arranging differently, we can write them in the form

$$(50.4) \qquad \bar{x} - \frac{st_\alpha}{\sqrt{n-1}} < m < \bar{x} + \frac{st_\alpha}{\sqrt{n-1}},$$

which gives the required confidence intervals for the expectation m.

EXAMPLE 3. We can also build confidence intervals for variance. If $x_1, \ldots, x_n$ is a sample from a normal population with expectation m and standard deviation σ, then Theorem 1 of § 45 asserts that the statistic ns^2/σ^2 (where, as usual, $s^2 = \frac{1}{n} \sum_{i=1}^{n} (x_i - \bar{x})^2$ and $\bar{x} = \frac{1}{n} \sum_{i=1}^{n} x_i$) has a chi-square distribution with $n-1$ degrees of freedom. Thus, we can find in the tables positive numbers λ_1 and λ_2 such that the inequality

$$(50.5) \qquad \lambda_1 < \frac{ns^2}{\sigma^2} < \lambda_2$$

holds with probability α. Rewriting it in the form

$$(50.6) \qquad \frac{ns^2}{\lambda_2} < \sigma^2 < \frac{ns^2}{\lambda_1}$$

or in the form

$$(50.7) \qquad s\sqrt{\frac{n}{\lambda_2}} < \sigma < s\sqrt{\frac{n}{\lambda_1}},$$

we obtain the desired confidence intervals on the confidence level α for variance σ^2 and standard deviation σ.

The construction of confidence intervals illustrated here by means of a few examples is very general, and under certain assumptions concerning the continuity of distributions it can be repeated for an arbitrary parameter. We can also apply it in the case of several parameters simultaneously, and construct confidence regions for them.

Indeed, let X be a multidimensional Euclidean space of points $x = (x_1, \ldots, x_n)$ representing the results of observations. Next, let Ω be the space od parameters θ (numerical or vector-valued) determining the distributions in X with densities $f_\theta(x)$. Let us fix α from the interval $0 < \alpha < 1$, and choose for each $\theta \in \Omega$ a set $S_\theta \subset X$ such that

$$\int_{S_\theta} f_\theta(x) dx = \alpha.$$

VIII. STATISTICAL INFERENCE

Consider now the space $X \times \Omega$ and the set D in it, consisting of all points (x, θ) for which we have $\theta \in \Omega$ and $x \in S_\theta$. Then (see Fig. 3) for a fixed $\theta \in \Omega$ the set

$$\{(x, \theta): x \in S_\theta\}$$

is an intersection of the set D parallel to the x-axis. The set D has the following property: no matter whether θ has a fixed value, or is a random variable with a certain probability distribution, the random point (x, θ) will belong to D with probability α. Let us now write the condition $(x, \theta) \in D$ in a different way, taking into consideration the intersections of D parallel to the θ-axis. Write

$$T_x = \{\theta: (x, \theta) \in D\}.$$

Then the three conditions

$$\theta \in \Omega, x \in S_\theta; \quad (x, \theta) \in D;$$
$$x \in X, \theta \in T_x$$

express in three different ways the same requirement $(x, \theta) \in D$. Thus, T_x are the desired confidence intervals with the confidence level α, having the property that, no matter whether θ is fixed or random, with a fixed probability α the random interval T_x corresponding to the observation x contains the value of the parameter θ, determining the distribution according to which the sample determining x has been selected.

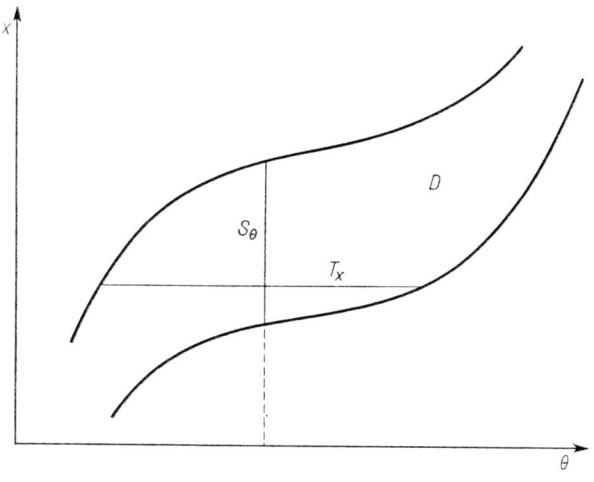

Fig. 3

It follows from the above construction that different confidence intervals at the same confidence level can be constructed for the same parameter. In choosing one of them, various additional considerations may be relevant. Intervals in Example 1 are symmetric and as short as possible. We could be interested, however, in one-sided intervals, and choose a number λ for which the inequality $m - \lambda\sigma/\sqrt{n} < \bar{x}$ would hold with the required probability. This leads to the confidence intervals $-\infty < m < \bar{x} + \lambda\sigma/\sqrt{n}$. Similarly, in Example 3 we could be interested in shortest intervals, or in one-sided intervals, bounding the parameter from below of from above. The reader can easily consider the details of the construction.

§ 51. MAXIMUM LIKELIHOOD ESTIMATORS

We shall now return to the discussion of estimators in the usual sense, where we want to give an approximate value of the parameter on the basis of the sample. One of the methods of constructing estimators, in which we do not utilize the prior distribution explicitly, is the construction of the so-called maximum likelihood estimators.

These estimators are constructed in the following manner. Let $f_\theta(x)$ be the probability density in the space X of the results of the experiment x, depending on the parameter θ assuming values in the space Ω. The *principle of maximum likelihood* tells us to take as an approximation value of θ, for a given value of x, that value $t(x) \in \Omega$ for which the probability density $f_\theta(x)$, for this x, attains its maximum with respect to θ. The function $f_\theta(x)$ considered for a fixed x as a function of θ is called the *likelihood function*.

We present some examples of constructing maximum likelihood estimators.

EXAMPLE 1. *Estimation of fraction of defective items.* Let X be the space of n-element sequences $x = \{i_1, \ldots, i_n\}$ consisting of zeros and ones. Next, let θ be a number from the interval $\Omega = \{\theta\colon 0 \leqslant \theta \leqslant 1\}$, and let

$$f_\theta(x) = \theta^m(1-\theta)^{n-m}$$

be the probability of obtaining a sequence $x = \{i_1, \ldots, i_n\}$ containing m ones in n independent random experiments, the probability of obtaining 1 in each experiment being equal to θ. Then, for a given sequence $x = \{i_1, \ldots, i_n\}$ containing m ones, the probability $f_\theta(x)$ (i.e., the probability density with respect to measure μ "counting" points $x = \{i_1, \ldots, i_n\}$) attains its maximum with respect to θ for $\theta = m/n$. Thus, the fraction m/n of 1's, i.e., the ratio of the number of successes to the number of trials, turns out to be the maximum likelihood estimator of the probability of success.

The reader will verify that we obtain the same results if we take as X the set $\{0, 1, \ldots, n\}$ and as $f_\theta(x)$ the binomial probability $\binom{n}{x}\theta^x(1-\theta)^{n-x}$.

VIII. STATISTICAL INFERENCE

EXAMPLE 2. *Estimating the parameter in a Poisson distribution.* Let X be the space of points $x = \{x_1, \ldots, x_n\}$ whose coordinates x_i are all non-negative integers. Let

$$f_\theta(x) \equiv \prod_{i=1}^{n} \frac{\theta^{x_i}}{x_i!} e^{-\theta} = \frac{1}{x_1! \ldots x_n!} \theta^{(x_1 + \ldots + x_n)} e^{-n\theta}$$

be the probability of obtaining the observation x when θ is the value of the parameter (the distribution in X with density $f_\theta(x)$ with respect to the counting measure is the probability distribution of n independent random variables $\xi_1, \ldots, \xi_n$, each having a Poisson distribution with the parameter θ). Clearly, $f_\theta(x)$ attains, for a given x, its maximum with respect to θ for $\theta = (x_1 + \ldots + x_n)/n = \bar{x}$. Thus, the sample mean is the maximum likelihood estimator of the parameter in a Poisson distribution.

EXAMPLE 3. *Normal distribution.* Let X be the Euclidean space of points $x = (x_1, \ldots, x_n)$, $\Omega = \{(\mu, \sigma): -\infty < \mu < \infty, 0 < \sigma < \infty\}$, $\theta = (\mu, \sigma)$, and

$$f_\theta(x) = \prod_{i=1}^{n} \frac{1}{\sigma\sqrt{2\pi}} \exp\left\{-\frac{(x_i-\mu)^2}{2\sigma^2}\right\} = \frac{1}{\sigma_n(2\pi)^{n/2}} \exp\left\{-\frac{1}{2\sigma^2}\sum_{i=1}^{n}(x_i-\mu)^2\right\}.$$

Here the probability distribution in X corresponding to the given value of the parameter θ is the probability distribution of n independent random variables having the same normal distribution with expectation μ and standard deviation σ. For a given x the density $f_\theta(x)$ assumes its maximum for

$$\mu = \bar{x} = \frac{1}{n}(x_1 + \ldots + x_n),$$

$$\sigma = s, \quad \text{where} \quad s^2 = \frac{1}{n}\sum_{i=1}^{n}(x_i - \bar{x})^2.$$

Indeed, for a fixed x and σ the density $f_\theta(x)$ assumes its maximum with respect to μ when the sum $\sum (x_i - \mu)^2$ is smallest, i.e., for $\mu = \bar{x}$. If we substitute $\bar{x}$ in place of μ, it remains to determine σ so as to maximize the expression

$$\frac{1}{\sigma^n(\sqrt{2\pi})^n} \exp\left\{-\frac{1}{2\sigma^2}\sum_{i=1}^{n}(x_i-\bar{x})^2\right\} = \frac{1}{\sigma^n(\sqrt{2\pi})^n} \exp\left\{-\frac{ns^2}{2\sigma^2}\right\}.$$

If we take the partial derivative of the right-hand side with respect to σ and find the value for which this derivative vanishes, we see that the above expression assumes its maximum for $\sigma = s$.

Thus the sample mean and the sample standard deviation are the maximum likelihood estimates of the parameters of a normal distribution.

If both X and Ω contained at most countably many points, and $f_\theta(x)$ denoted the probability of obtaining the observation x when the value of the parameter is θ, then the principle of maximum likelihood could be justified as follows: suppose that we want to guess, on the basis of the observation x, the real value of the parameter θ; if we guess the real value correctly, and this value is θ, we obtain a prize $a(\theta) > 0$; if we do not guess correctly, we obtain nothing. Then, for a given value of the parameter θ and a given observation x, the expected prize for claiming that θ is the value of the parameter when x is observed is equal to $a(\theta)f_\theta(x)$. This implies that we ought to connect the observation x with that decision concerning θ for which this expected prize is the greatest. If $a(\theta) = 1$, the choice reduces to finding the maximum of the probability $f_\theta(x)$. In a similar way, in the continuous case one should find the maximum of the probability density.

As can be seen from a comparison with (49.2), the principle asserting that we should connect with observation x that decision which gives the maximum expected prize is equivalent to the principle asserting that we should take as the estimate of the parameter, given the value x, the mode of the posterior distribution of the parameter θ corresponding to the observation x when the prior distribution is given by the density.

$$g(\theta) = \frac{a(\theta)}{\sum_{\theta \in \Omega} a(\theta)}.$$

If $a(\theta) \equiv 1$, the above prior distribution is uniform. This remark explains the connection between the principle of maximum likelihood and Bayes' rule. We also see that the maximum likelihood principle is more general than Bayes' rule, since it comprises cases where θ belongs to an infinite set, such as the real line. In such cases the principle of maximum likelihood remains meaningful, while the concept of uniform distribution has no sense.

In spite of the fact that the principle of maximum likelihood is based on rather vague considerations concerning optimality of estimators, and is closely related to the violently attacked Bayes' rule, it is nevertheless frequently used in constructing estimators, and maximum likelihood estimators, have under fairly general conditions, certain desirable limit properties, such as asymptotic normality and asymptotic maximum efficiency (see § 52).

We complete this section with an example showing that the principle of maximum likelihood may lead to bad estimators.

EXAMPLE 4. Let X be the space of points $x = (x_1, y_1; \ldots; x_n, y_n)$ representing the results of observations, and let $\theta = (\sigma, \mu_1, \ldots, \mu_n)$ be the parameter determining in X the probability distribution with the density

$$f_\theta(x) = \prod_{i=1}^n \frac{1}{2\pi\sigma^2} \exp\left\{-\frac{1}{2\sigma^2}[(x_i-\mu_i)^2+(y_i-\mu_i)^2]\right\},$$

$$\Omega = \{\theta: \sigma > 0, -\infty < \mu_i < \infty, i = 1, 2, ..., n\}.$$

The probability distribution in X with the above density corresponds to the probability distribution of $2n$ independent random variables $\xi_1, \eta_1, ..., \xi_n, \eta_n$ having a normal distribution with the same standard deviation and expectations changing from pair to pair. This corresponds to an empirical situation in which, in order to determine the accuracy of measurement, we make series of pairs of measurements, each pair on a different object.

We can easily see that for a given x the probability density $f_\theta(x)$ assumes its maximum for

$$\mu_i = \bar{x}_i = \frac{1}{2}(x_i+y_i), \quad \sigma^2 = \frac{1}{n}\sum_{i=1}^n s_i^2,$$

where

$$s_i^2 = \frac{1}{2}[(x_i-\bar{x}_i)^2+(y_i-\bar{x}_i)^2].$$

The expectation of the statistic s_i^2 equals $\frac{1}{2}\sigma^2$ and, consequently, the maximum likelihood estimator of the variance σ^2 is for large n heavily biased, i.e. it has an expectation differing considerably from the estimated parameter. Consequently, this estimator is not even consistent.

§ 52. UNBIASED ESTIMATORS WITH MINIMUM VARIANCE. INFORMATIONAL INEQUALITY

If we know that the distribution in the population is normal with a known standard deviation σ, and we want to estimate the expectation on the basis of an n-element sample $x = (x_1, ..., x_n)$, we can use as an estimator the sample mean $\bar{x}$. This estimator is *unbiased*, i.e., for every expectation in the population, the expectation of this estimator equals the parameter estimated, i.e., the expectation of the population. Clearly, this property of estimators is desirable. Unbiasedness is one of the possible formulations of the requirement that the estimator should be "unprejudiced" with respect to various values of the parameter estimated if we do not know anything about the value of the parameter. It excludes, for example, the estimator which would assume the same value irrespective of the results of the observation, in spite of the fact that such an estimator may in some conditions (when the true value of the parameter coincides with this constant value of the estimator) give accurate results. The second desirable property of an estimator is that its values be concentrated as closely as possible around the value of the estimated parameter if they are computed on the basis of samples from this population. This concentration may be measured by variance. The sample mean mentioned above has variance equal to σ^2/n. The question arises: is this the minimal possible variance of an unbiased estimators? We know that a sample mean will have such variance provided only that the population variance is σ^2. Can we somehow use the fact that we know the distribution to be normal? We shall now discuss briefly the problem of determining the minimal variance of unbiased estimators, or generally, estimators whose expectation depends in a known way on the parameter of the population. It turns out that such estimators cannot have a smaller variance than a certain constant depending on the manner in which the distribution of the population depends on the parameter. Here is the corresponding theorem for the case of one numerical parameter.

THEOREM 1 (Informational inequality). *Let X be a space of points x representing the results of observations, let μ be a fixed measure on X, and let $f(x, \theta)$ be the probability density with respect to measure μ, depending on a parameter θ which assumes values from an open interval. Further, let $t(x)$ be the statistic which we take as an estimator of θ. Suppose that the densities $f(x, \theta)$ are differentiable in θ, except perhaps on a set of points x of measure μ equal to zero, this set being independent of θ, and that we can differentiate under the integral sign the expressions*

(52.1) $$\int_X f(x; \theta) d\mu(x) = 1,$$

and

(52.2) $$E(t|\theta) \stackrel{df}{=} \int_X t(x) f(x; \theta) d\mu(x) = \theta + b(\theta).$$

Moreover, assume that we have the inequality

(52.3) $$I(\theta) \stackrel{df}{=} \int_X \left\{\frac{\partial}{\partial \theta} \log f(x; \theta)\right\}^2 f(x; \theta) d\mu(x) > 0.$$

Under these conditions,

(a) *the variance*

$$D^2(t|\theta) \stackrel{df}{=} \int (t(x) - \theta - b(\theta))^2 f(x; \theta) d\mu(x)$$

satisfies the inequality

(52.4) $$D^2(t|\theta) \geq \frac{[1+b'(\theta)]^2}{I(\theta)};$$

(b) *the equality sign in* (52.4) *holds if and only if the probability densities* $f(x;\theta)$ *are of the form*

(52.5) $$f(x;\theta) = C(\theta)e^{Q(\theta)t(x)}$$

where $C(\theta)$ *and* $Q(\theta)$ *depend only on* θ;

(c) *if the densities* $f(x;\theta)$ *are of form* (52.5) *and the estimator* $t(x)$ *is unbiased, i.e.* $E(t|\theta) = \theta$, *then*

$$\frac{\partial}{\partial \theta} Q(\theta) = I(\theta).$$

Inequality (52.4) is called the *Frechèt–Cramér–Rao inequality*, or *informational inequality*; the latter name is connected with the fact that the value $I(\theta)$ appearing in the denominator on the right-hand side of this inequality has some properties which qualify it as a measure of the amount of information about the parameter θ contained in the observation x. A family of probability distributions with densities of form (52.5) is called a *one-parameter exponential family with respect to the statistic* $t(x)$. Thus, assertion (b) of Theorem 1 may be formulated as follows: the equality sign in relation (52.4) holds if and only if the probability densities $f(x;\theta)$ constitute a one-parameter exponential family with respect to the statistic $t(x)$.

Unbiased estimators with a minimal variance will be called *most efficient*.

PROOF. Differentiating formulas (52.1) and (52.2) under the integral sign with respect to θ, we obtain the equalities

(52.6)
$$\int_X \frac{\partial}{\partial \theta} f(x;\theta) d\mu(x) = 0,$$
$$\int_X t(x) \frac{\partial}{\partial \theta} f(x;\theta) d\mu(x) = 1 + b'(\theta).$$

Multiplying the first of them by $\theta + b(\theta)$, subtracting it from the second, and using the equality

$$\frac{\partial}{\partial \theta} f(x;\theta) = f(x;\theta) \frac{\partial}{\partial \theta} \log f(x;\theta),$$

we obtain

(52.7) $$\int_X (t(x) - \theta - b(\theta)) f(x;\theta) \frac{\partial}{\partial \theta} \log f(x;\theta) d\mu(x) = 1 + b'(\theta).$$

Squaring both sides, we get

$$\left\{ \int_X (t(x)-\theta-b(\theta))f(x;\theta)\frac{\partial}{\partial\theta}\log f(x;\theta)\,d\mu(x) \right\}^2 = (1+b'(\theta))^2.$$

Now, we apply the Schwarz inequality to the left-hand side:

$$\int_X a^2(x)\,d\mu(x) \cdot \int_X b^2(x)\,d\mu(x) \geq \left\{ \int_X a(x)b(x)\,d\mu(x) \right\}^2,$$

where we put

$$a(x) = (t(x)-\theta-b(\theta))\sqrt{f(x;\theta)},$$

$$b(x) = \sqrt{f(x;\theta)}\,\frac{\partial}{\partial\theta}\log f(x;\theta).$$

As a result we obtain the inequality

$$\int_X (t(x)-\theta-b(\theta))^2 f(x;\theta)\,d\mu(x) \cdot \int_X \left\{ \frac{\partial}{\partial\theta}\log f(x;\theta) \right\}^2 f(x;\theta)\,d\mu(x) \geq (1+b'(\theta))^2$$

which implies inequality (52.4) and completes the proof of (a).

To prove (b) note that in the Schwarz inequality we have the equality sign if and only if the functions $a(x)$ and $b(x)$ are proportional, i.e., if there exists a constant $k(\theta)$ depending only on θ and such that

(52.8) $$\frac{\partial}{\partial\theta}\log f(x;\theta) = k(\theta)(t(x)-\theta-b(\theta)).$$

By integration we obtain

$$\log f(x;\theta) = Q(\theta)t(x)+\log C(\theta),$$

or

$$f(x;\theta) = C(\theta)e^{Q(\theta)t(x)},$$

where $Q(\theta)$ denotes the indefinite integral of the function $k(\theta)$, and $\log C(\theta)$ denotes the indefinite integral of the function $-k(\theta)[\theta+b(\theta)]$. Thus, if we have the equality sign in relation (52.4), then the densities $f(x;\theta)$ constitute the exponential family with respect to the statistic $t(x)$. Conversely, if the densities $f(x;\theta)$ are of form (52.5), by differentiation we check that (52.8) holds, which proves assertion (b).

To prove (c) note that if $f(x;\theta)$ is of form (52.5), then

$$\frac{\partial}{\partial\theta}\log f(x;\theta) = t(x)\frac{\partial}{\partial\theta}Q(\theta)+k(\theta)$$

where

$$k(\theta) = \frac{\partial}{\partial\theta}\log C(\theta).$$

VIII. STATISTICAL INFERENCE

In view of (52.6) we then have

$$0 = \int_X \frac{\partial}{\partial \theta} f(x;\theta)\,d\mu(x) = \int_X \left\{\frac{\partial}{\partial \theta} \log f(x;\theta)\right\} f(x;\theta)\,d\mu(x)$$

$$= \int_X \left\{t(x)\frac{\partial}{\partial \theta} Q(\theta) + k(\theta)\right\} f(x;\theta)\,d\mu(x),$$

which shows that

$$\frac{\partial}{\partial \theta} Q(\theta) E(t|\theta) = -k(\theta).$$

If $E(t|\theta) = \theta$, we obtain therefore

$$\frac{\partial}{\partial \theta} \log f(x;\theta) = \bigl(t(x)-\theta\bigr)\frac{\partial}{\partial \theta} Q(\theta).$$

Substituting this in (52.7) and using once more the fact that $t(x)$ is unbiased, we obtain the equality

$$\frac{\partial}{\partial \theta} Q(\theta) D^2(t|\theta) = 1,$$

and consequently

$$\frac{\partial}{\partial \theta} Q(\theta) = I(\theta),$$

which completes the proof of Theorem 1.

A number of important distributions depending on one parameter constitute an exponential family. For these families Theorem 1 allows us to find the bound for the variance of estimators. We give here some examples.

EXAMPLE 1. *Bernoulli trials.* Let X be the space of n-element sequences consisting of zeros and ones, $x = \{i_1, \ldots, i_n\}$, and let Ω be the interval $\{\theta: 0 < \theta < 1\}$. Next, let

$$f(x;\theta) = \theta^m(1-\theta)^{n-m}$$

(where m is the number of ones in the sequence x) be the density with respect to the counting measure μ on X.

Substituting $t(x) = m/n$, we can write this density in the form

$$f(x;\theta) = (1-\theta)^n \exp\left(t(x) n \log \frac{\theta}{1-\theta}\right).$$

Thus, these distributions turn out to constitute a one-parameter exponential family with respect to the statistic $t(x)$. As we know, this statistic is an unbiased estimator of the parameter θ. It follows from Theorem 1 that this statistic is an unbiased estimator of the parameter θ with minimal possible variance.

EXAMPLE 2. *Poisson distribution.* Let X be the space of all n-element sequences $x = \{k_1, \ldots, k_n\}$ with non-negative integer coordinates, and let Ω be the half-line $\{\theta: 0 < \theta < \infty\}$. Next, let

$$f(x; \theta) = \prod_{i=1}^{n} \frac{\theta^{k_i}}{k_i!} e^{-\theta}$$

be the probability density in X with respect to the counting measure μ. We can write the density $f(x; \theta)$ in the form

$$f(x; \theta) = \frac{1}{k_1! \ldots k_n!} \exp[-n\theta] \exp[-t(x) n \log \theta],$$

where $t(x) = (k_1 + \ldots + k_n)/n$. Thus, these densities determine a one parameter exponential family with respect to the above statistic $t(x)$. Here $n \log \theta$ plays the role of $Q(\theta)$, $e^{-n\theta}$ plays the role of $C(\theta)$ and a measure equal to $1/(k_1! \ldots k_n!)$ on the set containing the single point $x = \{k_1, \ldots, k_n\}$ plays the role of the measure μ. It follows from Theorem 1 that $t(x)$, i.e., the sample mean, is the unbiased estimator of θ with minimal variance.

EXAMPLE 3. *Estimating the expectation of normal distribution with a known variance.* Let us take for X the n-dimensional Euclidean space of points $x = \{x_1, \ldots, x_n\}$, and for Ω the real line $\Omega = \{\theta: -\infty < \theta < \infty\}$. Let

$$f(x; \theta) = \prod_{i=1}^{n} \frac{1}{\sigma \sqrt{2\pi}} \exp\left\{-\frac{1}{2\sigma^2}(x_i - \theta)^2\right\},$$

where σ is known.

The density $f(x; \theta)$ may be written in the form

$$f(x; \theta) = \exp\left\{-\frac{n\theta^2}{2\sigma^2}\right\} \exp\left\{\bar{x} \frac{\theta n}{\sigma^2}\right\} \frac{1}{(\sigma \sqrt{2\pi})^n} \exp\left\{-\frac{1}{2\sigma^2} \sum_{i=1}^{n} x_i^2\right\},$$

where $\bar{x} = (x_1 + \ldots + x_n)/n$. We thus have here a one-parameter exponential family of distributions with respect to the sample mean $t(x) = \bar{x}$. The function $n\theta/\sigma^2$ plays the role of $Q(\theta)$, the function $\exp(-n\theta^2/2\sigma^2)$ plays the role of $C(\theta)$ and the measure with density

$$\frac{1}{(\sigma \sqrt{2\pi})^n} \exp\left\{-\frac{1}{2\sigma^2} \sum_{i=1}^{n} x_i^2\right\}$$

plays the role of the measure μ. Since the sample mean is an unbiased estimator of the parameter θ, by Theorem 1 it has minimal variance.

VIII. STATISTICAL INFERENCE

The informational inequality has been investigated by many authors. In Theorem 1 we discussed the so-called *regular case of estimation*. It is known that in the regular case the minimal variance of unbiased estimators decreases as $1/n$ when the sample size n increases (see problem 1), while in certain non-regular cases this variance may decrease as $1/n^2$ (see problem 2). The informational inequality generalized to the case of several parameters shows that the covariance matrix of a system of unbiased estimators must satisfy certain conditions.

Problems

1. If $\xi_1, \ldots, \xi_n$ are independent random variables with the same probability distribution with density $f(x_1; \theta)$ depending on a numerical parameter θ, and $f(x; \theta) = f(x_1; \theta) \ldots f(x_n; \theta)$ is the density of their joint distribution, then

$$\int \cdots \int \left\{ \frac{\partial \log f(x; \theta)}{\partial \theta} \right\}^2 f(x; \theta) dx_1 \ldots dx_n = n \int_{-\infty}^{+\infty} \left(\frac{\partial \log f(x_1; \theta)}{\partial \theta} \right)^2 f(x_1; \theta) dx_1.$$

2. Let

$$f(x; \theta) = \begin{cases} 1 & \text{for } \theta - \tfrac{1}{2} < x < \theta + \tfrac{1}{2}, \\ 0 & \text{otherwise} \end{cases}$$

be the density of the uniform distribution concentrated on the interval of length 1 and centre θ. Let $\xi_1, \ldots, \xi_n$ be independent random variables having the same distribution with density $f(x; \theta)$, and let

$$\zeta = \frac{1}{2} \{\max[\xi_1, \ldots, \xi_n] + \min[\xi_1, \ldots, \xi_n]\}.$$

Prove that $E\zeta = \theta$ and $D^2\zeta = 2/(n+1)(n+2)$.

3. Let X be the n-dimensional Euclidean space of points x and let $f(x; \theta)$, $f(x; \theta+h)$ and $t(x)$ be probability densities and a statistic such that

$$\int_X t(x) f(x; \theta) dx = \theta,$$

$$\int_X t(x) f(x; \theta+h) dx = \theta+h.$$

Prove that the variance of the statistic t satisfies the inequality

$$D^2(t|\theta) \stackrel{\text{df}}{=} \int_X (t(x) - \theta)^2 f(x; \theta) dx \geq \frac{1}{\int_X \left\{ \frac{f(x; \theta+h) - f(x; \theta)}{h} \right\}^2 \frac{1}{f(x; \theta)} dx}.$$

§ 53. STATISTICAL DECISION FUNCTION. MINIMAX ESTIMATORS

In the forties, owing to the research of A. Wald, it was observed that the majority of important statistical problems may be treated uniformly as special cases of two-person zero-sum games in the sense of the modern theory of games, and that these problems belong to a wider class of problems of making decisions under uncertainty. This point of view comprises the problems of testing hypotheses, discrimination and estimation. Since this approach has led to the ordering of problems of statistics, and also to new discoveries, we shall discuss it briefly.

We start from presenting the concept of *two-person zero-sum game*.

There are two players. The game is played as follows: the first player chooses an element a from the set A, and the second chooses an element b from the set B; the choices are independent and made without the knowledge of the opponent's choice. After the choice is made, it is disclosed, and then the second player pays to the first player the sum $K(a, b)$. The sets A and B, and the function $K(a, b)$ are known to the players before the beginning of the game. The elements of the set A are called (pure) *strategies of the first player*, the elements of the set B are called (pure) *strategies of the second player*, and the function $K(a, b)$ is called the *payoff function*. Thus the formal structure of the game is determined by the triple $\langle A, B, K(a, b)\rangle$.

The question arises which strategies should be chosen by each of the players. The theory of games suggests choosing then according to the so-called *minimax principle*, which tells us to play against the most damaging choice of the opponent. The argument goes as follows:

The second player wants to pay as little as possible. If he chooses the strategy b, he must take into account a choice of his opponent that is worst for him. In other words, the upper bound

(53.1) $$\sup_{a\in A} K(a, b)$$

determines the payoff which he risks if he chooses the strategy b. The smaller is this upper bound, the better is the strategy, and the minimax principle tells the second player to choose a strategy b_0 (if it exists) such that

(53.2) $$\sup_{a\in A} K(a, b_0) = \inf_{b\in B}\sup_{a\in A} K(a, b).$$

Such a strategy b_0 is called the *minimax strategy for the second player*, and the right-hand side of equality (53.2) is called the *value of the game for the second player*.

In an analogous way, a strategy a_0 (if it exists) for which

(53.3) $$\inf_{b\in B} K(a, b)$$

attains its maximum, i.e. such that

(53.4) $$\inf_{b\in B} K(a_0, b) = \sup_{a\in A}\inf_{b\in B} K(a, b)$$

is called the *minimax strategy for the first player*, and the right-hand side of (53.4) is called the *value of the game for the first player*.

The value of the game for the second player tells him to what extent he may decrease his payoff by a suitable choice of his strategy, irrespective of the choice of the first player. Similarly, the value of the game for the first player tells him how much he can safeguard himself by a suitable choice, irrespective of the choice of the second player. It follows, of course, that the value of the game for the first player does not exceed the value of the game for the second player.

The situation when both values of the game are equal is considered as advantagous for both players. The advantage is that (if minimax strategies exist) none of the players can improve his situation when he learns that his opponent chooses his minimax strategy, i.e. the first player cannot increase his payoff when he knows that his opponent will choose a minimax strategy, and the second player cannot decrease the payoff when he learns that his opponent chooses a minimax strategy. Such a game is therefore called *closed* in the terminology suggested by H. Steinhaus* or *determined* in the terminology accepted in the West.

When the game is *open*, in the sense that the value of the game for the first player is smaller that the value of the game for the second player, at least one of the players may profit from the knowledge that his opponent chooses a minimax strategy. This is illustrated by the following matrices, which represent games; here the first player chooses the row and the second player chooses the column, and the term at the intersection of the chosen row and column represents the payoff:

$$\begin{bmatrix} 1 & 3 \\ 4 & 2 \end{bmatrix} \qquad \begin{bmatrix} 1 & 4 \\ 3 & 2 \end{bmatrix} \qquad \begin{bmatrix} 2 & 9 & 4 \\ 5 & 3 & 8 \\ 7 & 6 & 1 \end{bmatrix}$$

Advantage for first player Advantage for second player Advantage for both players

In proofs of the closeness of games frequent use is made of the following theorem:

THEOREM 1. *Given a two-person zero-sum game, suppose that there exist strategies $a_0 \in A$ and $b_0 \in B$ such that a_0 is the best answer of the first player against the strategy b_0 of the second player, i.e.*

(53.5) $$K(a_0, b_0) = \sup_{a \in A} K(a, b_0)$$

---

* H. Steinhaus, *Definitions needed in the theory of games and pursuit* (in Polish), Myśl Akademicka, 1 (1925), pp. 13–14.

and b_0 is the best answer of the second player against the strategy a_0 of the first player, i.e.
$$K(a_0, b_0) = \inf_{b \in B} K(a_0, b).$$

Then the game is closed, a_0 and b_0 are minimax strategies of the first and the second players, and $K(a_0, b_0)$ is the common value for both players.

The pair of strategies $a_0 \in A$ and $b_0 \in B$ optimal against one another is called a *saddle point of the game*. Thus, we can formulate Theorem 1 as follows: *if the game has a saddle point, then it is closed and the strategies of saddle point are minimax strategies*.

In studying statistical games one makes also use of the following condition of closeness of the game, which does not require the existence of a minimax strategy of the first player:

THEOREM 2. *Given a two-person zero-sum game, if there exists a strategy b_0 such that the payoff $K(a, b_0)$ does not depend on a, and there exists a sequence of strategies $a_1, a_2, \ldots$ such that*

(53.7) $$\liminf_{n \to \infty} \inf_{b \in B} K(a_n, b) = K(a_1, b_0),$$

then the game is closed and b_0 is a minimax strategy for the second player.

The statistical problems may be treated as games in the following sense: We are given the space X of sample points x and a family of probability distributions in X with densities $f(x; \theta)$, taken with respect to a measure μ on X, and depending on a parameter θ which assumes values in a given space Ω. Usually this parameter is a number or a vector. Next, we are given the space D of decisions d, and a function $L(\theta, d)$ defined on the Cartesian product $\Omega \times D$, interpreted as loss resulting from a decision d when the true value of the parameter is θ. The statistician does not know the value of the parameter θ, and knows only the value of the observation x. He is to take a decision $d \in D$ on the basis of the observation x, and he is interested in minimizing his loss. Such a situation may be interpreted as a game between the statistician and his opponent, whom we may conveniently call *Nature*. The strategies of Nature are values of the parameter θ, i.e., elements of the set Ω, and the strategies of the statistician are *decision functions* $\delta(x)$, i.e. functions defined on X with values in D. Denote this class of decision functions by $\mathscr{D}$. Now, we have defined the sets of strategies for both players, and it remains to define the payoff function of the game. We take as the payoff function the risk function, i.e. the expected loss: if the statistician chooses a decision function $\delta(x)$, the value observed was x, and the true value of the parameter is θ, then the statistician's loss is $L(\theta, \delta(x))$. The result of observations, however, appears randomly according to a probability distribution with density $f(x; \theta)$. The value of the decision function $\delta(x)$ must therefore be evaluated in terms of the expected loss, or the risk, defined as

$$(53.8) \qquad r(\theta, \delta) = \int_X L(\theta, \delta(x)) f(x; \theta) d\mu(x).$$

We have described here a game in a non-randomized, or pure, form. Various statistical problems may be interpreted in terms of the above scheme by specifying the elements X, Ω, D, $L(\theta, d)$ and $\mathscr{D}$.

For instance, we obtain a problem of testing hypotheses if we take as D a two-element set $D = \{d_0, d_1\}$, split Ω into two disjoint sets H and K, and define the loss $L(\theta, d)$ as follows:

$$L(\theta, d) = \begin{cases} 0 & \text{if } \theta \in H \text{ and } d = d_0, \\ 0 & \text{if } \theta \in K \text{ and } d = d_1, \\ 1 & \text{if } \theta \in H \text{ and } d = d_1, \\ 1 & \text{if } \theta \in K \text{ and } d = d_0. \end{cases}$$

Here H denotes the null hypothesis, K is the alternative hypothesis, d_0 means the decision "the null hypothesis is true", d_1 means the decision "the alternative hypothesis is true", and the risk expresses the probability of a decision inconsistent with the true state.

If Ω and D have the same finite number of elements, $\Omega = \{\theta_1, \ldots, \theta_n\}$, $D = \{d_1, \ldots, d_n\}$, and $L(\theta_i, d_j)$ is the loss resulting from accepting the decision d_j if θ_i is the true value of the parameter, we obtain the problem which we have discussed under the name of the discrimination problem.

If $\Omega = D$ and $L(\theta, d)$ is a measure of error, for instance, if Ω and D are real lines, and $L(\theta, d) = |\theta - d|$, or $L(\theta, d) = (\theta - d)^2$, we have the estimation problem.

The importance of the concept of a statistical game lies precisely in the fact that various statistical problems may be obtained by specializing its elements. This creates the possibility of a general discussion of certain fundamental problems of taking decisions based on statistical observations.

Treating statistical problems as games, which has led to a new theory of statistical decision functions, has led also to some new suggestions concerning the criteria of choice of decision functions. Such a suggestion is contained in the minimax principle, i.e., it is suggested that the statistician uses the minimax decision function in the sense of the theory of games. The question arises of finding these minimax decision functions in various statistical games. Some problems of this type were already discussed in § 48.

Finally, what is probably most important, the studies connected with statistical decision functions and, more precisely, with the concept of completeness of a class of strategies, have led to a sort of rehabilitation of Bayes' rule. We shall explain this below.

Besides non-randomized statistical games, one can also consider randomized games, or mixed extensions of pure games. A game is randomized from the point of view of Nature if the strategies of Nature are not the particular values of the parameter θ, but probability distributions $\varkappa$ on Ω. These distributions are simply prior distributions of the parameter θ. The risk is computed according to the formula

(53.9) $$r(\varkappa, \delta) = \int_\Omega r(\theta, \delta) d\varkappa(\theta)$$

where $r(\theta, \delta)$ is given by (53.8).

As regards minimax decision functions, it proves easier to determine them for games randomized from the point of view of Nature, since these games are usually closed and one can either determine the saddle point or apply Theorem 2. On the other hand, non-randomized games are usually open, and a direct determination of minimax decision functions is often rather complicated. It turns out, however, that we have

LEMMA 1. *If all probability distributions on Ω are admitted, then a minimax decision function in a game randomized from the point of view of Nature is at the same time a minimax decision function in a non-randomized game.*

Indeed, for a given decision function $\delta(x)$ we have the obvious relation

(53.10) $$\sup_\varkappa r(\varkappa, \delta) = \sup_{\theta \in \Omega} r(\theta, \delta),$$

where on the left-hand side sup is taken with respect to all probability distribution functions on Ω, including the degenerate distributions (i.e. those concentrated at one point). It follows that the value of the decision function, as measured by risk, is the same in randomized and in non-randomized games.

Thus, if the interpretation of the parameter θ as a random variable has no empirical meaning, considering a randomized game may be regarded as a method of determining a minimax decision function for a non-randomized game; if θ may be treated as a random variable, we obtain a minimax decision function against a larger class of strategies of Nature.

We can also discuss the problem of admitting randomness on the statistician's side. By a randomized decision function we may mean either a probability distribution on the class $\mathscr{D}$ of decision functions, or assigning to observation x a probability distribution $\delta(x)$ on D. Both approaches prove essentially equivalent;

in the sequel, we shall use the second approach. The risk $r(\theta, \delta)$ corresponding to the value θ of the parameter and the randomized decision function $\delta(x)$ is then computed as follows: first, we compute the expected loss for a given x, with respect to the distribution $\delta(x)$, and then we take the expectation of such conditional risks with respect to the distribution $f(x; \theta)$ according to the formula

(53.11) $$r(\theta, \delta) = \int_X \left\{ \int_D L(\theta, \delta(x)) \, d\delta \right\} f(x; \theta) \, d\mu(x).$$

Besides this change in computing risk $r(\theta, \delta)$ formula (53.9) for risk in a game randomized from the point of view of Nature remains the same. We find that this approach does not reduce to idle formalism, since sometimes the introduction of randomized decision functions improves the situation of the statistician (see problem 1). However, for a large class of problems this approach does not constitute any improvement. Theorem 3 below explains the situation to some extent.

Given the decision functions $\delta'(x)$ and $\delta''(x)$, randomized or not, the function $\delta'(x)$ is *no worse* that the function $\delta''(x)$, if for every $\theta \in \Omega$ we have

(53.12) $$r(\theta, \delta') \leq r(\theta, \delta'').$$

If this inequality is strict for at least one θ, the function δ' is *better* than δ''. The function δ is *admissible* in the class $\mathscr{D}$ if this class does not contain any function better than δ.

THEOREM 3. *If D is the real line, and for every θ the function $L(\theta, d)$ is a convex function of d such that*

(53.13) $$\lim_{d \to -\infty} L(\theta, d) = \lim_{d \to +\infty} L(\theta, d) = \infty,$$

then for every randomized decision function $\delta(x)$ there exists a non-randomized decision function no worse than the former.

The theorem below explains why it is advantageous to built decision functions based on sufficient statistics:

THEOREM 4. *Under the assumptions of Theorem 3, let $\delta(x)$ be an arbitrary decision function, let $t(x)$ be a sufficient statistic for the parameter θ, and let $\delta^*(x)$ be the decision function obtained from $\delta(x)$ as follows*: $\delta^*(x)$ *assumes on the orbit A_t of the statistic $t(x)$ a value equal to the average of the decision function $\delta(x)$ with respect to a conditional distribution on A_t. Then for every $\theta \in \Omega$ we have*

$$r(\theta, \delta^*) \leq r(\theta, \delta).$$

In the preceding paragraph we discussed Bayes' rule, Bayesian discrimination functions, Bayesian estimators, and the criticism of this approach. We also discussed the construction of tests, discrimination functions and estimators which do not rely on the prior distribution and Bayes' rule: the theory of most powerful tests, or the theory of unbiased estimators of minimal variance, may serve as examples of the latter approach.

The principle of maximum likelihood and the construction of confidence intervals do not rely on the concept of prior distribution either. Thus, we have discussed various methods of avoiding the objections directed against Bayes' rule: this rule was, in a sense, rejected by the multiplicity of these methods. On the other hand, however, the theory of statistical decision functions led to a sort of rehabilitation of Bayes' rule. There exist, as we have seen, a variety of criteria of optimality, and the theory of statistical decision functions has added a new one, namely the minimax criterion. It may happen that all these criteria lead to different decision procedures. Thus, since there was no general agreement as to which of these decision functions is the best, the question was asked whether it might not be easier to agree which decision functions are no good, so as to restrict the choice from the other side.

We have already defined what it means that a decision function δ' is better, or at least no worse than a decision function δ''. It seems desirable to reduce the statistical problem in such a way as to obtain, after reduction, a class $\mathscr{C}$ of decision functions such that for every function from outside $\mathscr{C}$ one may find a function in $\mathscr{C}$ no worse than the former. Such a class of decision functions is called *complete*. Clearly, the class of all decision functions is complete. If the reduction is to be efficient, one should try to obtain a complete class consisting of a possibly small number of decision functions. Let us agree to say that the complete class of decision function is *minimal* if it cannot be decreased, i.e., if it does not contain any proper subclass which is also complete. Now, under very general assumptions, which we shall not formulate here, the minimal complete class of decision functions coincides with the class of Bayesian decision function and their limits. By a *Bayesian decision function* we mean a decision function which constitutes the best answer against some prior distribution. Thus, the choice of statistical decision functions may, roughly speaking, be replaced by a choice of a prior distribution.

We now present some examples of statistical games and minimax estimators.

EXAMPLE 1. *Estimating the fraction of defective items on the basis of a sample of a fixed size.* As the sample space X we take here the set of integers $X = \{0, 1, \ldots \ldots, n\}$ representing the numbers of defective items in a sample of size n. As the parameter we take the fraction of defective items; we have, therefore $\Omega = \{\theta: 0 \leqslant \theta \leqslant 1\}$, and as the probability distribution on X corresponding to the parameter θ we take the binomial distribution

$$f(m; \theta) = \binom{n}{m} \theta^m (1-\theta)^{n-m}, \quad m = 0, 1, \ldots, n.$$

As the space of decisions we take $D = \{d: 0 \leqslant d \leqslant 1\}$. The decision functions, or estimators, are all functions $\delta(m)$ with values in D and arguments m from X. The strategies of Nature are probability distribution functions $G(\theta)$ determining the probability distributions on Ω, i.e., such that $G(0) = 0$ and $G(1+) = 1$.

If the loss function is

$$L(\theta, d) = (\theta - d)^2,$$

then the game is closed,

$$\delta(m) = \frac{m + \frac{1}{2}\sqrt{n}}{n + \sqrt{n}}, \quad m = 0, 1, \ldots, n$$

is the minimax estimator,

$$G(\theta) = \frac{\Gamma(\sqrt{n})}{\{\Gamma(\frac{1}{2}\sqrt{n})\}^2} \int_0^\theta [\theta(1-\theta)]^{\frac{1}{2}\sqrt{n}-1} d\theta, \quad 0 < \theta < 1$$

is the minimax prior distribution, and

$$\frac{1}{4(1+\sqrt{n})^2}$$

is the value of the game common to both players.

EXAMPLE 2. Under the same assumptions as in Example 1, if the loss function is

$$L(\theta, d) = \frac{(\theta - d)^2}{\theta(1-\theta)},$$

then the game is also closed,

$$\delta(m) = \frac{m}{n}, \quad m = 0, 1, \ldots, n$$

is the minimax estimator, and the uniform distribution

$$G(\theta) = \theta \quad \text{for} \quad 0 \leqslant \theta \leqslant 1$$

is the minimax prior distribution. The value of the game is $1/n$.

EXAMPLE 3. *Estimating the composition of an urn on the basis of sampling with replacement.* Let X be the space of systems $x = \{m_1, \ldots, m_k\}$ of non-negative integers $m_1, \ldots, m_k$ whose sum is n. Let Ω be the space of systems $\theta = (p_1, \ldots, p_k)$ of real numbers with sum 1, and let

$$f(x; \theta) = \frac{1}{m_1! \ldots m_k!} p_1^{m_1} \ldots p_k^{m_k}$$

be the family of probability densities on X with respect to the counting measure. Finally, let D be the space of real-valued vectors $d = (d_1, \ldots, d_k)$.

The statistical game in which all probability distributions on Ω are strategies of Nature and all functions $\delta(x)$ defined on X with values in D are the statistician's strategies and the loss function is

(53.15) $$L(\theta, d) = \sum_{i=1}^{k} (p_i - d_i)^2$$

is closed; the statistician's minimax strategy is given by the function $\delta(x) = (\delta_1(x), \ldots, \delta_k(x))$ such that

(53.16) $$\delta_i(x) = \frac{m_i + \frac{\sqrt{n}}{k}}{n + \sqrt{n}}, \quad i = 1, 2, \ldots, k,$$

and the value of the game is

$$\frac{n}{(n + \sqrt{n})^2} \cdot \frac{k-1}{k}.$$

If D is restricted to systems of vectors $d = (d_1, \ldots, d_k)$ with the sum equal to one, and $L(\theta, d)$ is replaced by the function

$$L(\theta, d) = \sum_{i=1}^{k} \frac{(p_i - d_i)^2}{p_i}$$

then the minimax estimator is given by the function $\delta(x) = (\delta_1(x), \ldots, \delta_k(x))$ such that

$$\delta_i(x) = \frac{m_i}{n}, \quad i = 1, 2, \ldots, k,$$

and $(n-1)/n$ is the value of the game.

Problems

1. The coin is either good and has head and tail, or is bad and has heads on both sides. We can toss it once and have to decide whether the coin is good or not. The loss is 1 if the decision is wrong, and 0 if it is correct. The decision rule is to say that the coin is good if it comes up tails and if it comes up heads, to say that it is good with probability ϱ and to say that it is not good with probability $1-\varrho$. Show that $1/2$ is the minimax value of probability ϱ.

2. In problem 3 of § 47 the statistic $2(x_1 + \ldots + x_n)/n$ is an unbiased estimator of the number N of cars in the town. Improve this estimator in the sense of mean square error according to Theorem 4, using the fact that $t = \max\{x_1, \ldots, x_n\}$ is a sufficient statistic for N. Carry on the numerical computations for $N = 2$.

References

1. D. Blackwell and M. A. Girshick, *Theory of Games and Statistical Decisions*, John Wiley, New York, London, 1954.
2. D. G. Chapman and H. Robbins, Minimum variance estimation without regularity conditions, *Annals of Mathematical Statistics*, 22 (1951), 581–586.
3. H. Cramér, *Mathematical Methods in Statistics*, Princeton University Press, Princeton, 1946.
4. W. Feller, *An Introduction to Probability Theory and its Applications*, John Wiley, New York, London, 1961.
5. B. de Finetti, La legge dei grandi numeri nel caso dei numeri aleatori equivalenti, *Atti R.A. Naz. Lincei* 18 (1933), 203–207.
6. M. Fisz, *Probability Theory and Mathematical Statistics*, John Wiley, New York, London, 1963.
7. V. I. Glivenko, *Probability Theory* (in Russian), Moscow–Leningrad.
8. B. W. Gnedenko, *Course in Probability Theory* (in Russian), Moscow, 1950.
9. P. R. Halmos, *Measure Theory*, Van Nostrand, New York, 1950.
10. E. L. Lehman, *Testing Statistical Hypotheses*, John Wiley, New York, London, 1959.
11. H. Steinhaus, The problem of estimation, *Annals of Mathematical Statistics*, 28 (1957), 633–648.
12. S. Trybuła, Some problems in simultaneous minimax estimation, *Annals of Mathematical Statistics*, 29 (1958), 245–253.
13. — On some loss functions, *Colloquium Mathematicum* 7 (1960), 297–305.
14. B. L. van der Waerden, *Mathematische Statistik*, Springer, Berlin–Göttingen–Heidelberg, 1957.

Index

Additive class of subsets, 8
Admissible decision function, 311
Alternative statistical hypothesis, 269
  composite, 269
  simple, 269
Alternatives of the null hypothesis, 240
Angular transformation, 236
A posteriori probability, 60
A priori probability, 60
Asymptotically normal random variables, 194

Bayes' rule, 291
Bayes' theorem, 59
Bayesian decision function, 312
Bayesian discrimination function, 285
Bayesian estimators, 290
Bernoulli (binomial) distribution, 76, 273
Bernoulli trials, 75, 303

Cartesian product
  of events, 66, 70
  of fields of events, 67, 70
  of probabilities, 68, 72, 73
  of spaces of elementary events, 66, 70
Cauchy distribution, 177
Causes, 60
Central moments, 120
Central term, 77
Characteristic function, 209
  of a random variable, 160
Chi-square distribution, 187
Class
  additive of subsets, 8
  countably additive (σ-additive) of subsets, 8
  of decision functions complete, 312
  of decision functions complete, minimal 312
  of subsets complementative, 8
Closed game, 307
Complementative class of subsets, 8
Complement of an event, 7

Complete class of decision function, 213
Composite statistical hypothesis, 240
Conditional probability, 54, 55
Confidence level, 293
Consistent estimators, 289
Continuous distribution, 53
Convergence
  fundamental, 165
  in probability, 131
  with probability one, 133
  weak, 165
Conversion formula, 162
Correlation axes, 207
Correlation coefficient, 205
Correlation ellipse, 208
Correlation hyperellipsoid, 209
Correlation hyperplanes, 208
Countably additive (σ-additive) class of sets, 8
Covariance, 199
Covariance matrix, 199
Critical region, 240
  of the test, 269
Čebyšev's inequality, 128

Decision function, 308
  admissible, 311
  Bayesian, 312
  δ' better than δ'', 311
  δ' no worse than δ'', 311
Degenerate distribution, 224
Density of the distribution, 53
Dependent events, 61
Determined game, 307
Difference of events, 7
Discrete distribution, 53
Disjoint events, 2, 7
Distribution
  Bernoulli, 76
  beta, 197
  Cauchy, 177
  chi-square, 187

continuous, 57
degenerate, 224
discrete, 53
exponential, 179
Fisher's F, 191
Fisher's z, 193
gamma, 176
Gaussian, 92
geometric, 158
Laplace, 177
log-normal, 183
marginal, 51, 67, 72
multinomial, 77
n-dimensional normal, 214
non-degenerate n-dimensional normal, 219
normal, 224, 297
Pascal, 158
Poisson, 80, 277
posterior of the parameter, 290
prior of the parameter, 290
Snedocor's, 193
Snedecor's T, 193
Student's, 188, 265
Distribution function, 41, 113
empirical, 129
Distribution of system of n random variables 115
Drawing balls from two urns, 57

Empirical distribution function, 129
Error
of the first kind, 268
of the second kind, 269
Estimation problems, 237, 289
Estimator, 289
Bayesian, 290
consistent, 289
most likely, 301
unbased, 299
Event, 2, 3, 5
impossible, 7
sure, 7
Events
dependent, 61
independent, 61, 62
independent *en bloc*, 64
Expectation, 117

Expectation of the random variable, 103, 107, 110
Exponential distribution, 179

Field, 8
of Borel sets, 24
of events, 43
σ-, 8
Fisher's F distribution, 191
Fisher's z distribution, 193
Formula for absolute probability, 57
Frechèt–Cramér–Rao inequality, 301
Frisch theorem, 202
Function
Bayesian discrimination, 285
Borel, 96
Borel of n variables, 98
characteristic, 160, 290
decision, 308
distribution, 113
generating, 155
Laplace, 91
likelyhood, 296
minimax discrimination, 285
moment generating, 159
randomized critical, 270
Fundamental convergence, 165
Fundamental lemma of Neyman–Pearson, 271

Galton's order test, 268
Gamma distribution, 176
Gaussian probability distribution, 92
Geometric distribution, 158
Generating function of the random variable, 155
Game
closed, 307
determined, 307
open, 307
saddle point of, 308
two-person zero-sum, 306
Glivenko Theorem, 144

Helly's theorem the first, 167
Helly's theorem generalized second, 167
Impossible event, 7
Independent events, 61, 62

INDEX

Indepenentd random variables, 122
Independent trials, 65
Informational inequality, 300, 301
Integral de Moivre–Laplace theorem, 85
Inversion, 265

Jump, 52

Kolmogorov–Smirnov test, 264
Kolmogorov test, 263
Kolmogorov theorem, 235
kth factorial moment, 152

Laplace distribution, 177
Laplace function, 91
Lapunov theorem, 183
Laws of de Morgan, 10
Level of significance of the test, 269
Likelihood function, 296
Lindberg–Feller theorem, 185
Lindberg–Lévy theorem, 180
Local de Moivre–Laplace theorem, 81
log-normal distribution, 183
Lower integral sum, 107
Lower quantile, 118

Marginal distribution, 51, 68, 72
Mean absolute deviation, 118
Mean square deviation, 119
Median, 118
Minimax discrimination function, 285
Minimax principle, 306
Minimax strategy for the first player, 307
Minimax strategy for the second player, 306
Mode of distribution, 290
Moment generating, 159
Moments, 120
Monotone likelihood ratio, 280
Most likely number of successes, 77
Most powerful test, 269
Most efficient estimators, 301
Multinomial distribution, 77
Multiple correlation coefficient, 205

Nature, 308
n-dimensional distribution function, 48
Negative part of random variable, 110
Non-degenerate n-dimensional normal distribution, 219

Normal distribution, 224
Normal (Gaussian) probability distribution, 92
nth difference of the function, 50
Null statistical hypothesis, 239, 269
  composite, 269
  simple, 269
n-vise independent system of events, 64

One-parameter exponential family with respect to the statistic $t(x)$, 301
Open game, 307
Orbit, 275
Orthogonal random variables, 140
Outer measure, 32

Paradox of d'Alembert, 19
Parametric statistical hypothesis, 239
Pascal distribution, 158
Payoff function, 306
Pearson's chi-square statistic, 256
Pearson's chi-square test, 256
Pearson's theorem, 256
Poisson distribution, 80
Poisson theorem, 78
Positive part of random variable, 110
Posterior distribution, 290
Power
  of a randomized test, 270
  of the test, 270
  of the test W against the alternative, 270
Principle of maximum likelihood, 296
Prior distribution, 290
Probability, 14
  a posteriori, 60
  a priori, 60
  conditional, 54, 55
Probability distribution, 22, 41
  of the random variable, 113
Probability distribution function, 48, 113
  of the system of n-random variables, 114
Probability distribution of the characteristic in the population, 116
Probability distribution of the waiting time for the rth success, 158
Problems
  of discrimination, 283
  of verification, 238

Product of events, 3, 5

Quadratic form
 positive definite, 199
 positive semidefinite, 199
Quantile
 lower, 118
 of order p, 117
 upper, 118

Random variable, 93, 94
 with a finite number of values, 103
Random variables
 asymptotically normal, 194
 independent, 122, 124
 orthogonal, 140
 truncated, 109
 uncorrelated, 140
Random vector, 198
Randomized critical function, 270
Randomized test, 270
Regression, 204
 coefficient, 205
 hyperplane, 205
 of the second kind, 204
Regular case of estimation, 305
Residual variance, 205
Results, 60
Retrospective plans, 291

Saddle point of the game, 308
Sample distribution function, 129
Sample expectation, 129
Sample mean, 129
Sample variance, 129
Sampling without replacement, 65
Second factorial moment, 152
Sequences of independent random variables, 126
Significance level of the test, 240, 270
Simple statistical hypothesis, 239, 240
Snedecor's distribution, 193
Snedecor's T distribution, 193
Space of elementary events, 5
Standard deviation, 119
Standard error of measurement, 248
Statistic, 242, 289

Pearson's chi-square, 256
 sufficient, 276
 sufficient minimal, 276
 Student's t, 243
Statistical hypothesis, 239
 alternative, 269
 composite, 240
 null, 269
 parametric, 239
 simple, 239, 240
Strategies
 of the first player, 306
 of the second player, 306
Strong law of large numbers, 135
Student's t distribution, 188
Student's test, 242, 245
Student's t statistic, 243
Sufficient statistic, 276
 minimal, 276
Sum of events, 3, 6
Sure event, 7

Taking a random sample, 116
Terms asymptotically negligible, 186
Test
 Galton's order, 268
 Kolmogorov, 263
 Kolmogorov–Smirnov, 264
 most powerful, 269
 of the hypothesis against the alternative, 269
 one-sided for the hypothesis of zero expectation in the normal distribution with a known variance, 238
 Pearson's chi-square, 256
 randomized (randomized critical function), 270
 Student's, 242, 245
 two-sided of the hypothesis concerning the expectation in a normal distribution with a known variance, 240
 uniformly most powerful, 275
 Wilcoxon, 265
Testing the hypothesis
 concerning variance in the normal distribution, 247
 concerning the value of the correlation coefficient, 252

INDEX 321

Testing the hypothesis
  concerning the value of the regression coefficient, 249
  of the equality of the means in normal populations with the same known variance, 241
  of zero expectation in a normal distribution with unknown variance, 242
  of the hypothesis of the equality of expectations in two normal populations with the same unknown standard deviation, 245
Theorem
  Bayes, 59
  Frisch, 202
  Glivenko, 144
  integral de Moivre–Laplace, 85
  Kolmogorov, 235
  Lapunov, 183
  Linderberg–Feller, 185
  Linderberg–Lévy, 180
  local de Moivre–Laplace, 85
  on extension of measures, 31
  on factorization, 276
  Pearson, 256
  Poisson, 78
  Smirnov, 263

Translation parameter, 264
Transformation to orthogonal coordinates, 207, 209
Truncated random variables, 109
Two-dimensional (bivariate) distribution function, 48
Two-person zero-sum game, 306

Unbiased estimator, 299
Uncorrelated random variables, 140
Uniformly most powerful test of composite alternative hypothesis, 275
Upper integral sum, 107
Upper quantile, 118

Value of the game
  for the first player, 307
  for the second player, 307
Variance
  in normal distribution with known expectation, 279
  of a random variable, 119

Waiting time for the first success, 158
Weak convergence, 165
Weak law of large number, 130
Wilcoxon test, 265